The Painlevé Handbook

Robert Conte Micheline Musette

The Painlevé Handbook

Springer

Robert Conte
Service de physique de l'état condensé
CEA-Saclay
F–91191 Gif-sur-Yvette Cedex
France

Micheline Musette
Dienst Theoretische Natuurkunde
Vrije Universiteit Brussel
Pleinlaan 2
B–1050 Brussels
Belgium

ISSN 1860-0832
ISBN 978-94-007-9627-0
ISBN 978-1-4020-8491-1 (eBook)

Published by Springer Science+Business Media B.V.
P.O. Box17, 3300 AA Dordrecht, The Netherlands
In association with
Canopus Publishing Limited,
27 Queen Square, Bristol BS1 4ND, UK

www.springer.com and www.canopusbooks.com

Les cas où l'on peut intégrer une équation différentielle sont extrêmement rares, et doivent être regardés comme des exceptions; mais on peut considérer une équation différentielle comme définissant une fonction, et se proposer d'étudier les propriétés de cette fonction sur l'équation différentielle elle-même.

Charles Briot et Jean-Claude Bouquet, 1859.

Preface

Nonlinear differential or difference equations are encountered not only in mathematics, but also in many areas of physics (evolution equations, propagation of a signal in an optical fiber), chemistry (reaction-diffusion systems) and biology (competition of species).

The purpose of this book is to introduce the reader to nonperturbative methods allowing one to build *explicit* solutions to these equations. A prerequisite task is to investigate whether the chances of success are high or low, and this can be achieved without any *a priori* knowledge of the solutions, with a powerful algorithm called the Painlevé test. If the equation under study passes the Painlevé test, the equation is presumed *integrable* in some sense, and one can try to build the explicit information displaying this integrability:

- for an ordinary differential equation, the closed form expression of the general solution;
- for a partial differential equation, the nonlinear superposition formula to build soliton solutions;

and similar elements in the discrete situation. If on the contrary the test fails, the system is nonintegrable or even chaotic, but it may still be possible to find solutions. Indeed, the methods developed for the integrable case still apply and may in principle produce all the available pieces of integrability, such as the *solitary waves* of evolution equations, or solutions describing the collision of solitary waves, or the first integrals of dynamical systems, etc.

The examples chosen to illustrate these methods are mostly taken from physics. These include on the integrable side the nonlinear Schrödinger equation (continuous and discrete), the Korteweg–de Vries equation, the Boussinesq equation, the Hénon–Heiles Hamiltonians, and on the nonintegrable side the complex Ginzburg–Landau equation (encountered in optical fibers, turbulence, etc), the Kuramoto–Sivashinsky equation (phase turbulence), the reaction-diffusion model of Kolmogorov–Petrovski–Piskunov (KPP), the Lorenz model of atmospheric circulation and the Bianchi IX cosmological model which are both chaotic.

Written at a graduate level, the book contains tutorial text as well as detailed examples and describes the state of the art in some current areas of research.

Brussels,
February 2008

Robert Conte
Micheline Musette

Outline

In Chap. 1, we insist that a nonlinear equation should *not* be considered as the perturbation of a linear equation. We illustrate using two simple examples the importance of taking account of the singularity structure in the complex plane to determine the general solution of nonlinear equations. We then present the point of view of the Painlevé school to *define new functions* from nonlinear ordinary differential equations (ODEs) possessing a general solution which can be made single valued in its domain of definition (*Painlevé property*, PP).

In Chap. 2, we present a local analysis, called the *Painlevé test*, in order to investigate the nature of the movable singularities (i.e. whose location depends on the initial conditions) of the general solution of a nonlinear differential equation. The simplest of the methods involved in this test was historically introduced by Sophie Kowalevski [257] and later turned into an algorithm by Bertrand Gambier [163]. For equations possessing the Painlevé property, the test is by construction satisfied, therefore we concentrate on equations which generically fail the test, in order to extract some constructive information on cases of partial integrability. We first choose four examples describing physical phenomena, for which the test selects cases which may admit closed form particular solutions[1] or first integrals.

This procedure is illustrated in several examples.

In the first example, the *Lorenz model* of atmospheric circulation [284]

$$\frac{\mathrm{d}x}{\mathrm{d}t} = \sigma(y-x),\ \frac{\mathrm{d}y}{\mathrm{d}t} = rx - y - xz,\ \frac{\mathrm{d}z}{\mathrm{d}t} = xy - bz,$$

the test isolates four sets of values of the parameters (b, σ, r).

We next consider the *Kuramoto–Sivashinsky equation* (KS),

$$u_t + \nu u_{xxxx} + b u_{xxx} + \mu u_{xx} + u u_x = 0,\ \nu \neq 0,$$

[1] By definition, a solution is called particular if it can be obtained from the general solution by setting some constants of integration to numerical values.

an equation which describes the propagation of flames on a vertical wall, and we analyze for simplicity the ODE for its stationary flow. The test first detects the presence of movable multivaluedness[2] in the general solution whatever the parameters (ν, b, μ), then it displays the possible existence of particular solutions without movable branching.

We then analyze the one-dimensional cubic *complex Ginzburg–Landau equation* (CGL3),

$$iA_t + pA_{xx} + q|A|^2 A - i\gamma A = 0,\ pq\gamma \neq 0,\ (A, p, q) \in \mathscr{C},\ \gamma \in \mathscr{R}.$$

This is a generic equation which describes many physical phenomena, such as the propagation of a signal in an optical fiber [10], or spatiotemporal intermittency in spatially extended dissipative systems [296]. The test first uncovers the generic nonintegrable nature of this PDE, then it selects as values of the parameters (p, q, γ) those $(q/p \in \mathscr{R}, \gamma = 0)$ of the *nonlinear Schrödinger equation* (NLS), an equation which is integrable in many acceptations. Finally it shows the possible existence of particular single valued solutions in the CGL3 case $\mathrm{Im}(q/p) \neq 0$.

The next example is the *Duffing–van der Pol oscillator*

$$E(u) \equiv u'' + (au^2 + b)u' - cu + \beta u^3 = 0.$$

It is chosen to illustrate a weaker form of the test (weak Painlevé test) in which the general solution is allowed to possess more than one determination around a movable singularity, but only a finite number (weak Painlevé property), like the square root function.

The last example is the two-degree of freedom Hamiltonian system

$$\begin{aligned} H &= \frac{1}{2}(p_1^2 + p_2^2 + \omega_1 q_1^2 + \omega_2 q_2^2) + \alpha q_1 q_2^2 - \frac{1}{3}\beta q_1^3 + \frac{c_3}{2q_2^2},\ \alpha \neq 0 \\ & q_1'' + \omega_1 q_1 - \beta q_1^2 + \alpha q_2^2 = 0, \\ & q_2'' + \omega_2 q_2 + 2\alpha q_1 q_2 - c_3 q_2^{-3} = 0, \end{aligned}$$

in which $\alpha, \beta, \omega_1, \omega_2, c_3$ are constants. In the case $c_3 = 0, \beta/\alpha = 1$, it was introduced by Hénon and Heiles to describe the chaotic motion of a star in the axisymmetric potential of a galaxy [198]. It is now known as the *cubic Hénon–Heiles Hamiltonian* system (HH3). The test selects only three values $\beta/\alpha = -1, -6, -16$.

The last two sections (2.2 and 2.3) deal with two fairly common situations when the test, as initiated by Sophie Kowalevski, is inconclusive, because of the insufficient number of arbitrary constants in the local representation of the general solution.

Chapter 3 is devoted to the explicit integration of nonlinear ODEs by methods based on singularities, mainly taking the examples of the previous chapter. We pro-

[2] A point where multivaluedness occurs is classically called a *critical point* or *ramification point* or *branch point*.

cess successively the integrable (Sect. 3.1) and partially integrable (Sect. 3.2) situations.

In Sect. 3.1.1, in the four cases when the Lorenz model passes the Painlevé test, we give a systematic method to compute the polynomial first integrals and we perform the full integration in terms of elliptic or Painlevé functions.

In Sects. 3.1.2 and 3.1.3, one looks for the traveling waves[3] of two important evolution equations, respectively the *Korteweg–de Vries equation* (KdV), which governs the propagation of waves in shallow water [39, 256],

$$bu_t + u_{xxx} - \frac{6}{a}uu_x = 0, \ (a,b) \text{ constant},$$

and the *nonlinear Schrödinger equation* (NLS),

$$iA_t + pA_{xx} + q|A|^2A = 0, \ pq \neq 0, \ A \in \mathscr{C}, \ (p,q) \in \mathscr{R}.$$

This is an easy task because the ODEs have the Painlevé property and, from their general traveling wave, which is an elliptic function, one defines the various physically relevant particular solutions (pulses, fronts).

In Sect. 3.2, the partially integrable situation is mainly illustrated through the two examples of the equations for the traveling waves of the KS equation and the CGL3 equation, which have been seen to fail the Painlevé test.

In Sect. 3.2.1.2, we introduce the concept of *general analytic solution* of a nonintegrable ODE, defined as the closed form particular single valued solution which depends on the maximum possible number of integration constants, and we count precisely this number. We then look for two classes of solutions which are not too difficult to obtain and which have a great physical interest, the doubly periodic ones (elliptic) and the simply periodic ones (trigonometric).

Those particular solutions which are doubly periodic (elliptic) are easy to find because of necessary conditions arising from a nice property of elliptic functions. These conditions and the associated solutions are established in Sect. 3.2.2.

Among the particular solutions which are simply periodic (trigonometric), some are also easy to find by representing the possible solution as a polynomial in one elementary variable τ or two elementary variables (σ, τ) which obey fundamental nonlinear first order ODEs. These *truncation methods* are described in Sects. 3.2.3 (for KS) and 3.2.4 (for CGL3).

In Sect. 3.2.5, in order to overcome the limitations of the truncation methods, by implementing an old theorem of Briot and Bouquet (1856), we introduce a method able to find *all* the doubly periodic or simply periodic solutions of a given ODE, while any truncation method can only find *some* of these. Instead of searching an expression for the solution, it builds an intermediate, equivalent information, namely the *first order* autonomous ODE satisfied by the unknown solution. For KS and CGL3, it provides no new result, this fact will be explained in Sect. 3.2.8 as an application of the Nevanlinna theory.

[3] A **traveling wave** of a given PDE $E(u,x,t) = 0$ is any solution of the reduction $\xi = x - ct$ if it exists.

Section 3.2.6 deals with the *Duffing–van der Pol* oscillator when it passes the weak Painlevé test. In a particular case when a first integral exists, the resulting ODE can be mapped by a special transformation (hodograph) to another equation possessing the Painlevé property.

In Sect. 3.2.7, we display an example (the *Bianchi IX cosmological model* in vacuum), in which the necessary conditions to pass the test are used in a constructive, unusual way, in order to isolate all possible single valued solutions. The perturbative method of Sect. 2.2 shows the probable existence of one additional solution to the known ones.

In Sect. 3.2.8, we briefly present additional results on the KS equation which are obtained by the Nevanlinna theory. This theory, which is not based on singularity analysis, gives a complementary insight on the analytic structure of the solutions.

Chapter 4 deals with the extension to nonlinear *partial differential equations* (PDEs) of the Painlevé property and Painlevé test previously introduced for ODEs. In Sect. 4.1, we mention solutions of a PDE which are also solutions of some ODEs, i.e. what is called a *reduction*. In Sect. 4.2, we introduce the quite important class of *soliton equations*, together with their main properties: existence of an N-soliton solution and of a remarkable transformation called the Bäcklund transformation (BT). In Sect. 4.3, we extrapolate to PDEs the notion of integrability and the definition of the Painlevé property. After defining in Sect. 4.4.1 the expansion variable χ which minimizes the computation of the Laurent series representing the local solution, we present in Sect. 4.4.2 the successive steps of the Painlevé test, on the example of the KdV equation in order to establish necessary conditions for the Painlevé property. Finally, in Sect. 4.4.3, we apply the test to the equation of *Kolmogorov–Petrovski–Piskunov* (KPP) [255, 383] to generate necessary conditions for the existence of closed form particular solutions.

The subject of Chap. 5 is the "integration" of nonlinear PDEs. Constructive algorithms must be devised to establish the Painlevé property and ultimately to find explicit solutions. Known as the *singular manifold method* (SMM), these algorithms are the natural extension of the truncation methods already encountered in Chap. 3.

In Sect. 5.1, we first extract from the numerous results of the Painlevé test some global information about the analytic property of the solutions. In Sect. 5.2, we recall the two main approaches to build the so called N-soliton solution and briefly introduce the main integrability tools of the soliton equations: Lax pair, Darboux transformation, Bäcklund transformation, nonlinear superposition formula and the Crum transformation. The precise definitions are then given in Sect. 5.3, with application to two physically important equations, the KdV and Boussinesq equations [39], which are integrable by the *inverse spectral transformation* method (IST) [1].

In order to establish the Painlevé property of the PDEs under consideration, the challenge is to derive these integrability items by using methods based only on the singularity structure of the equations.

In Sect. 5.5.1 we present the basic ideas of this singular manifold method mainly consisting in converting the local information provided by the Painlevé test into the above mentioned (global) integrability items. The next two sections are respectively devoted to the SMM in the case of equations possessing the PP (Sect. 5.6) and in the case of partially integrable equations, i.e. equations which fail the Painlevé test but nevertheless admit particular singlevalued solutions (Sect. 5.7).

More precisely, in Sect. 5.6.1 we process the Korteweg-de Vries and Boussinesq equations, which possess only one family of movable singularities. Their two nonlinear superposition formulae are found to be the same, the reason being that the KdV and Boussinesq equations are two different reductions of a 2+1-dimensional IST-integrable equation, the Kadomtsev–Petviashvili (KP) equation [243]. However, the two reductions induce two different solitonic behaviors: KdV only describes the overtaking interaction of solitary waves, while Boussinesq may also describe the head-on collision of solitary waves.

In Sect. 5.6.2, the SMM is applied to two IST-integrable equations (sine-Gordon, modified KdV) which possess two families of movable singularities, and again obtain for both equations the same form of the NLSF.

In Sect. 5.6.3, we apply the SMM to two other integrable PDEs which have a third order Lax pair, the Sawada–Kotera (SK) [387] and Kaup–Kupershmidt (KK) [246, 148] equations. The key ingredient is to consider, in the list of Gambier [163] of second order first degree nonlinear ODEs possessing the PP, the very few equations which are linearizable into a third order ODE, yielding simultaneously the Darboux transformation and the x-part of the Lax pair. In addition to the auto-Bäcklund transformation and the NLSF in each case, the SMM provides a BT between SK and KK.

We next apply the SMM to partially integrable PDEs. In Sect. 5.7.1, we handle the *Fisher equation* [140], which models the evolution of mutant genes or the propagation of flames. In this one-family equation, by finding a particular solution of the necessary conditions generated by the Painlevé test, one obtains two elliptic solutions [8].

In Sect. 5.7.2, we handle the KPP reaction-diffusion equation, possessing two opposite families. The output is two one-soliton solutions (one tanh and one sech), and a degenerate two-soliton without coupling factor.

In the last section (5.8), we examine what these integrability items become when an integrable PDE reduces to an ODE:

∘ Lax pair $\rightarrow$ isomonodromic deformation
∘ Bäcklund transformation $\rightarrow$ birational transformation
∘ nonlinear superposition formula $\rightarrow$ contiguity relation.

In Chap. 6, we give an illustration on the various ways to "integrate" a Hamiltonian system using two examples of Hamiltonian systems with two degrees of freedom: the cubic HH Hamiltonian introduced in Sect. 2.1.5, three cases of which pass the Painlevé test, and the quartic HH Hamiltonian (HH4),

$$\begin{aligned} H = & \frac{1}{2}(P_1^2 + P_2^2 + \Omega_1 Q_1^2 + \Omega_2 Q_2^2) + C Q_1^4 + B Q_1^2 Q_2^2 + A Q_2^4 \\ & + \frac{1}{2}\left(\frac{\alpha}{Q_1^2} + \frac{\beta}{Q_2^2}\right) + \gamma Q_1,\ B \neq 0, \end{aligned}$$

$$Q_1'' + \Omega_1 Q_1 + 4CQ_1^3 + 2BQ_1Q_2^2 - \alpha Q_1^{-3} + \gamma = 0,$$
$$Q_2'' + \Omega_2 Q_2 + 4AQ_2^3 + 2BQ_2Q_1^2 - \beta Q_2^{-3} = 0,$$

in which $A,B,C,\alpha,\beta,\gamma,\Omega_1,\Omega_2$ are constants. The Painlevé test selects four sets of values of these constants, $A:B:C=$ 1:2:1, 1:6:1, 1:6:8, 1:12:16 (the notation $A:B:C=p:q:r$ stands for $A/p=B/q=C/r=$ arbitrary).

These various ways to integrate are

∘ (Liouville integrability) to find a second invariant in involution with the Hamiltonian, which is however insufficient to perform a global integration; we recall the seven first integrals which establish this integrability for both HH3 (Sect. 6.2.1) and HH4 (Sect. 6.3.1);
∘ (Arnol'd–Liouville integrability) to find the variables which separate the Hamilton–Jacobi equation, thus leading to a global integration; this has been done for HH3 (Sect. 6.2.2), and nearly finished for HH4 (Sect. 6.3.2);
∘ (Painlevé property) to find an explicit closed form single valued expression for the general solution $q_j(t), Q_j(t)$; this has been done in all seven cases (Sects. 6.2.3 and 6.3.3), *via* birational transformations to fourth order ODEs isolated and integrated by Cosgrove.

Chapter 7 deals with discrete nonlinear equations. After some generalities, in Sect. 7.1 we consider the logistic map of Verhulst,

$$u_n = a u_{n-1}(1-u_{n-1}),$$

a paradigm of chaotic behavior [405, 139], which admits a continuum limit to the Riccati equation. From the point of view of integrability, the logistic map is a "bad" discretization of the Riccati equation, because it cannot be linearized, and it must be replaced by a "good" discrete equation, i.e. one which preserves the property of linearizability. More generally, the goal is to extend the Painlevé property to the discrete world.

Section 7.2 presents an outlook of the difficulty to give an undisputed definition for the *discrete Painlevé property*.

In Sects. 7.3.1, 7.3.2 and 7.3.3, we present the three main methods of the discrete Painlevé test: the *singularity confinement method* [184], the *criterium of polynomial growth* [206], and the *perturbation of the continuum limit* [88].

In order to prove the discrete Painlevé property, one can either linearize the discrete equation, or explicitly integrate or, as admitted by most researchers, exhibit a discrete Lax pair.

In Sect. 7.4, we return to the question of finding a "good" discretization of the Riccati equation; this results in the homographic map

$$u_n = \frac{a_1 u_{n-1} + a_2}{a_3 u_{n-1} + a_4}.$$

The notion of discrete Lax pair is introduced in Sect. 7.5.

We then describe two examples of exact discretizations, i.e. for which the analytic expression of the general solution is the same for the continuous and discrete equations.

In Sect. 7.6.1 we consider the question of discretizing the nonlinear ODE for the modulus $v = |\psi|$ of the linear Schrödinger equation, namely [133, 305, 358],

$$v'' + fv + c^2 v^{-3} = 0,$$

usually called the *Ermakov–Pinney equation*. Again, the property to be preserved is the linearizability, since the starting equation is linear.

In Sect. 7.6.2 we recall the remark by Baxter and Potts that the addition formula of the Weierstrass function $\wp$ can be identified to an exact discretization of the Weierstrass equation. This is the foundation for a family of special two-component rational maps [367, 368] which, like its continuous counterpart, is a starting point to isolate discrete equations which may possess the discrete PP.

In Sect. 7.7, we briefly review two related problems. The first problem, still open but of a very high physical interest in optical fibers, is to find exact solitary waves (dark and bright) for the *nonintegrable discrete nonlinear Schrödinger equation*,

$$iu_t + p\frac{u(x+h,t) + u(x-h,t) - 2u(x)}{h^2} + q|u|^2 u = 0,\ i^2 = -1,\ pq \neq 0.$$

In the context of optical fibers or Bose–Einstein condensation [10], this equation is not obtained as a discretization of NLS but it arises by a direct construction. The second one is to isolate discrete versions of the nonlinear Schrödinger equation which might possess the discrete Painlevé property, and one such equation is the Ablowitz and Ladik [4] discrete equation.

Finally, in Sect. 7.8, after setting up the natural problem to extend to the discrete world the six transcendents of Painlevé, we introduce the two methods which have been devised to handle it. In the analytic method (Sect. 7.8.1), the procedure starts from the addition formula of the elliptic function, takes some inspiration from the method of Painlevé and Gambier and produces a rather long list of discrete Pn equations, but no proof exists that the list is exhaustive. The geometric method (Sect. 7.8.2) first displays the importance of two groups describing the continuous Pn, then uses the theory of rational surfaces to build an object which admits the largest of the just mentioned groups, object interpreted as the master discrete Painlevé equation e − P6, whose coefficients have an elliptic dependence on the independent variable. The main properties of all these d − Pn are then summarized in Sect. 7.8.3.

After an FAQ chapter, a few appendices collect material too technical to be presented in the main text.

Contents

1 Introduction ... 1
1.1 Singularities in the Complex Plane ... 1
1.1.1 Perturbative Method ... 2
1.1.2 Nonperturbative Method ... 3
1.2 Painlevé Property and the Six Transcendents ... 6

2 Singularity Analysis: Painlevé Test ... 11
2.1 Kowalevski–Gambier Method ... 11
2.1.1 Lorenz Model ... 12
2.1.2 Kuramoto–Sivashinsky (KS) Equation ... 16
2.1.3 Cubic Complex Ginzburg–Landau (CGL3) Equation ... 19
2.1.4 Duffing–van der Pol Oscillator ... 21
2.1.5 Hénon–Heiles System ... 23
2.2 Fuchsian Perturbative Method ... 28
2.3 NonFuchsian Perturbative Method ... 31

3 Integrating Ordinary Differential Equations ... 33
3.1 Integrable Situation ... 34
3.1.1 First Integrals and Integration of the Lorenz Model ... 34
3.1.1.1 Case $(1,1/2,0)$ ... 35
3.1.1.2 Case $(2,1,1/9)$ ... 36
3.1.1.3 Case $(0,1/3,r)$ ... 37
3.1.1.4 Case $(1,0,r)$ ... 37
3.1.2 General Traveling Wave of KdV Equation ... 38
3.1.3 General Traveling Wave of NLS Equation ... 40
3.2 Partially Integrable Situation ... 44
3.2.1 Traveling Wave Reduction of KS and CGL3 ... 44
3.2.1.1 Nonexistence of a First Integral ... 45
3.2.1.2 Counting of the Arbitrary Constants ... 45
3.2.2 Elliptic Traveling Waves ... 47
3.2.2.1 Necessary Conditions for Elliptic Solutions ... 48

3.2.2.2 The Elliptic Solutions ... 49
3.2.3 Trigonometric Traveling Waves of KS ... 51
3.2.3.1 Gauge Transformation ... 51
3.2.3.2 Polynomials in tanh ... 52
3.2.4 Trigonometric Traveling Waves of CGL3 ... 55
3.2.4.1 Polynomials in tanh ... 57
3.2.4.2 Polynomials in tanh and sech ... 60
3.2.5 General Method to Find Elliptic Traveling Waves ... 63
3.2.5.1 Application to KdV ... 67
3.2.5.2 Application to KS ... 68
3.2.5.3 Application to CGL3 ... 69
3.2.6 First Integral of the Duffing–van der Pol Oscillator ... 71
3.2.7 Singlevalued Solutions of the Bianchi IX Cosmological Model ... 72
3.2.8 Results of the Nevanlinna Theory on KS and CGL3 ... 74

4 Partial Differential Equations: Painlevé Test ... 77
4.1 On Reductions ... 78
4.2 Soliton Equations ... 80
4.3 Painlevé Property for PDEs ... 82
4.4 Painlevé Test ... 85
4.4.1 Optimal Expansion Variable ... 85
4.4.2 Integrable Situation, Example of KdV ... 87
4.4.3 Partially Integrable Situation, Example of KPP ... 89

5 From the Test to Explicit Solutions of PDEs ... 93
5.1 Global Information from the Test ... 93
5.2 Building *N*-Soliton Solutions ... 94
5.3 Tools of Integrability ... 94
5.3.1 Lax Pair ... 95
5.3.2 Darboux Transformation ... 97
5.3.3 Crum Transformation ... 99
5.3.4 Singular Part Transformation ... 100
5.3.5 Nonlinear Superposition Formula ... 101
5.4 Choosing the Order of Lax Pairs ... 103
5.4.1 Second-Order Lax Pairs and Their Privilege ... 103
5.4.2 Third-Order Lax Pairs ... 104
5.5 Singular Manifold Method ... 105
5.5.1 Algorithm ... 105
5.5.2 Level of Truncation and Choice of Variable ... 107
5.6 Application to Integrable Equations ... 109
5.6.1 One-Family Cases: KdV and Boussinesq ... 110
5.6.1.1 The Case of KdV ... 110
5.6.1.2 The Case of Boussinesq ... 113
5.6.2 Two-Family Cases: Sine-Gordon and Modified KdV ... 116

5.6.2.1 The Sine-Gordon Equation 117
5.6.2.2 The Modified Korteweg–de Vries Equation 120
5.6.3 Third Order: Sawada–Kotera and Kaup–Kupershmidt 122
5.6.3.1 Help from the Gambier Classification 123
5.6.3.2 Singularity Structure of SK and KK Equations 124
5.6.3.3 Truncation with a Second Order Lax Pair 126
5.6.3.4 Truncation with a Third Order Lax Pair and G5 . . . 127
5.6.3.5 Truncation with a Third Order Lax Pair and G25 . . 127
5.6.3.6 Bäcklund Transformation 127
5.6.3.7 Nonlinear Superposition Formula 128
5.7 Application to Partially Integrable Equations 130
5.7.1 One-Family Case: Fisher Equation 130
5.7.2 Two-Family Case: KPP Equation 133
5.8 Reduction of the Singular Manifold Method to the ODE Case 136
5.8.1 From Lax Pair to Isomonodromic Deformation 137
5.8.2 From BT to Birational Transformation 139
5.8.3 From NLSF to Contiguity Relation 141
5.8.4 Reformulation of the Singular Manifold Method: an Additional Homography 143

6 Integration of Hamiltonian Systems 145
6.1 Various Integrations 145
6.2 Cubic Hénon–Heiles Hamiltonians 146
6.2.1 Second Invariants 146
6.2.2 Separation of Variables 147
6.2.2.1 Case $\beta/\alpha = -6$ (KdV5) 147
6.2.2.2 Cases $\beta/\alpha = -1$ (SK) and -16 (KK) 148
6.2.3 Direct Integration 151
6.3 Quartic Hénon–Heiles Hamiltonians 153
6.3.1 Second Invariants 153
6.3.2 Separation of Variables 154
6.3.2.1 Case 1:2:1 (Manakov System) 154
6.3.2.2 Cases 1:6:1 and 1:6:8 155
6.3.2.3 Case 1:12:16 158
6.3.3 Painlevé Property 160
6.4 Final Picture for HH3 and HH4 162

7 Discrete Nonlinear Equations 163
7.1 Generalities 163
7.2 Discrete Painlevé Property 166
7.3 Discrete Painlevé Test 167
7.3.1 Method of Singularity Confinement 167
7.3.2 Method of Polynomial Growth 170
7.3.3 Method of Perturbation of the Continuum Limit 172
7.4 Discrete Riccati Equation 174

7.5 Discrete Lax Pairs ... 175
7.6 Exact Discretizations ... 177
7.6.1 Ermakov–Pinney Equation ... 177
7.6.2 Elliptic Equation ... 180
7.7 Discrete Versions of NLS ... 181
7.8 A Sketch of Discrete Painlevé Equations ... 182
7.8.1 Analytic Approach ... 183
7.8.2 Geometric Approach ... 184
7.8.3 Summary of Discrete Painlevé Equations ... 185

8 FAQ (Frequently Asked Questions) ... 187

A The Classical Results of Painlevé and Followers ... 191
A.1 Groups of Invariance of the Painlevé Property ... 192
A.2 Irreducibility. Classical Solutions ... 193
A.3 Classifications ... 194
A.3.1 First Order Higher Degree ODEs ... 194
A.3.2 Second Order First Degree ODEs ... 195
A.3.3 Second Order Higher Degree ODEs ... 196
A.3.4 Third Order First Degree ODEs ... 196
A.3.5 Fourth Order First Degree ODEs ... 196
A.3.6 Higher Order First Degree ODEs ... 198
A.3.7 Second Order First Degree PDEs ... 198

B More on the Painlevé Transcendents ... 199
B.1 Coalescence Cascade ... 200
B.2 Invariance Under Homographies ... 203
B.3 Invariance Under Birational Transformations ... 204
B.3.1 Normal Sequence ... 204
B.3.2 Biased Sequence ... 206
B.4 Invariance Under Affine Weyl Groups ... 207
B.5 Invariance Under Nonbirational Transformations ... 209
B.6 Hamiltonian Structure ... 209
B.7 Lax Pairs ... 210
B.8 Classical Solutions ... 212

C Brief Presentation of the Elliptic Functions ... 215
C.1 The Notation of Jacobi and Weierstrass ... 216
C.2 The Symmetric Notation of Halphen ... 217

D Basic Introduction to the Nevanlinna Theory ... 219

E The Bilinear Formalism ... 223
E.1 Bilinear Representation of a PDE ... 223
E.2 Bilinear Representation of a Bäcklund Transformation ... 225

F Algorithm for Computing Laurent Series 229

References 234
Index 253

Acronyms

AKNS	Ablowitz, Kaup, Newell and Segur
BT	Bäcklund transformation
CGL3	Cubic complex Ginzburg–Landau equation
CGL5	Quintic complex Ginzburg–Landau equation
DT	Darboux transformation
FAQ	Frequently asked question
Gn	Gambier equation number n
HH3	Cubic Hénon–Heiles Hamiltonian
HH4	Quartic Hénon–Heiles Hamiltonian
IST	Inverse spectral transform
KdV	Korteweg and de Vries equation
KK	Kaup-Kupershmidt equation
KP	Kadomtsev and Petviashvili equation
KPP	Kolmogorov–Petrovski–Piskunov equation
KS	Kuramoto and Sivashinsky equation
NLS	Nonlinear Schrödinger equation
NLSF	Nonlinear superposition formula
ODE	Ordinary differential equation
PDE	Partial differential equation
Pn	Painlevé equation number n
PP	Painlevé property
SG	sine-Gordon equation
SH	Swift and Hohenberg equation
SK	Sawada–Kotera equation
SME	Singular manifold equation
SMM	Singular manifold method
WTC	Weiss, Tabor and Carnevale

Chapter 1
Introduction

Abstract A nonlinear equation should *not* be considered as the perturbation of a linear equation. We illustrate using two simple examples the importance of taking account of the singularity structure in the complex plane to determine the general solution of nonlinear equations. We then present the point of view of the Painlevé school to *define new functions* from nonlinear ordinary differential equations possessing a general solution which can be made single valued in its domain of definition (the Painlevé property).

1.1 Singularities in the Complex Plane

Given some nonlinear differential equation, an intuitive approach to find a solution is to split the equation into the sum of a so-called *linear part* and a *perturbation*. Let us explain, using two examples, why this should *not* be done.

Consider the following elementary nonlinear equations

$$u' = k(1-u^2),\ ' = \frac{\mathrm{d}}{\mathrm{d}x}, \tag{1.1}$$

$$v'^2 = k^2 v^2 (1-v^2), \tag{1.2}$$

with the aim of finding their general solution,

$$u = \tanh k(x-x_0), \tag{1.3}$$

$$v = \frac{1}{\cosh k(x-x_0)} = \mathrm{sech}\, k(x-x_0). \tag{1.4}$$

The arbitrary complex constant x_0 is linked to the initial condition $u(x_i) = u_i, v(x_i) = v_i$ by the relation

$$u_i = \tanh k(x_i - x_0),\ v_i = \mathrm{sech}\, k(x_i - x_0). \tag{1.5}$$

On the real axis, u and v have no singularities but, in the complex plane, the singularities of u and v are a countable number of simple poles, located at $x = x_0 + (2n+1)i\pi/(2k), n \in \mathscr{Z}$. Such singularities are by definition said to be *movable*, as opposed to *fixed*, because their location depends on the initial conditions, i.e. on the constants of integration. The general solution of any linear differential equation has no movable singularity because it depends linearly on the constants of integration.

1.1.1 Perturbative Method

In the *perturbative* method [199], one first determines the stationary points, which leads to $u_0 = \pm 1$ and $v_0 = 0, \pm 1$, then one perturbs the solution in the neighborhood of a stationary point by expanding it in series of a small parameter ε. Under this perturbation

$$u = \sum_{n=0}^{\infty} \varepsilon^n u_n,\ u_0 = 1, \tag{1.6}$$

Equation (1.1) splits accordingly into

$$\sum_{n=0}^{\infty} \varepsilon^n E_n = 0,\ E_0 \equiv 0. \tag{1.7}$$

$$E_1 \equiv -u_1' - 2ku_0u_1 = 0, \tag{1.8}$$

$$E_2 \equiv -u_2' - 2ku_0u_2 - ku_1^2 = 0, \cdots \tag{1.9}$$

Choosing $u_1 = c_1 e^{-2kx}$ with c_1 arbitrary, this infinite set of linear equations with the same homogeneous part admits the particular solution

$$u_n = 2^{-n+1} c_1^n e^{-2knx},\ n \ge 1, \tag{1.10}$$

which defines a geometric series, and its sum provides the general solution of (1.1)

$$u = 1 + \frac{\varepsilon c_1 e^{-2kx}}{1 - (\varepsilon c_1/2)e^{-2kx}} = \tanh k(x - x_0),\ x_0 = \frac{1}{2k} \log\left(-\frac{\varepsilon c_1}{2}\right). \tag{1.11}$$

Equation (1.2) is handled slightly differently because of its nonlinearity in the highest derivative. One first takes its derivative,

$$v'' = k^2 v(1 - 2v^2), \tag{1.12}$$

to make the first perturbed equation $E_1 = 0$ linear in the highest derivative. Then the computation is quite similar: the expansion $v = \sum_{n=0}^{\infty} \varepsilon^n v_n$ around $v_0 = 0$ generates an infinite set of linear equations with the same homogeneous part. Choosing $v_1 = c_1 e^{-kx}$, with c_1 arbitrary, the particular solutions are

$$v_{2n+1} = (-4)^{-n} \left(c_1 e^{-kx}\right)^{2n+1}, \; v_{2n} = 0, \; n = 1, 2, \cdots \tag{1.13}$$

Therefore the series for v is geometric and it sums into:

$$v = \frac{\varepsilon c_1 e^{-kx}}{1 + (\varepsilon c_1/2)^2 e^{-2kx}} \equiv \operatorname{sech} k(x - x_0), \; x_0 = \frac{1}{k} \log \frac{\varepsilon c_1}{2}, \tag{1.14}$$

which represents the general solution of (1.2).

Why is this perturbative method not efficient for equations more complicated than (1.1)–(1.2)? There are several reasons for this:

1. In order to obtain the general term u_n, one must solve a recurrence relation, a difficult task even for a linear recurrence relation.
2. The resummation must be performed in closed form[1] and this is generically impossible; indeed, any solution which is not in closed form is what Painlevé calls "illusoire", in a sense to be developed soon.
3. After performing the resummation, one must check whether the closed form expression is valid everywhere except at a few points, called *singularities*; the location of these singularities cannot be restricted to the real axis but must be extended to the whole complex plane $\mathscr{C}$; in the above example, the reason for the finite value of the radius of convergence is the presence of a simple pole on the imaginary axis at $x = x_0 \pm i\pi/(2k)$.

To summarize, the main reason for the generic inapplicability of this perturbative method is that the singularity structure has not been taken into account: the movable singularity which is present in the exact solution is absent at all orders of the perturbation.

1.1.2 Nonperturbative Method

Let us now present a *nonperturbative* method, which yields the same result in a *finite* number of steps because it takes the singularity structure into account from the beginning.

Since nonlinear ODEs generically possess movable singularities, let us first establish the behavior of the general solution of (1.1) near such a movable singularity $x = x_0$. Assuming this behavior to be algebraic, this amounts to computing the possible values of the *leading power* p and the *leading coefficient* u_0 defined by

$$u \underset{\chi \to 0}{\sim} u_0 \chi^p, \; u_0 \neq 0, \; \chi = x - x_0, \tag{1.15}$$

with p not a positive integer. Then

[1] This will be defined precisely later. For the moment, it is sufficient to know that an example of such a closed form is $u = \psi'/\psi$, with ψ the solution of any linear equation.

$$u' \sim p u_0 \chi^{p-1} + u_0' \chi^p \sim p u_0 \chi^{p-1}, \tag{1.16}$$

so u_0 can be assumed constant when determining the leading behavior. The various terms of (1.1) then contribute as

term	$-u'$	$+k$	$-ku^2$
leading power	$p-1$	0	$2p$
leading coefficient	$-pu_0$	k	$-ku_0^2$

and the l.h.s. of the ODE, which must vanish, evaluates to

$$\begin{aligned} E(u) &\equiv -u' + k(1-u^2) \\ &= \left(-pu_0\chi^{p-1} + O(\chi^p)\right) + k\chi^0 - k\left(u_0^2\chi^{2p} + O(\chi^{2p+1})\right) \qquad (1.17) \\ &= E_0\chi^q + O(\chi^{q+1}) = 0. \qquad (1.18) \end{aligned}$$

The condition $u_0 \neq 0$ implies the equality of at least two of the three leading powers, the two equal powers being lower than or equal to the third one. As to the condition $E_0 = 0$, it expresses the vanishing of the sum of the two corresponding leading coefficients. Out of the three possibilities

$$(q = p - 1 = 0 \le 2p) \text{ and } (-pu_0 + k = 0), \tag{1.19}$$
$$(q = 0 = 2p \le p - 1) \text{ and } (k - ku_0^2 = 0), \tag{1.20}$$
$$(q = p - 1 = 2p \le 0) \text{ and } (-pu_0 - ku_0^2 = 0), \tag{1.21}$$

only the third one defines a solution,

$$p = -1,\ q = -2,\ u_0 = 1/k. \tag{1.22}$$

To summarize, the local behavior of u is that of a simple pole,

$$u \underset{\chi \to 0}{\sim} k^{-1}\chi^{-1},\ \chi = x - x_0. \tag{1.23}$$

In order to turn this local information into a global one, one then establishes a parallel with a well known generator of simple poles, namely the *logarithmic derivative* operator. If some function $\psi(x)$ has an algebraic behavior $\psi \sim \psi_0(x - x_0)^p$ near x_0 (with ψ_0 and p any complex numbers), under action of the logarithmic derivative operator,

$$\mathscr{D} = \frac{\mathrm{d}}{\mathrm{d}x}\log, \tag{1.24}$$

this behavior (whatever it is, regular or singular, multivalued or singlevalued) becomes that of a simple pole of residue p,

$$\frac{\mathrm{d}}{\mathrm{d}x}\log\psi \sim \frac{p}{x - x_0}. \tag{1.25}$$

The crucial point is then to match (1.23) and (1.25), by introducing the transformation from u to ψ *defined* by

$$u = k^{-1} \frac{\mathrm{d}}{\mathrm{d}x} \log \psi. \tag{1.26}$$

This transformation, called the *singular part transformation*, maps the Riccati ODE (1.1) to the second order ODE

$$\psi'' - k^2 \psi = 0, \tag{1.27}$$

which has no more movable singularities since it is linear. Therefore its general solution is known,

$$\psi = c \cosh k(x - x_1),\ (c, x_1) \text{ arbitrary}, \tag{1.28}$$

and this provides the closed form single valued expression (1.3) for the general solution of the Riccati ODE (1.1).

With our second example (1.2), one similarly obtains the two local behaviors

$$v \underset{\chi \to 0}{\sim} \pm i k^{-1} \chi^{-1},\ \chi = x - x_0. \tag{1.29}$$

This complex value $\pm ik^{-1}$ for the residue should be no surprise, since it is the root of an algebraic equation with real coefficients. The map from v to ψ must now involve two functions ψ_1, ψ_2, and indeed, if one defines the singular part transformation as

$$v = ik^{-1} (\log \psi_1)' - ik^{-1} (\log \psi_2)', \tag{1.30}$$

in which ψ_1 and ψ_2 are two different solutions of the same second order linear equation

$$\psi'' - \frac{k^2}{4} \psi = 0, \tag{1.31}$$

which can be chosen as

$$\psi_1 = c_1 \cosh \frac{k}{2}(x - x_1),\ \psi_2 = c_2 \cosh \frac{k}{2}(x - x_2), \tag{1.32}$$

the expression (1.30) satisfies the ODE (1.2), provided x_1, x_2, k obey the relation

$$k(x_1 - x_2) = i\pi + 2mi\pi,\ m \in \mathscr{Z}, \tag{1.33}$$

with the correspondence of notation $x_0 = (x_1 + x_2)/2$.

Therefore, the fact of taking account of the singularity structure (one family of simple poles, two families of simple poles with opposite residues, etc) allows one to establish an explicit closed form link towards another ODE (in our examples a linear ODE) which has no movable singularities, *ipso facto* performing the explicit integration of the nonlinear ODE.

The purpose of this book is to explain how to *explicitly* build analytic solutions of nonlinear differential equations, whether ordinary or partial, by nonperturbative methods such as the simple one presented above.

Since all the exact solutions one can derive by any method necessarily obey the singularity structure of the equation in the complex plane, it is therefore a prerequisite to study these singularities. For instance, the solutions (1.3)–(1.4) have respectively one family and two families of movable simple poles, therefore one must be able to detect, directly on their ODEs without knowing the solutions in advance, respectively one family and two families of movable simple poles.

1.2 Painlevé Property and the Six Transcendents

How can this be generalized? This is the whole problem of the explicit integration of ODEs. *To integrate an ODE*, according to a definition attributed to Poincaré, is to express its *general solution* as a finitely many term explicit expression, possibly multivalued, built from elementary objects called functions. A *function* in turn is defined as a map which can be made singlevalued in its whole domain of definition. Any linear ODE defines a function because its general solution can be made singlevalued, by classical uniformization procedures such as cuts in the complex plane. Typical examples are all the "special functions" of mathematical physics defined by some linear equation (exponential and trigonometric functions, functions of Bessel, Hermite, Legendre, Gauss, ...). Therefore, with the above definition, a large class of ODEs are considered as integrated[2]: linear ODEs, linearizable ODEs, ODEs whose general solution is rational in the solution of a linear equation, ...

In order to extend the class of available functions, L. Fuchs and Poincaré stated the problem of defining new functions from algebraic nonlinear differential equations. One such function had already been discovered by Jacobi when he solved the motion of the pendulum. In this Hamiltonian system

$$H = \frac{1}{2} m l^2 \left(\frac{\mathrm{d}\theta}{\mathrm{d}t}\right)^2 + mgl(1-\cos\theta), \tag{1.34}$$

the problem is to find the position (characterized by the angle θ of the pendulum of length l and mass m with the vertical axis) as a function of the time t. After equating H to its constant value E, one obtains a first order second degree equation (the degree is by definition the polynomial degree in the highest derivative) which is often "integrated by separation of variables" as

[2] For instance, the stationary Schrödinger equation of quantum mechanics, called the Sturm–Liouville equation by mathematicians, is considered as integrated. To solve the spectral problem is outside the scope of this volume.

$$u = \tan\frac{\theta}{2},\ t = t_0 + \int_{u_0}^{u} \frac{\sqrt{2ml^2}\mathrm{d}u}{\sqrt{(1+u^2)(E(1+u^2)-2mglu^2)}}, \tag{1.35}$$

which expresses the time t as an *elliptic integral* of the position u. However, this does not answer the question, which was to express the position as a function of time. Indeed, the above answer is as bad as would be a multivalued expression like

$$t = t_0 + \int_{u_0}^{u} \frac{\mathrm{d}u}{1+u^2} = t_0 + \mathrm{Arctan}\, u - \mathrm{Arctan}\, u_0, \tag{1.36}$$

instead of the singlevalued answer

$$u = \frac{u_0 + \tan(t-t_0)}{1 - u_0 \tan(t-t_0)}. \tag{1.37}$$

This classical problem, called inversion of the elliptic integral, was solved by Abel and Jacobi, who proved that, for the pendulum, the coordinates $(l\cos\theta, l\sin\theta)$ of the position are singlevalued expressions of the time,

$$\sin\frac{\theta}{2} = k\,\mathrm{sn}\left(\sqrt{\frac{g}{l}}(t-t_0), k\right),\ k = \sqrt{\frac{E}{mgl}}, \tag{1.38}$$

$$\cos\frac{\theta}{2} = \mathrm{dn}\left(\sqrt{\frac{g}{l}}(t-t_0), k\right). \tag{1.39}$$

The symbols $\mathrm{sn}(x,k)$ and $\mathrm{dn}(x,k)$, in which k is a constant, denote two of the twelve Jacobi *elliptic functions* (Appendix C), which all satisfy equations of the type

$$\left(\frac{\mathrm{d}u}{\mathrm{d}x}\right)^2 - P(u) = 0, \tag{1.40}$$

with P a polynomial independent of x of degree four with complex coefficients. The general solution of (1.40) is singlevalued not only on the real x axis but in the whole complex plane. Considering the complex plane is mandatory to unveil the beautiful property of this function, which is to be a doubly periodic meromorphic function, a characteristic property of elliptic functions. This equation is form invariant under a transformation which plays a fundamental role in the present theories, the *homographic transformation* or homography,

$$u \mapsto \frac{\alpha u + \beta}{\gamma u + \delta},\ (\alpha,\beta,\gamma,\delta) \text{ complex constants},\ \alpha\delta - \beta\gamma \neq 0. \tag{1.41}$$

The canonical representative in this equivalence class is the *Weierstrass equation*

$$u'^2 = 4u^3 - g_2 u - g_3 = 4(u-e_1)(u-e_2)(u-e_3), \tag{1.42}$$

in which g_2, g_3, e_1, e_2, e_3 are complex constants and one zero of the polynomial P has been moved to infinity by choosing $-\delta/\gamma$ equal to the affix of that zero. The peculiarity of the homographic group is to be the *unique* bijection of the complex plane (to which one has added the point at infinity) to itself, this is why this group does not alter the singularity structure of the elliptic equation.

Another characteristic property of the elliptic equation, much more important in our context than the previous one, is to be the unique first order algebraic ODE able to define a "new" function in the above sense, i.e. from a nonlinear ODE.

This question (of defining new functions) has been investigated at higher orders (up to six for special classes) by the Painlevé school (Painlevé, Gambier, Chazy, Garnier) and its followers (Bureau, Exton, Martynov, Cosgrove). Its mathematical formulation [349, p. 2]

Déterminer toutes les équations différentielles algébriques du premier ordre, puis du second ordre, puis du troisième ordre, etc., dont l'intégrale a ses points critiques fixes.[3]

naturally leads to the definition of a property of differential equations.

Definition 1.1. If the general solution of an ODE can be made singlevalued, one says that such an ODE possesses the **Painlevé property** (PP).

A class of transformations which leaves invariant the singularity structure of u and therefore the PP of the ODE for u is the *homographic group* (also called Möbius group and denoted $\mathrm{PSL}(2,\mathscr{C})$)

$$\begin{aligned}&(u,x) \mapsto (U,X),\ u(x) = \frac{\alpha(x)U(X)+\beta(x)}{\gamma(x)U(X)+\delta(x)},\ X=\xi(x),\\&(\alpha,\beta,\gamma,\delta,\xi) \text{ functions},\ \alpha\delta-\beta\gamma\neq 0,\end{aligned} \tag{1.43}$$

which depends on four arbitrary functions and generalizes the group (1.41).

At present time, only second order nonlinear equations have defined additional functions, the six ones discovered by Painlevé and Gambier, called *Painlevé transcendents* $\mathrm{P}n, n=1,\dots,6$[4]

$$\begin{aligned}
\mathrm{P1}\ &:\ u'' = 6u^2+x,\\
\mathrm{P2}\ &:\ u'' = \delta(2u^3+xu)+\alpha,\\
\mathrm{P3}\ &:\ u'' = \frac{u'^2}{u}-\frac{u'}{x}+\frac{\alpha u^2+\gamma u^3}{4x^2}+\frac{\beta}{4x}+\frac{\delta}{4u},\\
\mathrm{P4}\ &:\ u'' = \frac{u'^2}{2u}+\gamma\left(\frac{3}{2}u^3+4xu^2+2x^2u\right)-2\alpha u+\frac{\beta}{u},\\
\mathrm{P5}\ &:\ u'' = \left[\frac{1}{2u}+\frac{1}{u-1}\right]u'^2-\frac{u'}{x}+\frac{(u-1)^2}{x^2}\left[\alpha u+\frac{\beta}{u}\right]+\gamma\frac{u}{x}+\delta\frac{u(u+1)}{u-1},
\end{aligned}$$

[3] To determine all the algebraic differential equations of first order, then second order, then third order, etc., whose general solution has no movable critical points.

[4] We adopt for P3 the choice made by Painlevé in 1906 [350] to replace his original choice of 1900 [348].

$$\text{P6} : u'' = \frac{1}{2}\left[\frac{1}{u} + \frac{1}{u-1} + \frac{1}{u-x}\right] u'^2 - \left[\frac{1}{x} + \frac{1}{x-1} + \frac{1}{u-x}\right] u'$$
$$+ \frac{u(u-1)(u-x)}{x^2(x-1)^2}\left[\alpha + \beta\frac{x}{u^2} + \gamma\frac{x-1}{(u-1)^2} + \delta\frac{x(x-1)}{(u-x)^2}\right],$$

in which $\alpha, \beta, \gamma, \delta$ are arbitrary complex parameters[5]. Their only singularities (in the complex e^x plane for P3, P5, in the x plane for the four others) are movable poles, with in addition three fixed critical singularities for P6, conveniently located at points $x = \infty, 0, 1$. Their main properties are briefly presented in the Appendix B.

We will not detail how these six equations were discovered, and we will only show, in Chap. 3, how to integrate given nonlinear ODEs with: linear ODEs, the elliptic equation, the six Painlevé equations.

A frequent question is (see the FAQ section), by which miracle these six functions, discovered from purely mathematical considerations, occur so widely in physics. The answer should now be evident: they are just functions defined by an ODE (like the exponential function is defined by the ODE $u' - u = 0$) and, as soon as the ODE governing the physical system possesses some singlevaluedness, the elliptic and Painlevé functions are likely to contribute to the corresponding singlevalued expression. A famous example is the two-dimensional Ising model, in which the two-point correlation function is a P6 function of the temperature [239] for parameter values of $\alpha, \beta, \gamma, \delta$ identical to those of the case of Picard (Appendix formula (B.3)) *modulo* transformations (B.15).

There is nothing more behind all this, and nothing less, than the explicit integration of nonlinear ODEs.

[5] Without loss of generality, one can assume $\delta(\delta-1) = 0$ in P5, $\gamma(\gamma-1) = 0, \delta(\delta-1) = 0$ in P3. See Appendix B.4.

Chapter 2
Singularity Analysis: Painlevé Test

Abstract In this chapter, we present the Painlevé test on various examples of nonlinear ODEs,

$$E(x,u^{(N)},...,u',u) = 0, \; ' = \frac{\mathrm{d}}{\mathrm{d}x}, \tag{2.1}$$

This is a local analysis which can be implemented as an algorithm to provide necessary conditions for the Painlevé property.
This method is historically due to Sophie (Sonya) Kowalevski and Bertrand Gambier, with many elements already in Hoyer [227][1]. In the motion of a rigid body around a fixed point, Kowalevski required the general solution to be a single valued function of time not only on the real axis but also in the complex plane and, by applying only a subset of the method which we are going to describe, she isolated a fourth case of possible singlevaluedness (the "Kowalevski case") which she succeeded to explicitly integrate [257], thus constructively proving the singlevaluedness. Gambier developed a direction initiated by Appelrot and made the method algorithmic. This method was later rediscovered by Ablowitz, Ramani and Segur [5]. As to Painlevé, he himself never used "le procédé connu de Madame Kowalevski ... dont le caractère nécessaire n'était pas établi" [349, pp. 10,83] [351, pp. 196,269]. We will not present in this book the α method of Painlevé [351] because it is more difficult to implement.

2.1 Kowalevski–Gambier Method

We present here its main subset, which consists in checking the existence of all possible local representations, near a movable singularity $x - x_0 = 0$, of the general solution as a locally single valued expression, e.g. the Laurent series

[1] Paul Hoyer and Sophie Kowalevski were both students of Weierstrass.

$$u = \sum_{j=0}^{+\infty} u_j \chi^{j+p},\ -p \in \mathcal{N},\ E = \sum_{j=0}^{+\infty} E_j \chi^{j+q},\ -q \in \mathcal{N}, \tag{2.2}$$

with coefficients u_j, E_j independent of the expansion variable χ. In case the ODE explicitly depends on x, the computed coefficients will depend on x_0; as noticed by Gambier [163, p. 50], this can be avoided by defining χ by the property $\chi' = 1$ rather than $\chi = x - x_0$, and the computed coefficients then depend on x. The reader interested in the full, detailed version of the test can refer to [76, p. 160].

2.1.1 Lorenz Model

As a first example, let us consider the system[2],

$$\frac{\mathrm{d}x}{\mathrm{d}t} = \sigma(y-x),\ \frac{\mathrm{d}y}{\mathrm{d}t} = rx - y - xz,\ \frac{\mathrm{d}z}{\mathrm{d}t} = xy - bz. \tag{2.3}$$

In the *first step*, as already explained in Sect. 1.1.2, one has to determine all the possible families of movable singularities, i.e. all the leading behaviors

$$\chi \to 0:\ x \sim x_0 \chi^{p_1}, y \sim y_0 \chi^{p_2}, z \sim z_0 \chi^{p_3},\ (x_0, y_0, z_0) \neq (0,0,0),\ \chi' = 1, \tag{2.4}$$

with at least one p_j not a positive integer. The various terms contribute as

$$\left|\begin{matrix} -x' \\ p_1 - 1 \end{matrix}\right|\begin{matrix} \sigma y \\ p_2 \end{matrix}\left|\begin{matrix} -\sigma x \\ p_1 \end{matrix}\right|\left|\begin{matrix} -y' \\ p_2 - 1 \end{matrix}\right|\begin{matrix} rx \\ p_1 \end{matrix}\left|\begin{matrix} -y \\ p_2 \end{matrix}\right|\begin{matrix} -xz \\ p_1 + p_3 \end{matrix}\left|\left|\begin{matrix} -z' \\ p_3 - 1 \end{matrix}\right|\begin{matrix} xy \\ p_1 + p_2 \end{matrix}\right|\begin{matrix} -bz \\ p_3 \end{matrix}\right|$$

the terms $-\sigma x$, $-y$ and $-bz$ are less singular than, respectively, x', y', z', and one obtains the unique possibility

$$\begin{cases} q_1 = p_1 - 1 = p_2, \\ q_2 = p_2 - 1 = p_1 + p_3, \\ q_3 = p_3 - 1 = p_1 + p_2, \end{cases} \mathrm{E}_0 = \begin{cases} -p_1 x_0 + \sigma y_0 = 0, \\ -p_2 y_0 - x_0 z_0 = 0, \\ -p_3 z_0 + x_0 y_0 = 0, \end{cases} \tag{2.5}$$

This defines two families,

$$\mathbf{p} = (-1, -2, -2),\ \mathbf{q} = (-2, -3, -3), \tag{2.6}$$

$$x_0 = 2i,\ y_0 = -2i\sigma^{-1},\ z_0 = -2\sigma^{-1},\ i^2 = -1. \tag{2.7}$$

A first necessary condition for the PP is that all components of $\mathbf{p}$ be integer, and in the present example it is satisfied.

After completion of this first step, one must check, for each family, whether it is possible to compute all the coefficients $u_j, j \geq 1$ of the Laurent series (2.2). The

[2] One could equivalently process the scalar third order equation (2.24), and this would yield the same results. However, as a general rule, any elimination increases the volume of expressions.

recursion relation for computing the j-th coefficient (x_j, y_j, z_j) of the Laurent series takes the form, in matrix notation,

$$\mathbf{P}_j \begin{pmatrix} x_j \\ y_j \\ z_j \end{pmatrix} + \mathbf{Q}_j = 0, \tag{2.8}$$

in which $\mathbf{P}_j$ is a square matrix depending only on j and the leading behavior $\mathbf{p}, x_0, y_0, z_0$,

$$\mathbf{P}_j = \begin{pmatrix} -j-p_1 & \sigma & 0 \\ -z_0 & -j-p_2 & -x_0 \\ y_0 & x_0 & -j-p_3 \end{pmatrix} = \begin{pmatrix} -j+1 & \sigma & 0 \\ 2\sigma^{-1} & -j+2 & -2i \\ -2i\sigma^{-1} & 2i & -j+2 \end{pmatrix}, \tag{2.9}$$

and the column vector $\mathbf{Q}_j$ depends on all the previously computed coefficients $(x_l, y_l, z_l), l = 0, \ldots, j-1$,

$$\mathbf{Q}_j = \begin{pmatrix} -\sigma x_{j-1} \\ -\sum_{k=1}^{j-1} x_k z_{j-k} + r x_{j-1} - y_{j-1} \\ \sum_{k=1}^{j-1} x_k y_{j-k} - b z_{j-1} \end{pmatrix}. \tag{2.10}$$

Therefore solving the recursion relation is a *linear* algebra problem, and three possibilities can occur.

1. If j is not a zero of the determinant of $\mathbf{P}_j$, the value of (x_j, y_j, z_j) is uniquely determined;
2. If j is a zero of the determinant of $\mathbf{P}_j$ and if the vector $\mathbf{Q}_j$ is orthogonal to the kernel of the transpose of $\mathbf{P}_j$, the solution (x_j, y_j, z_j) exists and it depends on as many (at least one) arbitrary coefficients as the dimension of the kernel;
3. If j is a zero of the determinant of $\mathbf{P}_j$ and if the vector $\mathbf{Q}_j$ is not orthogonal to the kernel of the transpose of $\mathbf{P}_j$, the Laurent series does not exist and the test fails.

The *second step* is to directly determine the locations j at which arbitrary coefficients may enter the Laurent series. These values j, the zeros of the determinant of $\mathbf{P}_j$, are computed as follows. Considering the *dominant terms* (subset of the terms which define the dominant behavior),

$$\hat{E}(x,y,z) \equiv \begin{cases} -x' + \sigma y \\ -y' - xz \\ -z' + xy, \end{cases} \tag{2.11}$$

one computes their derivative (i.e. the linearized system near the point $(x = x_0\chi^{p_1}, y = y_0\chi^{p_2}, z = z_0\chi^{p_3})$)

$$\lim_{\varepsilon \to 0} \frac{\hat{E}(x+\varepsilon X, y+\varepsilon Y, z+\varepsilon Z) - \hat{E}(x,y,z)}{\varepsilon} \equiv \begin{cases} -X' + \sigma Y \\ -Y' - zX - xZ \\ -Z' + yX + xY. \end{cases} \tag{2.12}$$

Near $\chi = 0$, this weighted homogeneous linear system for (X, Y, Z) admits a power law solution

$$X \sim X_j \chi^{j+p_1},\ Y \sim Y_j \chi^{j+p_2},\ Z \sim Z_j \chi^{j+p_3}, \tag{2.13}$$

and the coefficients of the lowest powers of χ in the linear system (2.12) evaluate to

$$\begin{cases} -(j+p_1)X_j + \sigma Y_j = 0, \\ -(j+p_2)Y_j - z_0 X_j - x_0 Z_j = 0, \\ -(j+p_3)Z_j + y_0 X_j + x_0 Y_j = 0. \end{cases} \tag{2.14}$$

Expressing that the solution (X_j, Y_j, Z_j) is nonzero results in the vanishing of the determinant of the matrix $\mathbf{P}_j$ of the system, see (2.9),

$$\det \mathbf{P}_j = -(j+1)(j-2)(j-4) = 0. \tag{2.15}$$

This equation (2.15) is called the *indicial equation* [232, Chap. XV], its roots are called *Fuchs indices* of the linear system (2.12) and, by a slight abuse of language, of the nonlinear system (2.3). For any ODE, the number of Fuchs indices is at most equal to the differential order of the ODE.

The value $j = -1$ seems unable to introduce an arbitrary constant since it is not a positive integer, but it represents in fact the arbitrary constant t_0. This results from a theorem of Poincaré: given a solution of the nonlinear system defined by the dominant terms (2.11), in the present case

$$(x, y, z) = (2i\chi^{-1}, -2i\sigma^{-1}\chi^{-2}, -2\sigma^{-1}\chi^{-2}),\ \chi = t - t_0, \tag{2.16}$$

its derivative with respect to any arbitrary constant in it (here t_0) is a solution of the linearized system (2.12), here

$$(X, Y, Z) = \partial_{t_0}(x, y, z) = (-2i\chi^{-2}, 4i\sigma^{-1}\chi^{-3}, 4\sigma^{-1}\chi^{-3}), \tag{2.17}$$

which indeed corresponds to $j = -1$ in (2.13). Therefore, for any family of any nonlinear ODE (and even PDE), the value -1 is always a Fuchs index.

The *third step* is to check whether or not, at each Fuchs index, the orthogonality condition is satisfied.

At $j = 1$, one finds

$$x_1 = (3\sigma - 2b - 1)i/3,\ y_1 = 2i,\ z_1 = 2\sigma^{-1}(3\sigma - b + 1), \tag{2.18}$$

At $j = 2$, the recursion relation admits a solution only under the condition

$$Q_2 \equiv (b - 2\sigma)(b + 3\sigma - 1) = 0, \tag{2.19}$$

in which case this solution is

$$x_2 = \text{ arbitrary},\ y_2 = \sigma^{-1}x_2 + (3\sigma - 2b - 1)i/3,$$

$$z_2 = -i\sigma^{-1}x_2 - \sigma^{-1}(3\sigma - 2b - 1)(3\sigma - b + 1)/9 + r - 1. \tag{2.20}$$

At $j = 4$, one similarly introduces an arbitrary coefficient x_4 under the single condition

$$\begin{aligned} Q_4 \equiv\ & -4i(b - \sigma - 1)(b - 6\sigma + 2)x_2 - (4/3)(b - 3\sigma + 5)b\sigma r \\ & +(-4 + 10b + 30b^2 - 20b^3 - 16b^4)/27 \\ & +(-38b - 56b^2 - (28/3)b^3 + 88\sigma + 86b^2\sigma)\sigma/3 \\ & -32\sigma/9 + 70b\sigma^2 - 64\sigma^3 - 58b\sigma^3 + 36\sigma^4 = 0, \end{aligned} \tag{2.21}$$

which splits into two conditions since x_2 must remain arbitrary. The third step terminates here, since no obstruction can occur above the highest Fuchs index $j = 4$, and this terminates the Painlevé test.

What is the conclusion of the test? Only if the *general solution* has been represented can some conclusion be drawn.

If the three generated conditions on (b, σ, r) are satisfied, there exists a Laurent series depending on three arbitrary coefficients (t_0, x_2, x_4), so one has represented the general solution by a locally singlevalued expression, and one says that the Lorenz model passes the Painlevé test. However, this does *not* imply that the model has the Painlevé property.

The three generated conditions on (b, σ, r) are called for brevity *no-log conditions*, because, if at least one of them is not satisfied, a local representation of the general solution still exists, which is a double series in χ and $\log \chi$ (see [280] for the Lorenz model), called a *psi-series*.

Unless the three generated conditions on (b, σ, r) are satisfied, the Laurent series with the leading behavior (2.7) does not exist and the test fails.

The three conditions admit only four solutions

$$(b, \sigma, r) = (1, 1/2, 0), (2, 1, 1/9), (0, 1/3, r), (1, 0, r), \tag{2.22}$$

which will be further examined in Sect. 3.1.1 from the point of view of their explicit integration.

The fourth solution ($\sigma = 0$) should not be rejected on the ground that the system (2.3) becomes linear. Indeed, there exists a scaling law in the model, and the system rewritten for $(x, \sigma y, \sigma z)$ has no such restriction. An easy way to remove it is to eliminate y and z and to consider the single third order ODE for $x(t)$ [391],

$$y = x + x'/\sigma,\ z = r - 1 - [(\sigma + 1)x' + x'']/(\sigma x), \tag{2.23}$$

$$\begin{aligned} & xx''' - x'x'' + x^3x' + \sigma x^4 + (b + \sigma + 1)xx'' + (\sigma + 1)(bxx' - x'^2) \\ & +b(1 - r)\sigma x^2 = 0. \end{aligned} \tag{2.24}$$

This ODE has exactly the same no-log conditions as the dynamical system.

2.1.2 Kuramoto–Sivashinsky (KS) Equation

Since the KS equation,

$$u_t + \nu u_{xxxx} + b u_{xxx} + \mu u_{xx} + u u_x = 0,\ \nu \neq 0, \tag{2.25}$$

is invariant under the Galilean transformation $(u,x,t) \to (u+c, x-ct, t)$, its traveling wave reduction is defined as

$$\begin{aligned} &u(x,t) = c + U(\xi),\ \xi = x - ct,\\ &\nu U'''' + bU''' + \mu U'' + UU' = 0,\ (\nu, b, \mu) \in \mathscr{R},\ \nu \neq 0, \end{aligned} \tag{2.26}$$

which integrates once as (renaming U as u for consistency with (2.2))

$$\nu u''' + b u'' + \mu u' + \frac{u^2}{2} + A = 0, \tag{2.27}$$

in which A is the integration constant. It has a chaotic behavior [296], and it depends on two dimensionless parameters, $b^2/(\mu\nu)$ and $\nu A/\mu^3$.

We follow the steps in Sect. 2.1.1. In the *first step*, one looks for all the singular dominant behaviors

$$\chi \to 0:\ u \sim u_0 \chi^p,\ u_0 \neq 0,\ \chi' = 1, \tag{2.28}$$

with p not a positive integer. The terms u'' and u' are less singular than u''', the term A is regular and cannot contribute, and so the dominant behavior is governed by the *dominant terms* $\hat{E}$

$$\hat{E}(u) \equiv \nu u''' + u^2/2, \tag{2.29}$$

which contribute the terms

$$\nu p(p-1)(p-2) u_0 \chi^{p-3} + (1/2)(u_0 \chi^p)^2, \tag{2.30}$$

and, since u_0 is nonzero, generate the two conditions

$$p - 3 = 2p,\ \nu p(p-1)(p-2) u_0 + (1/2) u_0^2 = 0,\ u_0 \neq 0. \tag{2.31}$$

Their unique solution is $p = -3, u_0 = 120\nu$, and the common value of the two powers is $q = p - 3 = 2p = -6$.

In the *second step*, for each dominant behavior, the indicial equation is computed as already indicated in Sect. 2.1.1 One builds the linearized equation near the solution which behaves like $u_0 \chi^p$

$$\lim_{\varepsilon \to 0} \frac{\hat{E}(u + \varepsilon w) - \hat{E}(u)}{\varepsilon} = (\nu \partial_x^3 + (u_0 \chi^p + \ldots)) w = 0, \tag{2.32}$$

then one requires this linear ODE for w to have a nonidentically zero power law solution $w = W\chi^{j+p}$ near $\chi = 0$,

$$\lim_{\chi\to 0} \chi^{-j-q}(\nu\partial_x^3 + u_0\chi^p)\chi^{j+p} \tag{2.33}$$

$$= \nu(j-3)(j-4)(j-5) + 120\nu = \nu(j+1)(j^2-13j+60) \tag{2.34}$$

$$= \nu(j+1)\left(j - \frac{13+i\sqrt{71}}{2}\right)\left(j - \frac{13-i\sqrt{71}}{2}\right) = 0. \tag{2.35}$$

The *third step* (to check the possibility to compute the sequence of coefficients u_j of the Laurent series for u) concludes without computation to the existence of the series, in which however two of the three arbitrary constants are missing.

In this example, the test as presented in Sect. 2.1.1 is indecisive since the obtained Laurent series does not represent the general solution, just a particular solution. In order to test the missing part of the solution for possible multivaluedness, one must undertake a perturbation [82], as follows. Denoting $u^{(0)}$ the particular solution computed above,

$$u^{(0)} = 120\nu\chi^{-3} - 15b\chi^{-2} + \frac{15(16\mu\nu - b^2)}{4\times 19\nu}\chi^{-1} + \frac{13(4\mu\nu - b^2)b}{32\times 19\nu^2} + O(\chi), \tag{2.36}$$

one defines another solution u by the Taylor series

$$u = u^{(0)} + \varepsilon u^{(1)} + \varepsilon^2 u^{(2)} + \dots, \tag{2.37}$$

in which the small parameter ε is not in the ODE (2.27). This splits the nonlinear ODE (2.27) according to the powers of ε into an infinite sequence

$$E(u) = \sum_{n=0}^{+\infty} \varepsilon^n E_n(u^{(0)}, \dots, u^{(n)}), \tag{2.38}$$

$$\forall n: \ E_n = 0, \tag{2.39}$$

and the additional requirement of this perturbative test is that each $u^{(n)}$ should be single valued near $\chi = 0$. The equation E_0 is just the nonlinear equation (2.27), so one must proceed to $n = 1$. Equation $E_1 = 0$ is nothing but the linearized equation around $u^{(0)}$

$$\left(\nu\frac{\mathrm{d}^3}{\mathrm{d}x^3} + b\frac{\mathrm{d}^2}{\mathrm{d}x^2} + \mu\frac{\mathrm{d}}{\mathrm{d}x} + u^{(0)}\right)u^{(1)} = 0, \tag{2.40}$$

therefore its indicial equation is identical to that already computed, see (2.35) and, according to classical results on linear ODEs of the Fuchsian type [232, Chap. XV], it admits three linearly independent solutions

$$u^{(1)} = c_j\chi^{j+p}\mathrm{Regular}(\chi), \tag{2.41}$$

in which j is a Fuchs index and "Regular" denotes a converging Taylor series of χ, so the local (near $\varepsilon = 0$ and $\chi = 0$) representation of u is now

$$\begin{aligned} u(x_0, \varepsilon c_+, \varepsilon c_-) = 120\nu\chi^{-3}\{&\text{Regular}(\chi) \\ &+\varepsilon[c_{-1}\chi^{-1}\text{Regular}(\chi) \\ &+ c_+\chi^{(13+i\sqrt{71})/2}\text{Regular}(\chi) \\ &+ c_-\chi^{(13-i\sqrt{71})/2}\text{Regular}(\chi)] + \mathcal{O}(\varepsilon^2)\}. \end{aligned} \tag{2.42}$$

The contribution c_{-1} is irrelevant, since, according to a result of Poincaré, the Fuchs index -1 only represents the constant x_0, already present in the unperturbed solution $u^{(0)}$, and at this perturbation order $n = 1$ the obtained solution contains three arbitrary constants $(x_0, \varepsilon c_+, \varepsilon c_-)$ and now locally represents the general solution. The test terminates here, with a conclusion of failure due to the presence of irrational Fuchs indices.

In fact, the test could have been stopped with the same conclusion immediately after having established the indices, formula (2.35), because all indices but -1 are irrational.

An important result, to be used in Sect. 3.2.5.2, is the existence of the converging Laurent series (2.36) depending on the movable constant x_0.

The ODE (2.27) admits other Laurent series in the variable $(u - \sqrt{-2A})^{-1}$, but they provide no additional information.

Remark. If one would consider the Taylor series whose existence is stated by the famous theorem of existence of Cauchy,

$$U = U_0 + \chi U_0' + \frac{1}{2}\chi^2 U_0'' + \frac{1}{6}\chi^3\left(-bU_0'' - \mu U_0' - \frac{U_0^2}{2} - A\right) + \dots, \tag{2.43}$$

one could never distinguish between chaos and integrability ("les méthodes dérivées de la doctrine de Cauchy ne semblaient pas susceptibles de répondre" [349, p. 6]). This feature displays the superiority of the Laurent series (which has a singularity) over the Taylor one (which has no singularity inside the disk of convergence).

Finally, although this is not part of the Painlevé test, let us introduce here the quite useful notion of *singular part operator* $\mathcal{D}$, which generalizes the logarithmic derivative operator already presented in Sect. 1.1. This operator is defined as the linear differential operator which represents all the singular terms of the Laurent series (2.36),

$$\begin{aligned} &u - \mathcal{D}\log(\xi - \xi_0) = \mathcal{O}(1), \\ &\mathcal{D} = 60\nu\partial_\xi^3 + 15b\partial_\xi^2 + \frac{15(16\mu\nu - b^2)}{76\nu}\partial_\xi. \end{aligned} \tag{2.44}$$

2.1.3 Cubic Complex Ginzburg–Landau (CGL3) Equation

The complex equation[3]

$$iA_t + pA_{xx} + q|A|^2A - i\gamma A = 0,\ pq\gamma \neq 0,\ (A,p,q) \in \mathscr{C},\ \gamma \in \mathscr{R}. \quad (2.45)$$

admits a scaling limit [362, 275] under which the variable $u = \arg A$ obeys the KS PDE (2.25), this is the reason why the KS equation is often called the *phase equation*. We will consider the two-component system [64]

$$\begin{cases} iA_t + pA_{xx} + qA^2B - i\gamma A = 0, \\ -iB_t + \bar{p}B_{xx} + \bar{q}B^2A + i\gamma B = 0, \end{cases} \quad (2.46)$$

in which B is not assumed to be the complex conjugate of A. Indeed, for $\gamma = 0$ and p,q real, this sytem (2.46), then known as the AKNS system [3], is IST-integrable whatever be (A,B), and the nonlinear Schrödinger equation (NLS),

$$iA_t + pA_{xx} + q|A|^2A = 0,\ pq \neq 0,\ A \in \mathscr{C},\ (p,q) \in \mathscr{R}. \quad (2.47)$$

is just one of its reductions, namely the complex conjugate reduction $B = \overline{A}$.

Since the Painlevé test for PDEs will only be defined in Chap. 4, let us process the traveling wave reduction of the system (2.46),

$$A(x,t) \to e^{i\omega t}A(\xi),\ B(x,t) \to e^{-i\omega t}B(\xi),\ \xi = x - ct, \quad (2.48)$$

and rename ξ as x, which defines an ODE system for $A(x), B(x)$. Fortunately, this will make no difference to the result of the test.

Let us denote the dominant behavior as

$$A \sim A_0\chi^{p_1},\ B \sim B_0\chi^{p_2},\ \chi' = 1,\ ' = \frac{\mathrm{d}}{\mathrm{d}x} \quad (2.49)$$

in which (A_0, B_0, p_1, p_2) are complex constants to be determined. Since the terms $iA_t, -i\gamma A, -iB_t, i\gamma B$ are less singular, the dominant terms $\hat{E}$ are

$$\hat{E}(A,B) \equiv \left(pA_{xx} + qA^2B,\ \bar{p}B_{xx} + \bar{q}B^2A\right), \quad (2.50)$$

they contribute to the following powers,

pA_{xx}	qA^2B	$\overline{p}B_{xx}$	$+\overline{q}B^2A$
$p_1 - 2$	$2p_1 + p_2$	$p_2 - 2$	$p_1 + 2p_2$

and generate the set of equations

[3] Only the imaginary part of γ can be absorbed in the definition of A, not the real part. As to the contribution of a group velocity term $-ivA_x$, it can be absorbed in the definition of x since v is real.

$$\begin{cases} p_1-2=2p_1+p_2, \; pp_1(p_1-1)+qA_0B_0=0, \\ p_2-2=p_1+2p_2, \; \overline{p}p_2(p_2-1)+\overline{q}A_0B_0=0. \end{cases} \quad (2.51)$$

It is convenient to introduce the six real parameters $d_r, d_i, s_r, s_i, g_r, g_i$,

$$d_r+id_i=\frac{q}{p}, \; s_r-is_i=\frac{1}{p}, \; g_r+ig_i=\frac{\gamma+i\omega}{p}+\frac{1}{2}c^2s_rs_i+\frac{i}{4}c^2s_r^2. \quad (2.52)$$

In the CGL3 case $d_i \neq 0$, the equations (2.51) admit two solutions for (p_1, p_2, A_0B_0) [53]

$$\begin{aligned} &p_1=-1+i\alpha, \; p_2=-1-i\alpha, \\ &A_0B_0=\frac{3(3d_r+\varepsilon_1\Delta)}{2d_i^2}, \; \alpha=\frac{3d_r+\varepsilon_1\Delta}{2d_i}, \; \Delta=\sqrt{9d_r^2+8d_i^2}, \; \varepsilon_1^2=1. \end{aligned} \quad (2.53)$$

In the case $d_i=0$, they admit the unique solution

$$p_1=-1, \; p_2=-1, \; A_0B_0=-2\frac{p}{q}, \quad (2.54)$$

which corresponds to the limit $d_i \to 0, \alpha \to 0, \varepsilon_1=-\operatorname{sign}(d_r)$ in (2.53).

The CGL3 equation therefore admits two families for AB (and four for $\sqrt{AB}$), and the noninteger values for the leading powers p_1, p_2 imply the failure of the test for the CGL3 equation when written in the variables (A, B). However, when written in other variables, e.g. $(AB, \operatorname{grad}\log(A/B))$, the CGL3 equation is still algebraic and the leading powers become integer ($-2, -1$ in the just mentioned variables). Therefore, at this stage of the leading powers, one should declare failure only if there exists no algebraic writing with integer leading powers.

Let us proceed to the second step. The indicial equation is the determinant of the second order matrix

$$\begin{aligned} \mathbf{P}_j = &\begin{pmatrix} 2qA_0B_0 & qA_0^2 \\ \overline{q}B_0^2 & 2\overline{q}A_0B_0 \end{pmatrix} \\ &+\operatorname{diag}(p(j-1+i\alpha)(j-2+i\alpha), \overline{p}(j-1-i\alpha)(j-2-i\alpha). \end{aligned} \quad (2.55)$$

If one replaced in this determinant A_0B_0 and α with their algebraic values (2.53), the resulting algebraic expression in $j, d_r, d_i, \varepsilon_i$ would look quite ugly and the roots j would be difficult to obtain. What should be done instead [53, 86, 90] is to solve the system (2.51) as a *linear* system on $\mathscr{C}$, which is the case for the unknowns $(A_0B_0, \overline{p}, \overline{q})$,

$$\begin{cases} A_0B_0=-\dfrac{p}{q}(1-i\alpha)(2-i\alpha), \\ \overline{p}=Kp(1-i\alpha)(2-i\alpha), \; \overline{q}=Kq(1+i\alpha)(2+i\alpha), \end{cases} \quad (2.56)$$

in which K is an irrelevant arbitrary nonzero complex constant. With such a resolution, in the determinant the variables (p, q, K) are factored out and the result only depends on j and the real variable α,

$$|p|^{-2}\det \mathbf{P}_j = (j+1)j(j^2-7j+6\alpha^2+12) = 0, \tag{2.57}$$

yielding the set of Fuchs indices [53],

$$j = -1, 0, \frac{7+\sqrt{1-24\alpha^2}}{2}, \frac{7-\sqrt{1-24\alpha^2}}{2}. \tag{2.58}$$

For each of the two families, and for generic values of (p,q), two of the four indices are irrational, and this fact cannot be changed by considering other variables like $AB, \mathrm{grad}\log(A/B)$. The conclusion is that, for the same reason as in the KS equation, the stationary reduction of the CGL3 PDE fails the test.

The condition that the four irrational Fuchs indices be integer (such a condition, that numbers should be integer, is called a *diophantine condition*) constrain p,q by the condition $\mathrm{Im}(p/q) = 0$, i.e. $d_i = 0, \alpha = 0$. Let us then compute the no-log conditions arising at the indices $0,3,4$. With a rescaling of A and B, one can assume p and q to be real. At $j = 0$, one arbitrary constant is introduced, the ratio A_0/B_0. At $j = 3$, one can similarly introduce an arbitrary constant $A_3 = -B_3$, but at $j = 4$, the two recursion relations are [249],

$$A_4 - B_4 = 0, p^{-2}\gamma^2 = 0, \tag{2.59}$$

and only when γ vanishes can the general solution be represented as a locally singlevalued expression, namely

$$A = \sqrt{-2\frac{p}{q}}e^{i\varphi_0}\chi^{-1}\left(1+c_3\chi^3+c_4\chi^4+O(\chi^5)\right),\ \chi = x - x_0, \tag{2.60}$$

$$B = \sqrt{-2\frac{p}{q}}e^{-i\varphi_0}\chi^{-1}\left(1-c_3\chi^3+c_4\chi^4+O(\chi^5)\right), \tag{2.61}$$

the four arbitrary constants being x_0, φ_0, c_3, c_4, associated to the respective indices $-1,0,3,4$.

The conclusion is that, for the stationary reduction $A_t = 0, B_t = 0$, the CGL3 equation for (A,B) passes the Painlevé test if and only if p/q is real and γ vanishes. As will be seen in Chap. 4, this conclusion would be unchanged by applying the test directly to the CGL3 PDE, therefore the CGL3 PDE passes the test iff it reduces to the NLS equation.

2.1.4 Duffing–van der Pol Oscillator

The van der Pol oscillator

$$u'' + (au^2+b)u' - cu = 0 \tag{2.62}$$

was designed by van der Pol in 1922 to explain the triode oscillator, the friction term au^2 making the energy finite when b is negative. This model is famous [28] because, for $a > 0, b < 0, c < 0$, one can prove the existence, uniqueness and stability of a special closed trajectory called a limit cycle by Poincaré.

If one includes the next nonlinearities, the equation becomes

$$E(u) \equiv u'' + (au^2 + b)u' - cu + du^2 + \beta u^3 = 0, \tag{2.63}$$

and, like (2.62), it admits the two families [38]

$$u \sim u_0 \chi^{-1/2},\ u_0^2 = \frac{3}{2a},\ \text{Fuchs indices} = -1, \frac{3}{2}. \tag{2.64}$$

The Fuchs index $3/2$ is not integer but it is not an irrational number like for KS or CGL3. In such a case (positive rational number), a Laurent series, which has only integer powers, cannot represent the general solution. Such a representation is achieved [369] by a generalization of the Laurent series called the *Puiseux series* [207] containing integer powers of $\chi^{1/2}$,

$$u = \sum_{2j=0}^{+\infty} u_j \chi^{j-1/2} = u_0 \chi^{-1/2} + u_{1/2} + u_1 \chi^{1/2} + u_{3/2} \chi^1 + \dots, \tag{2.65}$$

$$E(u) = \sum_{2j=0}^{+\infty} E_j \chi^{j-5/2},\ \forall j:\ E_j = 0, \tag{2.66}$$

in which one must check whether the coefficient $u_{3/2}$ is arbitrary or not. The algebraic equations for u_j are

$$\begin{cases} E_0 \equiv \dfrac{u_0}{4}(3 - 2au_0^2) = 0, \\ E_{1/2} \equiv -au_0^2 u_{1/2} = 0, \\ E_1 \equiv -\dfrac{1 + 2au_0^2}{4} u_1 - \dfrac{a}{2} u_0 u_{1/2}^2 - \dfrac{b}{2} u_0 + \beta u_0^3 = 0, \\ E_{3/2} \equiv 3\beta u_0^2 u_{1/2} + du_0^2 = 0, \end{cases} \tag{2.67}$$

their resolution is quite similar to the Laurent case, i.e. it must be done by increasing values of j. At the Fuchs index $j = 3/2$, under penalty of a movable logarithm, the coefficient d must be set to zero, resulting in the series

$$u = \chi^{-1/2} \left(u_0 + \frac{3\beta - ab}{2a} u_0 \chi + u_{3/2} \chi^{3/2} + \dots \right), \tag{2.68}$$

in which the coefficient $u_{3/2}$ is arbitrary. The ODE (2.63) with $d = 0$ (usually called the Duffing–van der Pol oscillator) passes a weaker form of the Painlevé test called the *weak Painlevé test* [369, 180], and the associated global property which the ODE may possess is a weaker form of the Painlevé property: around any movable singularity, the general solution is required to take only a finite number of determinations, this number being prescribed in advance or not. This weaker form was studied in detail by Painlevé in [346, Leçons 5–10,13,19].

Remark. Despite the invariance of the Duffing–van der Pol oscillator under $u \to -u$, the ODE for u^2 still involves a Puiseux series instead of a Laurent series, because the values of the Fuchs indices $(-1, 3/2)$ are invariant under the change $u \to u^2$.

The integration of this Duffing–van der Pol oscillator is examined in Sect. 3.2.6.

2.1.5 Hénon–Heiles System

This cubic Hamiltonian system is defined by

$$H = \frac{1}{2}(p_1^2 + p_2^2 + \omega_1 q_1^2 + \omega_2 q_2^2) + \alpha q_1 q_2^2 - \frac{1}{3}\beta q_1^3 + \frac{c_3}{2q_2^2}, \ \alpha \neq 0, \tag{2.69}$$

$$q_1'' + \omega_1 q_1 - \beta q_1^2 + \alpha q_2^2 = 0, \tag{2.70}$$

$$q_2'' + \omega_2 q_2 + 2\alpha q_1 q_2 - \frac{c_3}{q_2^3} = 0. \tag{2.71}$$

Before analyzing the system (2.70)–(2.71), let us give some background. Taking the most general two-degree of freedom, classical (as opposed to quantum), time-independent Hamiltonian of the physical type (i.e. the sum of a kinetic energy and a potential energy),

$$H = \frac{1}{2}(p_1^2 + p_2^2) + V(q_1, q_2), \tag{2.72}$$

let us determine all the potentials $V(q_1, q_2)$ such that some unspecified integer power $q_j^{n_j}(t)$ of the general solution is a single valued function of the complex time t.

In the case of one degree of freedom, with the additional requirement that V is rational, this problem only admits two solutions,

$$n = \pm 1, \ H = \frac{p^2}{2} + \sum_{j=1}^{4} c_j q^j, \tag{2.73}$$

$$n = \pm 2, \ H = \frac{p^2}{2} + a q^{-2} + c_2 q^2 + c_4 q^4 + c_6 q^6. \tag{2.74}$$

In both cases, q^n is an elliptic function and, in the second case, q is generically multivalued.

The equations of motion for $q_j(t)$ are obtained by eliminating the momenta p_1, p_2 between the Hamilton's equations of motion,

$$\frac{\mathrm{d}p_j}{\mathrm{d}t} = -\frac{\partial V(q_1, q_2)}{\partial q_j}, \ \frac{\mathrm{d}q_j}{\mathrm{d}t} = q_j' = p_j, \ j = 1, 2 \tag{2.75}$$

which results in the system of two coupled second order ordinary differential equations

$$q_j'' + \frac{\partial V(q_1,q_2)}{\partial q_j} = 0,\ j = 1,2 \tag{2.76}$$

together with the first integral

$$H \equiv \frac{q_1'^2}{2} + \frac{q_2'^2}{2} + V(q_1,q_2) = E. \tag{2.77}$$

Following basic principles stated by Painlevé [348] which are outside the scope of this volume, the Painlevé test preselects [97] a finite number of admissible potentials V, depending on a finite number of arbitrary constants. Two of these preselected potentials are in addition rational in q_1,q_2, these are the *cubic Hénon–Heiles Hamiltonians* (HH3) (2.69) and the *quartic Hénon–Heiles Hamiltonians* (HH4) (2.78),

$$\begin{aligned} H = {} & \frac{1}{2}(P_1^2 + P_2^2 + \Omega_1 Q_1^2 + \Omega_2 Q_2^2) + CQ_1^4 + BQ_1^2 Q_2^2 + AQ_2^4 \\ & + \frac{1}{2}\left(\frac{\alpha}{Q_1^2} + \frac{\beta}{Q_2^2}\right) + \gamma Q_1,\ B \neq 0, \end{aligned} \tag{2.78}$$

$$Q_1'' + \Omega_1 Q_1 + 4CQ_1^3 + 2BQ_1 Q_2^2 - \alpha Q_1^{-3} + \gamma = 0, \tag{2.79}$$

$$Q_2'' + \Omega_2 Q_2 + 4AQ_2^3 + 2BQ_2 Q_1^2 - \beta Q_2^{-3} = 0, \tag{2.80}$$

so named according to their global degree in (q_1,q_2).

Finally, the "usual" Painlevé test (i.e. the subset introduced in previous sections) selects seven potentials, three cubic [58, 147, 82] and four quartic [369, 179].

These seven potentials were initially isolated by the condition that a special second integral of the motion should exist [204]. The difference between the two approaches is worth being emphasized: requiring singlevaluedness generates *necessary* conditions on $V(q_1,q_2)$, while requiring the existence of a second integral of motion only results in *sufficient* conditions since one must give as an input [204] the form of the additional first integral (e.g. quartic in the momenta).

In the present section, we perform the Painlevé test on the cubic Hamiltonian (2.69), more precisely on the system made of the two equations of the motion (2.70)–(2.71), in order to select the admissible values of the constants in (2.69).

Analyzing a system like (2.70)–(2.71) requires the introduction of several leading behaviors, here one for q_1 and one for q_2. This typical situation for systems may lead to the useless consideration of apparently different cases, implying a waste of time and effort. This is indeed the case here since, as the interested reader can check, q_1 admits the only leading power -2, while q_2 admits the three leading powers $2, -3/2, 4$ (the fractional power is acceptable since the system is still algebraic in the variables (q_1, q_2^2) which only admit integer leading powers). It is elementary to eliminate either q_2^2 from (2.70) or q_1 from (2.71), but the resulting fourth order ODE for q_1 can be written in a much simpler form [147].

The elimination of c_3 between (2.69) and (2.71) first provides

$$q_2 q_2'' + q_2'^2 + 2\omega_2 q_2^2 + 4\alpha q_1 q_2^2 + \omega_1 q_1^2 - (2\beta/3) q_1^3 - 2E = 0,\ H = E,$$

then the elimination of q_2^2 with (2.70) yields

$$\begin{aligned} & q_1'''' + (8\alpha - 2\beta) q_1 q_1'' - 2(\alpha+\beta) q_1'^2 - \frac{20}{3}\alpha\beta q_1^3 \\ & + (\omega_1 + 4\omega_2) q_1'' + (6\alpha\omega_1 - 4\beta\omega_2) q_1^2 + 4\omega_1\omega_2 q_1 + 4\alpha E = 0, \end{aligned} \tag{2.81}$$

and this ODE is much simpler to process, because q_1 has only one admissible leading power, namely movable double poles.

This elimination of q_2 and c_3 establishes the identification [147] of the HH3 Hamiltonian system to the stationary flow $u_t = 0$ of the fifth order conservative PDE

$$u_t + \left(u_{xxxx} + (8\alpha - 2\beta) u u_{xx} - 2(\alpha+\beta) u_x^2 - \frac{20}{3}\alpha\beta u^3 \right)_x = 0, \tag{2.82}$$

and the result which we are going to establish is that the only values of β/α which make (2.81) pass the Painlevé test are precisely the only values for which the PDE (2.82) is a soliton equation.

The leading behavior of (2.81) $q_1 \sim q_{1,0}\chi^{-2}$ does not involve ω_1, ω_2, E,

$$120 q_{1,0} + 6(8\alpha - 2\beta) q_{1,0}^2 - 8(\alpha+\beta) q_{1,0}^2 - (20/3)\alpha\beta q_{1,0}^3 = 0, \tag{2.83}$$

and it defines two families, $q_{1,0} = 3/\alpha$ and $q_{1,0} = -6/\beta$. Then, the indicial equation

$$\begin{aligned} & (i-2)(i-3)(i-4)(i-5) + (8\alpha - 2\beta)((i-2)(i-3)+6) q_{1,0} \\ & + 8(\alpha+\beta)(i-2) q_{1,0} - 20\alpha\beta q_{1,0}^2 = 0, \end{aligned} \tag{2.84}$$

yields the following Fuchs indices

$$p = -2,\ q_{1,0} = -\frac{3}{\alpha},\ \text{indices } (-1, 10, r_1, r_2),\ r_1 + r_2 = 5, \tag{2.85}$$

$$p = -2,\ q_{1,0} = \frac{6}{\beta},\ \text{indices } (-1, 5, s_1, s_2),\ s_1 + s_2 = 10, \tag{2.86}$$

in which r_i and s_i are the roots of the equations

$$r^2 - 5r + 12 + 6\gamma = 0,\ s^2 - 10s + 24 + 48\gamma^{-1} = 0,\ \gamma = \frac{\beta}{\alpha}. \tag{2.87}$$

Thus, one first has to solve the diophantine conditions that, for every family of (2.81), the four indices are distinct integers.

The general method to solve these conditions has been developed by classical authors [59, 46, 104]. For the class

$$q_1'''' = A q_1 q_1'' + B q_1'^2 + C q_1^3. \tag{2.88}$$

the leading coefficient $q_{1,0}$ obeys the second degree equation

$$Cq_{1,0}{}^2 + 2(3A + 2B)q_{1,0} - 120 = 0, \tag{2.89}$$

and the indicial equation, after division by $i+1$, is

$$i^3 - 15i^2 + (86 - Aq_{1,0})i - 240 + 2(3A + 2B)q_{1,0} = 0. \tag{2.90}$$

The method [46] is to establish, for each of the two roots $q_{1,0} = s_1, s_2$ of (2.89), a lower bound and an upper bound for the product $N_j = 240 - 2(3A + 2B)s_j$ of the three Fuchs indices. One eliminates as many coefficients A, B, C as possible by replacing (2.89) by the equivalent information

$$C = -\frac{120}{s_1 s_2},\ 3A + 2B = 60\frac{s_1 + s_2}{s_1 s_2}, \tag{2.91}$$

which makes the indicial equation to depend on s_1, s_2 and only one fixed coefficient

$$i^3 - 15i^2 + (86 - As_1)i - 120\left(1 - \frac{s_1}{s_2}\right) = 0, \tag{2.92}$$

$$i^3 - 15i^2 + (86 - As_2)i - 120\left(1 - \frac{s_2}{s_1}\right) = 0. \tag{2.93}$$

The remaining coefficient A then does not contribute to the condition that each product $N_j, j = 1,2$ is integer. Since the two products N_1, N_2 only depend on one variable,

$$N_1 = 120\left(1 - \frac{s_1}{s_2}\right),\ N_2 = 120\left(1 - \frac{s_2}{s_1}\right), \tag{2.94}$$

an elimination is possible which yields a condition involving only N_1 and N_2, which is the desired diophantine condition [46, (2.15) p. 80]

$$\frac{1}{N_1} + \frac{1}{N_2} = \frac{1}{120}. \tag{2.95}$$

This equation admits 63 solutions [104, (2.27)].

Here, the diophantine conditions are simple enough to be handled directly. They can be parametrized as follows [82],

$$r_2 - r_1 = 2k - 1, s_2 - s_1 = 2l,\ r_1 < r_2,\ s_1 < s_2, \tag{2.96}$$

$$kl(r_i + 1)(r_i - 10)(s_i + 1)(s_i - 5) \neq 0, \tag{2.97}$$

with k and l two positive integers. A straightforward elimination yields

$$\gamma = \frac{48}{1 - l^2},\ l^2 = 1 + \frac{1152}{23 + (2k-1)^2}, \tag{2.98}$$

and this diophantine equation for (k, l) has only four solutions, listed in Table 2.1.

To finish the Painlevé test, one still has to compute the no-log conditions for each of the two families and each of the four selected values of β/α. The result [58, 82] is

Table 2.1 Cubic Hénon–Heiles system. Solution of the diophantine conditions. The last column is the name of the soliton equation (2.82) for the value of β/α (SK for Sawada–Kotera, KdV5 for fifth order Korteweg–de Vries, KK for Kaup–Kupershmidt).

β/α	k,l	r_1,r_2	s_1,s_2	Name
-1	$1,7$	$2,3$	$-2,12$	SK
-2	$3,5$	$0,5$	$0,10$	
-6	$6,3$	$-3,8$	$2,8$	KdV5
-16	$10,2$	$-7,12$	$3,7$	KK

summarized in Table 2.2. In the case $\beta/\alpha = -6$, the test succeeds unconditionally. In the case $\beta/\alpha = -2$, for which the two families coincide, the test fails unconditionally at the index 10, because the coefficient $q_{1,5}$ must remain arbitrary. In the cases $\beta/\alpha = -1$ and -16, the test selects one additional condition involving ω_1 and ω_2.

Table 2.2 Cubic Hénon–Heiles system. Necessary conditions for the absence of movable logarithms in (2.81), arising from the positive Fuchs indices, for the four cases of Table 2.1 and the two families $q_{1,0} = 3/\alpha$ and $q_{1,0} = -6/\beta$. The identically satisfied conditions are not listed, the Laurent coefficients are denoted $q_{1,j}$, and the symbol P denotes a polynomial of its arguments.

β/α	Family	Fuchs index	No-log condition
-1	$-3/\alpha$	2	$\omega_1 - \omega_2 = 0,$
-1	$-3/\alpha$	10	$(\omega_1 - \omega_2)P(q_{1,2}, q_{1,3}, \omega_1, \omega_2, E) = 0,$
-1	$6/\beta$	12	$(\omega_1 - \omega_2)q_{1,5}^2 = 0,$
-2		0	undefined
-2		10	$q_{1,5}^2 = 0,$
-16	$-3/\alpha$	12	$(\omega_1 - 16\omega_2)(q_{1,10} + P(\omega_1, \omega_2, E)) = 0,$
-16	$6/\beta$	5	$(\omega_1 - 16\omega_2)q_{1,3} = 0,$
-16	$6/\beta$	7	$(\omega_1 - 16\omega_2)q_{1,5} = 0.$

Therefore, the necessary conditions that the cubic Hénon–Heiles system passes the Painlevé test result in three sets of values for the parameters,

$$\text{(SK)}: \quad \beta/\alpha = -1, \omega_1 = \omega_2, \tag{2.99}$$

$$\text{(KdV5)}: \quad \beta/\alpha = -6, \tag{2.100}$$

$$\text{(KK)}: \quad \beta/\alpha = -16, \omega_1 = 16\omega_2. \tag{2.101}$$

These conditions are shown to be sufficient in Sect. 6.2.2, where the explicit integration is performed.

2.2 Fuchsian Perturbative Method

There exist two situations making the method of Kowalevski and Gambier inconclusive:

1. presence of negative integers among the set of Fuchs indices of the linearized equation,
2. insufficient number of Fuchs indices in the linearized equation.

These are dealt with in Sects. 2.2 and 2.3 respectively.

If one sets aside the ever present Fuchs index -1, it may happen that some family possesses at least one Fuchs index which is a negative integer. In such a case, the Laurent series (2.2) fails to locally represent the general solution, and the missing part of the solution may contain multivaluedness, therefore the Painlevé test must be extended to test this possibility.

From the presence of a negative integer Fuchs index, one should not assume the existence of a movable essential singularity like $e^{1/(x-x_0)}$. For instance, the ODE with a meromorphic general solution [348, 82]

$$u'' + 3uu' + u^3 = 0,\ u = \frac{1}{x-a} + \frac{1}{x-b}, \tag{2.102}$$

with a and b arbitrary, is linearizable into a third order equation,

$$u = \frac{\psi'}{\psi},\ \psi''' = 0, \tag{2.103}$$

and, despite the meromorphy of its general solution u, one of its two families,

(F1) $p = -1, u_0 = 1$, indices $(-1, 1)$,
(F2) $p = -1, u_0 = 2$, indices $(-2, -1)$,

admits the negative Fuchs index -2.

The method [149, 82] to unveil the information contained in the negative integer indices is to define a perturbation close to the identity

$$x \text{ unchanged},\ \mathbf{u} = \sum_{n=0}^{+\infty} \varepsilon^n \mathbf{u}^{(n)} : \mathbf{E} = \sum_{n=0}^{+\infty} \varepsilon^n \mathbf{E}^{(n)} = 0, \tag{2.104}$$

where, like for the α-method of Painlevé, the small parameter ε is not in the original equation.

Then, the single equation (2.1) is equivalent to the infinite sequence

$$n = 0\ \mathbf{E}^{(0)} \equiv \mathbf{E}(x, \mathbf{u}^{(0)}) = 0, \tag{2.105}$$

$$\forall n \geq 1\ \mathbf{E}^{(n)} \equiv \mathbf{E}'(x, \mathbf{u}^{(0)})\mathbf{u}^{(n)} + \mathbf{R}^{(n)}(x, \mathbf{u}^{(0)}, \ldots, \mathbf{u}^{(n-1)}) = 0, \tag{2.106}$$

in which $\mathbf{E}'$ is the derivative of the equation, like in (2.40), and $\mathbf{R}^{(1)}$ is identically zero. From a basic theorem of Poincaré [76, Theorem II, §5.3], necessary conditions for the PP are

- the general solution $\mathbf{u}^{(0)}$ of (2.105) has no movable critical points,
- the general solution $\mathbf{u}^{(1)}$ of (2.106) has no movable critical points,
- for every $n \geq 2$ there exists a particular solution of (2.106) without movable critical points.

Order $n = 0$ is not different from the original equation (2.1) for the unknown $\mathbf{u}^{(0)}$, so one takes for $\mathbf{u}^{(0)}$ the already computed (particular) Laurent series (2.2).

Order $n = 1$ is identical to the linearized equation

$$\mathbf{E}^{(1)} \equiv \mathbf{E}'(x, \mathbf{u}^{(0)})\mathbf{u}^{(1)} = 0, \tag{2.107}$$

and one must check the existence of N independent solutions $\mathbf{u}^{(1)}$ locally singlevalued near $\chi = 0$, where N is the order of (2.1).

The two main implementations of this perturbation are the Fuchsian perturbative method [149, 82], described hereafter, and the nonFuchsian perturbative method [318], described in the next section.

The Fuchsian perturbative method is adapted to the presence of negative integer indices in addition to the ever present value -1, and it generates additional no-log conditions. Denoting by ρ the lowest integer Fuchs index, $\rho \leq -1$, the Laurent series for $\mathbf{u}^{(1)}$

$$\mathbf{u}^{(1)} = \sum_{j=\rho}^{+\infty} \mathbf{u}_j^{(1)} \chi^{j+\mathbf{p}}, \tag{2.108}$$

represents a particular solution containing a number of arbitrary coefficients equal to the number of Fuchs indices, counting their multiplicity. If this number equals N, it represents the general solution of (2.107). Let us illustrate the method with two examples [76, §5.7.3].

The first example is the equation

$$u'' + 4uu' + 2u^3 = 0, \tag{2.109}$$

which possesses the single family

$$p = -1,\ q = -3,\ E_0^{(0)} = u_0^{(0)}(u_0^{(0)} - 1)^2 = 0, \text{ indices } (-1, 0), \tag{2.110}$$

with the puzzling fact that $u_0 \equiv u_0^{(0)}$ should be at the same time equal to 1 according to the equation $E_0^{(0)} = 0$, and arbitrary according to the index 0. The necessity of performing a perturbation arises from the multiple root of the equation for $u_0^{(0)}$, responsible for the insufficient number of arbitrary parameters in the zeroth order series $u^{(0)}$. The application of the method provides

$$u^{(0)} = \chi^{-1} \text{ (the series terminates)}, \tag{2.111}$$

$$E'(x, u^{(0)}) = \partial_x^2 + 4\chi^{-1}\partial_x + 2\chi^{-2}, \tag{2.112}$$

$$u^{(1)} = u_0^{(1)} \chi^{-1},\ u_0^{(1)} \text{ arbitrary}, \tag{2.113}$$

$$E^{(2)} = E'(x,u^{(0)})u^{(2)} + 6u^{(0)}u^{(1)^2} + 4u^{(1)}u^{(1)'}$$

$$= \chi^{-2}(\chi^2 u^{(2)})'' + 2u_0^{(1)^2} \chi^{-3} = 0, \tag{2.114}$$

$$u^{(2)} = -2u_0^{(1)^2} \chi^{-1}(\log \chi - 1). \tag{2.115}$$

The movable logarithmic branch point is therefore detected in a systematic way at order $n = 2$ and index $i = 0$. This result was of course found long ago by the α-method [348, §13, p 221].

Quite similarly, the no-log condition labelled "undefined" in Table 2.2 is in fact the failed condition $Q_0^{(2)} \equiv -40\alpha((u_{-1}^{(1)})' + u_0^{(1)})^2 = 0$ [82], found at perturbation order $n = 2$ from the Fuchs index 0, i.e. without the need for a lengthy computation to test the index 10.

The second example is a fourth order ODE isolated by Bureau [46, p. 79]

$$u'''' + 3uu'' - 4u'^2 = 0, \tag{2.116}$$

which possesses the two families

$$p = -2, u_0^{(0)} = -60, \text{ ind. } (-3,-2,-1,20), \hat{E} = u'''' + 3uu'' - 4u'^2, \tag{2.117}$$

$$p = -3, u_0^{(0)} \text{ arbitrary, indices } (-1,0), \hat{E} = 3uu'' - 4u'^2. \tag{2.118}$$

The second family has a Laurent series $(p : +\infty)$ which happens to terminate [82]

$$u^{(0)} = c(x - x_0)^{-3} - 60(x - x_0)^{-2},\ (c, x_0) \text{ arbitrary}. \tag{2.119}$$

For this family, the Fuchsian perturbative method is then useless, because the two arbitrary coefficients corresponding to the two Fuchs indices are already present at zeroth order.

The first family provides, at zeroth order, only a two-parameter expansion and, when one checks the existence of the perturbed solution

$$u = \sum_{n=0}^{+\infty} \varepsilon^n \left[\sum_{j=0}^{+\infty} u_j^{(n)} \chi^{j-2-3n} \right], \tag{2.120}$$

one finds that coefficients $u_{20}^{(0)}, u_{-3}^{(1)}, u_{-2}^{(1)}, u_{-1}^{(1)}$ can be chosen arbitrarily, and, at order $n = 7$, one finds two violations [82]

$$Q_{-1}^{(7)} \equiv u_{20}^{(0)} u_{-3}^{(1)^7} = 0, Q_{20}^{(7)} \equiv u_{20}^{(0)^2} u_{-3}^{(1)^6} u_{-2}^{(1)} = 0, \tag{2.121}$$

implying the existence of a movable logarithmic branch point. This method is easily computerizable.

2.3 NonFuchsian Perturbative Method

Whenever the number of indices is less than the differential order of the equation, the Fuchsian perturbative method fails to build a representation of the general solution, thus possibly missing some no-log conditions. The missing solutions of the linearized equation (2.107) are of the nonFuchsian type near $\chi = 0$.

In Sect. 2.2, the fourth order equation (2.116) has been shown to fail the test after a computation which is practically untractable without a computer. Let us prove the same result directly [318]. The linearized equation

$$E^{(1)} = E'(x, u^{(0)})u^{(1)} \equiv [\partial_x^4 + 3u^{(0)}\partial_x^2 - 8u_x^{(0)}\partial_x + 3u_{xx}^{(0)}]u^{(1)} = 0, \qquad (2.122)$$

is known *globally* for the second family because the two-parameter solution (2.119) is closed form, therefore one can test all the singular points χ of (2.122). These are $\chi = 0$ (nonFuchsian) and $\chi = \infty$ (Fuchsian), and the key to the method is the information obtainable from $\chi = \infty$. Let us first lower by two units the order of the linearized equation (2.122), by taking advantage of the knowledge of the two global single valued solutions $u^{(1)} = \partial_{x_0} u^{(0)}$ and $\partial_c u^{(0)}$, i.e. $u^{(1)} = \chi^{-4}, \chi^{-3}$,

$$u^{(1)} = \chi^{-4} v : \ [\partial_x^2 - 16\chi^{-1}\partial_x + 3c\chi^{-3} - 60\chi^{-2}]v'' = 0, \qquad (2.123)$$

Then the local study of $\chi = \infty$ is unnecessary, since one recognizes the Bessel equation. The two other solutions in global form are

$$c \neq 0 : \ v_1'' = \chi^{-3} {}_0F_1(24; -3c/\chi) = \chi^{17/2} J_{23}(\sqrt{12c/\chi}), \qquad (2.124)$$

$$v_2'' = \chi^{17/2} N_{23}(\sqrt{12c/\chi}), \qquad (2.125)$$

where the hypergeometric function ${}_0F_1(24; -3c/\chi)$ is single valued and possesses an isolated essential singularity at $\chi = 0$, while the function N_{23} of Neumann is multivalued because of a $\log \chi$ term.

Chapter 3
Integrating Ordinary Differential Equations

Abstract The problem addressed in this chapter is, given an ODE which may admit a singlevalued solution, to find it explicitly in closed form. Two kinds of ODE are considered.
The first kind is made of ODEs which pass the Painlevé test, in which case the goal is to find their *general solution*, and our three main examples will be: the four cases of the Lorenz model isolated in Sect. 2.1.1, the traveling wave reduction of the Korteweg–de Vries equation, and the traveling wave reduction of the nonlinear Schrödinger equation. In these three examples, the general solution can indeed be found and is represented either by elliptic functions or by Painlevé functions.
The second kind is made of ODEs which fail the Painlevé test, but which do not fail it too badly, leaving open the possibility of a particular singlevalued solution. This second kind of ODE is much more difficult to handle, because of the nonexistence of a closed form general solution, making inapplicable most of the powerful methods available for the ODEs of the first kind. One must first count the maximal number of integration constants in the largest particular singlevalued solution, which we call for convenience the *general analytic solution*.
The methods of integration can then be divided into two classes, one "sufficient" and one "necessary".

In the first class, one looks for solutions u in a given class of expressions (usually polynomials) in some intermediate variable which obeys a given first order ODE (e.g. Riccati, Weierstrass, projective Riccati). Then, by a direct computation, one checks whether there indeed exist solutions in the given class. The solutions with a simple profile (such as tanh for a front, sech for a pulse), are easily found by this class of methods. The work involved can be done by hand but all solutions outside the given class are surely missed.

In the second class of methods [320], presented in Sect. 3.2.5, rather than directly looking for the unknown solution

$$u = f(\xi - \xi_0), \tag{3.1}$$

in which ξ_0 is a constant of integration, one looks as an intermediate piece of information for the first order nonlinear ODE

$$F(u,u')=0, \tag{3.2}$$

obtained by eliminating ξ_0 between (3.1) and its derivative, in which F is as unknown as f. Indeed, provided that f is singlevalued, by a classical theorem there is equivalence between the knowledge of the solution f and that of the subequation F which it satisfies.

3.1 Integrable Situation

Using six examples of ODEs which pass the Painlevé test, we explain how to find the first integrals if they exist, and then to perform the explicit integration.

3.1.1 First Integrals and Integration of the Lorenz Model

A first integral of an N-th order ODE is by definition a function of $x,u(x),u'(x), \dots, u^{(N-1)}$ which takes a constant value at any x, *including* the movable singularities of u.

Since the Lorenz model (2.3) is autonomous, first integrals can be defined as the product of $e^{\lambda t}$ by an expression only depending on x,y,z, for instance $P(x,y,z)e^{\lambda t}$, with P polynomial and λ constant. One should not search for first integrals in this class simply by putting an upper bound on the global degree of the polynomial P in the three variables [388]. Indeed, P must have no movable singularities, therefore the admissible polynomials P are provided by a generating function built from the singularity degrees $(-1,-2,-2)$ of (x,y,z) (see (2.7)) [280]

$$\begin{aligned}\frac{1}{(1-\alpha x)(1-\alpha^2 y)(1-\alpha^2 z)} &= 1+\alpha x+\alpha^2\left(x^2+y+z\right)+\alpha^3\left(x^3+xy+xz\right)\\ &+\alpha^4\left(x^4+x^2y+y^2+x^2z+yz+z^2\right)+\dots\end{aligned} \tag{3.3}$$

and this defines the basis, ordered by singularity degrees,

$$(1),\ (x),\ (x^2,y,z),\ (x^3,xy,xz),\ (x^4,x^2y,y^2,x^2z,yz,z^2),\ \dots \tag{3.4}$$

The existence of the two families (2.7) splits the possible first integrals into two disjoint classes according to their parity under the mapping $(x,y,z)\mapsto(-x,-y,z)$. The odd first integrals are generated by the basis

$$(x),\ (y),\ (x^3,xz),\ (x^2y,yz),\ \dots \tag{3.5}$$

and the even ones by

$$(1),\ (x^2,z),\ (xy),\ (x^4,y^2,x^2z,z^2),\ (x^3y,xyz),\ldots \tag{3.6}$$

The smallest admissible singularity degree is two and the associated candidate is the even linear combination

$$K=(c_0+c_1x^2+c_2z)e^{\lambda t}, \tag{3.7}$$

which indeed provides one first integral [390]

$$K_1=(x^2-2\sigma z)e^{2\sigma t},\ b=2\sigma. \tag{3.8}$$

Six polynomial first integrals are known [390, 264] with a singularity degree equal to two or four (in accordance with the Fuchs indices (2.15)), they all have an even parity[1] and it can be proven [283] that they are the only ones in the polynomial class. The case $b=2\sigma$ involved in (3.8) is one of the two solutions of $Q_2=0$ (2.19), and an interesting open problem is to find a first integral, if any, associated to the second solution $b=1-3\sigma$ of $Q_2=0$.

In order to find first integrals of dynamical systems, an original approach has been developed [40], which has no relation with singularities but which takes advantage of basic properties in linear algebra. This approach is explained well in [177].

For the four values of (b,σ,r) for which the test succeeds, the explicit integration happens to be possible, and it has been performed in [390] for the first three cases and in [438] for the fourth case. Each case admits at least one of the following first integrals [390]

$$(2\sigma,\sigma,r):\ K_1=(x^2-2\sigma z)e^{2\sigma t}, \tag{3.9}$$
$$(0,1/3,r):\ K_2=\left[-9x^4+12x^2(z-r)+4(y^2+2xy)\right]e^{4t/3}/12, \tag{3.10}$$
$$(1,\sigma,0):\ K_3=(y^2+z^2)e^{2t}, \tag{3.11}$$

therefore the elimination of y and z from (2.23) defines a second order ODE for $x(t)$ in which K_j represents the origin of t, itself associated with the Fuchs index -1.

3.1.1.1 Case $(1,1/2,0)$

Since K_1 and K_3 are both first integrals, $x(t)$ obeys a first order ODE,

$$\left(\frac{dx}{dt}\right)^2+x\frac{dx}{dt}+\frac{1}{4}\left(x^2-K_1e^{-t}\right)^2+\frac{x^2}{4}-\frac{K_3}{2}e^{-2t}=0, \tag{3.12}$$

[1] This is not a rule. The ODE with two opposite families $u'''-6u^2u'+au''-2au^3=0$ admits the odd first integral $K=(u''-2u^3)e^{ax}$.

and there exists a simultaneous change of dependent and independent variables $(x,t) \to (X,T)$ in the class (1.43),

$$x = e^{-t/2}X, \ T = e^{-t/2}, \tag{3.13}$$

which maps it to the autonomous, elliptic equation

$$\left(\frac{\mathrm{d}X}{\mathrm{d}T}\right)^2 + (X^2 - K_1)^2 - K_3 = 0, \tag{3.14}$$

whose general solution is single valued.

3.1.1.2 Case $(2,1,1/9)$

The second order ODE for $x(t)$

$$\frac{\mathrm{d}^2x}{\mathrm{d}t^2} + 2\frac{\mathrm{d}x}{\mathrm{d}t} + \frac{x^3}{2} + \left(\frac{8}{9} - \frac{K_1}{2}e^{-2t}\right)x = 0, \tag{3.15}$$

cannot be made autonomous by a transformation (1.43), therefore the strategy is to map it to one nonautonomous equation among the fifty equations of the list of Gambier [163, 232, 116] (see Appendix A). In order to decide which equation is suitable in this list, the method [348] is to write (3.15) in the canonical form

$$\frac{\mathrm{d}^2x}{\mathrm{d}t^2} = A_2(x,t)\left(\frac{\mathrm{d}x}{\mathrm{d}t}\right)^2 + A_1(x,t)\frac{\mathrm{d}x}{\mathrm{d}t} + A_0(x,t). \tag{3.16}$$

Since the ODE passes the Painlevé test, as shown by Painlevé [348, p. 258] [349, p. 74], the coefficient A_2 is the sum of at most four simple poles in x (possibly including $x = \infty$), whose sum of residues is equal to two

$$A_2(x,t) = \sum_j \frac{r_j}{x - a_j(t)}, \ \sum_j r_j = 2, \tag{3.17}$$

and the number of poles and the set of residues $\{r_j\}$ is invariant under the homographic transformations (1.43). In the present case, the value $A_2 = 0$ means the single pole $x = \infty$ with residue 2 (in order to check it, change $x \to 1/x$). The method is then to select the few Gambier equations which have the same number of poles and the same set of residues $\{r_j\}$, here one pole with residue two, finally to map the ODE (3.15) to one of these Gambier equations by a transformation (1.43). Here the suitable Gambier equation is either P2 or P1[2]. To choose between P2 and P1, one also matches the structure of singularities (two movable simple poles with opposite residues for P2 and $x(t)$, one movable double pole for P1), which now selects P2 as

[2] As noticed by Painlevé [347], the ODE $u'' = \delta(2u^3 + xu) + \gamma(6u^2 + x) + \beta u + \alpha$, which includes P2 and P1, has the PP, therefore, at least for the criterium of residues, P2 and P1 are not different.

the unique possible match. Indeed, the particular transformation (1.43)

$$x = ae^{-2t/3}X,\ T = \frac{i}{2}a^{-3/2}e^{-2t/3},\ K_1 = \frac{3}{8}ia^3, \tag{3.18}$$

maps (3.15) to

$$\frac{\mathrm{d}^2X}{\mathrm{d}T^2} = 2X^3 + TX + \alpha,\ \alpha = 0. \tag{3.19}$$

3.1.1.3 Case $(0, 1/3, r)$

The second-order equation for $x(t)$ is

$$\frac{\mathrm{d}^2x}{\mathrm{d}t^2} - \frac{1}{x}\left(\frac{\mathrm{d}x}{\mathrm{d}t}\right)^2 + \frac{x^3}{4} + \frac{K_2}{3x}e^{-4t/3} = 0. \tag{3.20}$$

For $K_2 = 0$ this equation admits the first integral

$$\frac{1}{x^2}\left(\frac{\mathrm{d}x}{\mathrm{d}t}\right)^2 + \frac{x^2}{4} = A^2, \tag{3.21}$$

and the general solution $x = \pm 2iA/\sinh A(t - t_0)$. For $K_2 \neq 0$, the coefficient A_2 defined by (3.16) decomposes as

$$A_2(x,t) = \frac{1}{x} + \frac{1}{x - \infty}, \tag{3.22}$$

therefore the unique possible nonautonomous match in the list of Gambier is P3. Indeed, the transformation

$$x = c_1X,\ T = ce^{-2t/3},\ c_1^2 = -\frac{4\gamma}{9},\ c^2 = \frac{27K_2}{4\gamma\delta}, \tag{3.23}$$

realizes the mapping to

$$X'' = \frac{X'^2}{X} - \frac{X'}{T} + \frac{\alpha X^2 + \gamma X^3}{4T^2} + \frac{\beta}{4T} + \frac{\delta}{4X},\ \alpha = \beta = 0. \tag{3.24}$$

3.1.1.4 Case $(1, 0, r)$

Because of the scaling invariance, the first integral (3.11) is also a first integral [438] of the third order equation (2.24) for $(b, \sigma, b\sigma r) = (1, \sigma, 0)$, which includes the particular case of interest to us $(b,\sigma,b\sigma r)=(1,0,0)$,

$$(b,\sigma,b\sigma r)=(1,0,0):\ K^2=\lim_{\sigma\to 0}\sigma^2 K_3=\left[\left(\frac{x''+x'}{x}\right)^2+x'^2\right]e^{2t}. \quad (3.25)$$

For $K=0$, the general solution is

$$x=ik\tanh\frac{k}{2}(t-t_0)-i,\ i^2=-1,\ (k,t_0)\text{ arbitrary}. \quad (3.26)$$

For $K\neq 0$, the ODE has second degree, therefore the classification of Gambier cannot be used. Instead of looking in higher classifications, one can take the usual parametric representation

$$\frac{x''+x'}{x}=Ke^{-t}\cos\lambda,\ x'=Ke^{-t}\sin\lambda, \quad (3.27)$$

and build the second order ODE for $\lambda(t)$. This ODE,

$$\lambda''-Ke^{-t}\sin\lambda=0,\ x(t)=\lambda'(t) \quad (3.28)$$

is not algebraic, but the variable $e^{i\lambda}$ obeys an algebraic second order first degree ODE which is easy to map, as done in previous cases, to the P3 equation. The overall result is the general solution

$$x=i+2i\frac{\mathrm{d}}{\mathrm{d}t}\log w(\xi(t)),\ i^2=-1,\ \xi=ae^{-t}, \quad (3.29)$$

in which $w(\xi)$ is the particular third Painlevé function defined by

$$\frac{\mathrm{d}^2w}{\mathrm{d}\xi^2}=\frac{1}{w}\left(\frac{\mathrm{d}w}{\mathrm{d}\xi}\right)^2-\frac{\mathrm{d}w}{\xi\mathrm{d}\xi}+\frac{\alpha w^2+\gamma w^3}{4\xi^2}+\frac{\beta}{4\xi}+\frac{\delta}{4w}, \quad (3.30)$$

$$\alpha=0,\ \beta=0,\ \gamma\delta=-(K/a)^2. \quad (3.31)$$

To conclude this example of the Lorenz model, it represents a good illustration of the power of the Painlevé test: the four values of (b,σ,r) selected by the test indeed correspond to dynamical systems having the Painlevé property.

3.1.2 General Traveling Wave of KdV Equation

The KdV PDE[3]

[3] Although the constants (a,b) can be scaled out, retaining them makes the conversion between different conventions easier.

$$bu_t + u_{xxx} - \frac{6}{a} u u_x = 0,\ (a,b) \text{ constant}, \tag{3.32}$$

is invariant under a Galilean transformation $(u,x,t) \to (u-(ab/6)c, x-ct, t)$, therefore it is convenient to define its traveling wave reduction as

$$u = -\frac{ab}{6}c + U,\ \xi = x - ct, \tag{3.33}$$

which leads to the ODE

$$U''' - \frac{6}{a} U U' = 0. \tag{3.34}$$

The singularities of U are a movable double pole, with the Fuchs indices $-1,4,6$, therefore the possible first integrals can only have the singularity degrees 4 and 6.

As a consequence of the conservative form of the KdV equation, the third order ODE (3.34) is a total derivative. Its primitive,

$$U'' - \frac{3}{a} U^2 + a g_2 = 0, \tag{3.35}$$

with g_2 an integration constant, admits the integrating factor U', which yields

$$\frac{1}{2} U'^2 - \frac{1}{a} U^3 + a g_2 U + 2a^2 g_3 = 0, \tag{3.36}$$

in which g_3 is another integration constant. The two first integrals g_2, g_3 have the respective singularity degrees 4 and 6, in agreement with the prediction from the Fuchs indices.

The equation (3.36) is identical to the Weierstrass equation (1.42), under the rescaling

$$U = 2au,\ X = x, \tag{3.37}$$

which belongs to the homographic group (1.43), therefore its general solution is the elliptic function

$$U(\xi) = 2a\wp(\xi - \xi_0, g_2, g_3), \tag{3.38}$$

a singlevalued expression which depends as expected on three arbitrary complex constants g_2, g_3, ξ_0.

The physically interesting traveling wave u is obtained by selecting the three complex constants g_2, g_3, ξ_0 so as to ensure the decreasing of u to some background value B when $\xi \to \pm\infty$, which requires

$$-\frac{3}{a}(B-\kappa)^2 + a g_2 = 0,\ \kappa = -\frac{abc}{6}, \tag{3.39}$$

$$-\frac{1}{a}(B-\kappa)^3 + a g_2 (B-\kappa) + 2a^2 g_3 = 0, \tag{3.40}$$

i.e.

$$g_2 = 3\left(\frac{B-\kappa}{a}\right)^2,\ g_3 = -\left(\frac{B-\kappa}{a}\right)^3. \tag{3.41}$$

For such values of g_2, g_3, the doubly periodic elliptic function $\wp$ degenerates to a hyperbolic trigonometric function, according to the classical formula

$$\forall x, d:\ \wp(x, 3d^2, -d^3) = 2d - \frac{3d}{2}\coth^2\sqrt{\frac{3d}{2}}x. \tag{3.42}$$

In the resulting complex solution,

$$u = -\frac{ab}{6}c + 2a\left(2\frac{B-\kappa}{a} - \frac{3(B-\kappa)}{2a}\coth^2\sqrt{\frac{3(B-\kappa)}{2a}}(\xi-\xi_0)\right), \tag{3.43}$$

the singularities, which are movable double poles located at

$$\xi = \xi_0 + in\pi/k, n \in \mathscr{Z},\ k^2 = 3\frac{B-\kappa}{2a}, \tag{3.44}$$

are moved outside the real axis of ξ by a shift on ξ_0, equivalent to the change of coth for tanh, resulting in the final, bell-shaped, solitary wave solution (Fig. 3.1),

$$u = B - 2ak^2\operatorname{sech}^2\left(kx - 4k^3t/b + 6k(B/a)t/b - k\xi_0\right). \tag{3.45}$$

3.1.3 General Traveling Wave of NLS Equation

Given the *nonlinear Schrödinger equation* (NLS),

$$iA_t + pA_{xx} + q|A|^2A = 0,\ pq \neq 0,\ (p,q) \in \mathscr{R}, \tag{3.46}$$

its traveling wave is defined by

$$A(x,t) = \sqrt{M(\xi)}e^{i(-\omega t + \varphi(\xi))},\ \xi = x - ct, \tag{3.47}$$

which generates the complex equation

$$\frac{M''}{2M} - \frac{M'^2}{4M^2} + i\varphi'' - \varphi'^2 + i\varphi'\frac{M'}{M} - \frac{i}{p}\frac{cM'}{2M} + \frac{1}{p}\left(c\varphi' + \omega\right) + \frac{q}{p}M = 0, \tag{3.48}$$

whose real and imaginary parts define the third order system in (M, φ'),

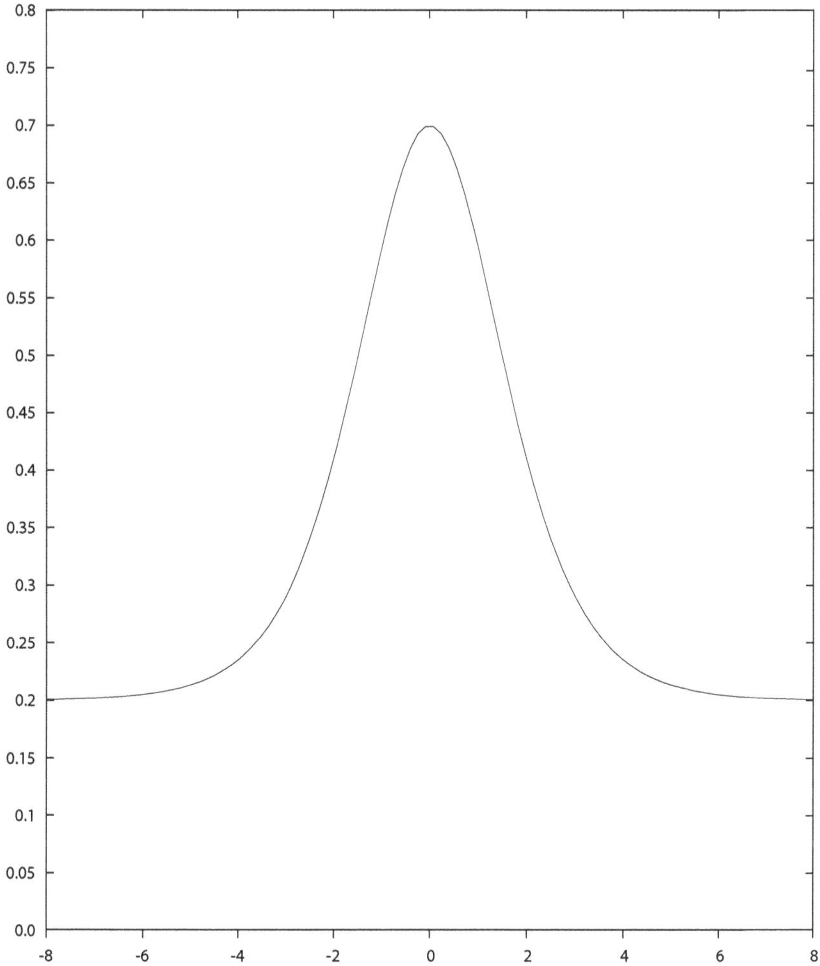

Fig. 3.1 KdV bell-shaped solitary wave (3.45), for the values $a = -1, b = 1, \xi_0 = 0, B = 0.2, t = 0, k = 0.5$.

$$\begin{cases} \dfrac{M''}{2M} - \dfrac{M'^2}{4M^2} - \left(\varphi' - \dfrac{c}{2p}\right)^2 + \dfrac{q}{p}M + \dfrac{\omega}{p} + \dfrac{c^2}{4p^2} = 0, \\ \varphi'' + \left(\varphi' - \dfrac{c}{2p}\right)\dfrac{M'}{M} = 0. \end{cases} \tag{3.49}$$

According to the results in Sect. 2.1.3, the field M has movable double poles, with the Fuchs indices $(-1,0,3,4)$, see (2.58). The indices -1 and 0 correspond to the two "angles" ξ_0, φ_0 (the irrelevant origins of ξ and φ), and the two others $(3,4)$ to two "actions" to be found.

The second equation of the system (3.49) admits the integrating factor M, resulting in the first integral

$$M\varphi' - \frac{c}{2p}M = K_1. \tag{3.50}$$

Since this relation is homographic, it is equivalent to consider either φ' or M. After elimination of φ', the first equation of the system (3.49) admits the integrating factor M', resulting in another first integral

$$\frac{M'^2}{M} + \frac{2q}{p}M^2 + \left(\frac{c^2 - 4\omega p}{p^2}\right)M + 4\frac{K_1^2}{M} = K_2. \tag{3.51}$$

The two first integrals K_1, K_2 have the respective singularity degrees 3 and 4, in accordance with the Fuchs indices. Up to an affine transformation, the elliptic equation (3.51) is identical to the Weierstrass equation (1.42), and the general solution (M, φ') is elliptic,

$$M = -2\frac{p}{q}\left(\wp(\xi - \xi_0, g_2, g_3) - e_0\right),\ \varphi' = \frac{c}{2p} + \frac{K_1}{M},\ e_0 = \frac{4\omega p - c^2}{12p^2}, \tag{3.52}$$

with the notation

$$K_2 = -4\frac{p}{q}\left(6e_0^2 - \frac{g_2}{2}\right),\ K_1^2 = -\left(\frac{p}{q}\right)^2 (4e_0^3 - g_2 e_0 - g_3). \tag{3.53}$$

It depends on four fixed constants, p, q, c, ω (fixed because they appear in the definition of the ODE), and three movable constants ξ_0, g_2, g_3. Out of these three, one (ξ_0) represents the translational invariance, and the two important ones are (g_2, g_3) or equivalently (K_1, K_2). Introducing a by $\wp(a) = e_0$, the solution can also be written as

$$\begin{cases} M = -2\frac{p}{q}\left(\wp(\xi - \xi_0) - \wp(a)\right), \\ \varphi' = \frac{c}{2p} + \frac{j}{2}\frac{\wp'(a)}{\wp(\xi - \xi_0) - \wp(a)},\ j^2 = -1, \\ \wp(a) = (4\omega p - c^2)/(12p^2), \end{cases} \tag{3.54}$$

The modulus $|A| = \sqrt{M}$ is single valued if $\wp'(a) = 0$, otherwise it is multivalued and behaves like $(\xi \pm a)^{1/2}$ near $\xi = \mp a$. Since the variables $e^{\pm i \arg A}$ display the same kind of branching, a compensation occurs making the two fields A and $\bar{A}$ single valued. Indeed, the quadrature for φ is classical [9, §18.7.3],

$$\wp'(a)\int \frac{\mathrm{d}\xi}{\wp(\xi) - \wp(a)} = 2\zeta(a)\xi + \log\sigma(\xi - a) - \log\sigma(\xi + a), \tag{3.55}$$

in which the meromorphic function ζ is the primitive of $-\wp$, the odd entire function $\sigma(z)$ behaves like z near $z = 0$, and the overall expressions of $e^{i\omega t}A$ and $e^{-i\omega t}\bar{A}$ in terms of ξ are indeed globally singlevalued (but not elliptic) [83]

$$\begin{cases} e^{i\omega t} A = \sqrt{-\frac{2p}{q}} \sqrt{\wp(\xi) - \wp(a)}\, e^{ij\zeta(a)\xi} \left(\frac{\sigma(\xi - a)}{\sigma(\xi + a)}\right)^{ij/2} e^{ic\xi/(2p)},\ j^2 = -1, \\ e^{-i\omega t} \bar{A} = \sqrt{-\frac{2p}{q}} \sqrt{\wp(\xi) - \wp(a)}\, e^{-ij\zeta(a)\xi} \left(\frac{\sigma(\xi - a)}{\sigma(\xi + a)}\right)^{-ij/2} e^{-ic\xi/(2p)}. \end{cases} \tag{3.56}$$

Remark. This multivaluedness of the modulus $\sqrt{M}$ also occurs in the stationary linear Schrödinger (Sturm–Liouville) equation of quantum mechanics

$$\psi'' + f(x)\psi = 0, \tag{3.57}$$

with ψ complex and f real, see details in Sect. 7.6.1.

Those traveling waves which decay exponentially fast at $\xi = \pm\infty$ are obtained by requiring that the elliptic functions (doubly periodic in $\mathscr{C}$) degenerate to trigonometric functions (simply periodic in $\mathscr{C}$), according to the formula (3.42). One may alternatively use the correspondence with the Jacobi functions

$$\wp\left(\frac{x}{\sqrt{e_1 - e_3}}, g_2, g_3\right) = e_1 + (e_1 - e_3)\,\mathrm{cs}^2(x,k),\ k^2 = \frac{e_2 - e_3}{e_1 - e_3}, \tag{3.58}$$

followed by the well known degeneracy of the Jacobi functions to trigonometric functions. One thus finds

1. the bright solitary wave [442], which only exists in the focusing regime $pq > 0$,

$$pq > 0:\ A = \sqrt{2\frac{p}{q}}\, k\, \mathrm{sech}(k(x - ct)) \times e^{i\frac{c}{2p}x + ip[k^2 - (\frac{c}{2p})^2]t}. \tag{3.59}$$

2. the dark solitary wave [444], which only exists in the defocusing regime $pq < 0$,

$$pq < 0:\ A = \sqrt{-2\frac{p}{q}} \left[\frac{k}{2} \tanh(\frac{k}{2}(x - ct)) - i(K - \frac{c}{2p})\right] \times e^{iKx - 2ip[\frac{k^2}{4} + (K - \frac{c}{2p})^2 + \frac{1}{2}K^2]t}. \tag{3.60}$$

The reason why the explicit integration of the traveling wave reduction of both KdV and NLS has been a quite easy task is their Painlevé property. For instance, the first integrals (g_2, g_3 for KdV, K_1, K_2 for NLS) can be generated by a systematic method involving the Lax pair [274, 147] and not by inspection as we have presented it.

For ODEs which do not possess the Painlevé property, finding all the possible particular single valued solutions is not at all easy. Let us examine this question.

3.2 Partially Integrable Situation

We now introduce the main methods to find particular single valued solutions of ODEs which fail the Painlevé test. These include truncation methods and a subequation method, and the output is typically all those solutions which are elliptic or degenerate of elliptic, i.e. rational in one exponential e^{kx} or rational in x.

3.2.1 Traveling Wave Reduction of KS and CGL3

The ODEs considered in the remaining part of this chapter are those which fail the Painlevé test but which do not fail it too much, leaving open the possibility for closed form single valued particular solutions to exist. The goal is then to find *all* such solutions.

The two main examples handled will be the traveling wave reduction of the Kuramoto–Sivashinsky equation (2.25) defined by (2.27)

$$\nu U''' + bU'' + \mu U' + \frac{U^2}{2} + A = 0,\ \nu \neq 0. \tag{3.61}$$

and of the CGL3 equation (2.45) defined by (3.47), resulting in the complex equation

$$\frac{M''}{2M} - \frac{M'^2}{4M^2} + i\varphi'' - \varphi'^2 + i\varphi'\frac{M'}{M} - \frac{i}{p}\frac{cM'}{2M} + \frac{1}{p}(c\varphi' + \omega) + \frac{q}{p}M - \frac{i\gamma}{p} = 0,$$

whose real and imaginary parts define the third order system in (M, φ'),

$$\begin{cases} \dfrac{M''}{2M} - \dfrac{M'^2}{4M^2} - \varphi'^2 - s_i\left(\dfrac{cM'}{2M} + \gamma\right) + s_r(c\varphi' + \omega) + d_r M = 0, \\ \varphi'' + \varphi'\dfrac{M'}{M} - s_r\left(\dfrac{cM'}{2M} + \gamma\right) - s_i(c\varphi' + \omega) + d_i M = 0, \end{cases} \tag{3.62}$$

with the notation (2.52).

These two examples are not really independent since there exists a perturbation, in the spirit of Sect. B.1, which maps the CGL3 PDE to the KS PDE [362, 276].

3.2.1.1 Nonexistence of a First Integral

As opposed to the KdV reduction (3.35), the ODE (2.27), whose local singularity analysis has been performed in Sect. 2.1.2, admits no integrating factor. Let us first prove this quite important point.

When performed on the family $U \sim 2a\chi^{-2}$ of the KdV reduction (3.34) and its successive first integrals (3.35), (3.36), the local singularity analysis yields the sequence of embedded Fuchs indices

$$(-1,4,6) \to (-1,6) \to (-1). \tag{3.63}$$

Realizing a similar sequence for KS

$$\left(-1,\frac{13+i\sqrt{71}}{2},\frac{13-i\sqrt{71}}{2}\right) \to \left(-1,\frac{13-i\sqrt{71}}{2}\right) \to (-1), \tag{3.64}$$

would dissociate a complex index from its complex conjugate, which is impossible, therefore the ODE (2.27) admits no first integral.

A similar result holds for the CGL3 system (3.62).

3.2.1.2 Counting of the Arbitrary Constants

When the ODE (2.1) is assumed nonintegrable, like (2.27) or the system (3.62), the number of integration constants which can be present in any closed form solution is strictly smaller than the differential order of the ODE. This difference, which is an indicator of the amount of integrability of the ODE, can be precisely computed from a local analysis.

Two local representations of the general solution of (2.1) exist. The first one, also the most well known, is useless for our purpose. This is the Taylor series near a regular point, whose existence, unicity, convergence, holomorphy, etc is stated by the famous existence theorem of Cauchy, e.g. (2.43) for the KS ODE (2.27). The reason why it is useless is its inability to make a distinction between chaotic ODEs such as (2.27) and integrable ODEs such as $u''' - 12uu' - 1 = 0$ (P1).

The second one is the Laurent series (or more generally psi-series and/or Puiseux series) near a movable singular point x_0, which has been considered in various examples in Chap. 2. This one does provide the expected information.

In the Kuramoto–Sivashinsky ODE (2.27), it has been argued essentially with physical considerations [402] that the two irrational indices $(13 \pm i\sqrt{71})/2$ are responsible for the dense movable branching of the solution in the complex plane and consequently for the chaotic behavior. The only way to remove such a behavior is to require $\varepsilon c_+ = \varepsilon c_- = 0$, i.e. $\varepsilon = 0$ in (2.42), thus restricting to a single arbitrary constant the analytic part of the solution.

A similar reasoning can be made for the CGL3 equation and its two irrational indices $(7 \pm \sqrt{1-24\alpha^2})/2$, (2.58).

The question is then to turn this local information into a global one, i.e. to find the closed form singlevalued expression depending on the maximal number (here one) of movable constants.

We will call *unreachable* any constant of integration which cannot participate in any closed form solution. The KS ODE (2.27) has two unreachable integration constants, the third one, x_0, being irrelevant since it reflects the invariance of (2.27) under a translation of x.

We will also call the *general analytic solution* the closed form solution which depends on the maximal possible number of reachable integration constants, and our goal is precisely to exhibit a closed form expression for this general analytic solution, whose local representation is a Laurent series like (2.36).

The above notions (irrelevant, unreachable) belong to an equation, not to a solution. Let us introduce another integer number, attached to a solution, allowing one to measure its distance to the general analytic solution.

The *distance* of a closed form solution to the general analytic solution is defined as the number of constraints between the fixed constants and the reachable relevant movable constants.

For the KS ODE (2.27), the fixed constants are ν, b, μ, A, the movable constant x_0 is irrelevant, the movable constants $c_1 = \varepsilon c_+, c_2 = \varepsilon c_-$ are unreachable, so the distance d is the number of constraints among the fixed constants.

For the CGL3 system (3.62), the movable constants x_0, φ_0 are irrelevant, the movable constants associated with the indices $(7 \pm \sqrt{1-24\alpha^2})/2$ are unreachable, so the distance d is also the number of constraints among the fixed constants. In [197], another counting, based on the various possible topological structures (fronts, sources, sinks, etc) is made for CGL3 and provides the same results.

Table 3.1 summarizes this counting for the two nonintegrable ODEs considered in this section. For convenience, and also because of their high physical interest, we include in this table the traveling wave reduction of the following related PDEs,

1. the one-dimensional quintic complex Ginzburg–Landau equation (CGL5),

$$\begin{aligned} & iA_t + pA_{xx} + q|A|^2 A + r|A|^4 A - i\gamma A = 0,\ pr \neq 0,\ \mathrm{Im}(p/r) \neq 0, \\ & (A, p, q, r) \in \mathscr{C},\ \gamma \in \mathscr{R}, \end{aligned} \tag{3.65}$$

2. the Swift–Hohenberg equation [398, 277] (SH)

$$\begin{aligned} & iA_t + bA_{xxxx} + pA_{xx} + q|A|^2 A + r|A|^4 A - i\gamma A = 0,\ br \neq 0, \\ & (A, b, p, q, r,) \in \mathscr{C},\ \gamma \in \mathscr{R}, \end{aligned} \tag{3.66}$$

   in which b, p, q, r, γ are constants,
3. two coupled one-dimensional CGL3 equations

$$E \equiv \begin{cases} iA_t + ivA_x + pA_{xx} + q(|A|^2 + \delta|B|^2)A - i\gamma A = 0, \\ iB_t - ivB_x + pB_{xx} + q(|B|^2 + \delta|A|^2)B - i\gamma B = 0, \end{cases} \tag{3.67}$$

in which the coupling parameter δ is a complex constant and v is the (real) group velocity. These describe for instance the amplitudes of two lasers [380], or the spatiotemporal intermittency [11, 202], or hydrothermal waves [48].

Table 3.1 Integer numbers involved in the characterization of the general analytic solution of a nonintegrable ODE. The vocabulary (irrelevant, unreachable) is defined in Sect. 3.2.1.2. The column "Available" indicates the number of relevant, reachable integration constants, this is the algebraic sum "Order" – "Irrelevant" – "Unreachable".

Equation	Order	Irrelevant	Unreachable	Available
CGL3	4	$2 = \xi_0, \varphi_0$	2	0
2 CGL3	8	$3 = \xi_0, \varphi_{A,0}, \varphi_{B,0}$	5	0
CGL5	4	$2 = \xi_0, \varphi_0$	2	0
KS	3	$1 = \xi_0$	2	0
SH	8	$2 = \xi_0, \varphi_0$	6	0

For both the KS equation (2.25) and the CGL3 equation (2.45), there is some experimental evidence (this is not a proof) for the existence of the general analytic solution. Indeed, computer simulations as well as real experiments (for a review, see [376]) sometimes display regular patterns in the (x,t) plane, and some of them are described by some analytic solution. For the remaining patterns, the guess is that there should exist analytic expressions, to be found, corresponding to these patterns. For the KS equation (2.25), one thus observes a homoclinic[4] solitary wave [403, Fig. 7] $u = f(\xi), \xi = x - ct$, while all solutions known to date are heteroclinic. For the CGL3 equation (2.45) the existence of a fourth physically interesting solution has been predicted [196], which is a distance-one homoclinic hole solution with an arbitrary velocity c.

Let us now turn to the methods which may be able to provide the general analytic solution, of course in closed form.

3.2.2 Elliptic Traveling Waves

Having found an elliptic function for the general traveling wave of KdV and NLS, one might wonder whether the general analytic solution of either KS or CGL3 can also be elliptic. It appears that this question (to characterize all the elliptic solutions) is much easier to answer [224] than the one to characterize all the trigonometric solutions.

Indeed, elliptic functions (i.e. doubly periodic functions in the complex plane) enjoy the following nice property. Inside a period parallelogram, any elliptic func-

[4] When the real variable $\xi = x - ct$ goes to $\pm\infty$, the solution is said to be homoclinic if the two limits of the solution are the same, otherwise it is called heteroclinic.

tion possesses as many zeroes as poles (counting multiplicity), and the sum of the residues at the poles is necessarily zero. For instance, the function $\wp$ has two simple zeroes and one double pole, the Laurent expansion at this pole is

$$\wp(x,g_2,g_3) = \frac{1}{x^2} + \frac{g_2}{20}x^2 + \frac{g_3}{28}x^4 + \mathcal{O}(x^6), \tag{3.68}$$

and the coefficient of x^{-1} is zero. Similarly, any one of the twelve Jacobi functions pq has two simple zeroes and two simple poles, with a polar expansion

$$\mathrm{pq}(x,k) = \pm\frac{c}{x-x_0} + \mathcal{O}(x-x_0),\ c \neq 0, \tag{3.69}$$

so the sum of the two residues is again zero.

3.2.2.1 Necessary Conditions for Elliptic Solutions

Let us now assume that either the KS ODE (2.27) or the CGL3 system (3.62) possesses an elliptic solution. For the KS ODE (2.27), which admits only one family of movable triple poles, this implies a period parallelogram containing one triple pole and three simple zeroes, and the residue criterium from the Laurent series (2.36) is the necessary condition [224]

$$b^2 - 16\mu\nu = 0. \tag{3.70}$$

For the CGL3 system (3.62), as seen from the NLS limit handled in Sect. 3.1.3, the relevant variable to consider is M or φ'; these variables admit two families of movable singularities represented by the Laurent series

$$d_i \neq 0: \begin{cases} M = \dfrac{9d_r \pm 3\Delta}{2d_i^2}\chi^{-2}\left(1 + \dfrac{cs_i}{3}\chi + \mathcal{O}(\chi^2)\right), \\ \varphi' = \dfrac{c}{2p_r} + \mathcal{O}(\chi), \end{cases} \tag{3.71}$$

with the degeneracy $d_i = 0$ to the single family

$$d_i = 0: \begin{cases} M = -\dfrac{2}{d_r}\chi^{-2}\left(1 + \dfrac{cs_i}{3}\chi + \mathcal{O}(\chi^2)\right), \\ \varphi' = \dfrac{c}{2p_r} + \mathcal{O}(\chi), \end{cases} \tag{3.72}$$

If the solution were elliptic, the period parallelogram would contain two double poles and four simple zeroes, with the necessary condition of vanishing of the sum of the two residues in (3.71),

$$d_i \neq 0:\ d_r c s_i = 0. \tag{3.73}$$

Under the degeneracy $d_i = 0$ to one Laurent series (3.72), i.e. one double pole and two simple zeroes, the necessary condition becomes

$$d_i = 0: \; cs_i = 0, \tag{3.74}$$

therefore the overall necessary condition is

$$\forall d_i: \; d_r cs_i = 0. \tag{3.75}$$

Infinitely many such necessary conditions exist, by requiring a similar property from any rational function of M and its derivatives, e.g. M^n or φ'^n with n integer, and the final necessary conditions are [224, 416]

$$\begin{cases} d_i \neq 0, \forall d_r: \; cs_i = 0, \; g_r = 0, \; g_i = 0, \\ d_i = 0: \; cs_i = 0. \end{cases} \tag{3.76}$$

Under the degeneracy $g_2^3 - 27g_3^2 = 0$ of elliptic to trigonometric, this residue criterium unfortunately does not survive, and more elaborate methods are necessary to characterize *all* the trigonometric solutions of an ODE, this will be the subject of Sects. 3.2.5 and 3.2.8.

3.2.2.2 The Elliptic Solutions

Knowing the necessary conditions (3.70) and (3.76) for the KS ODE (2.27) or the CGL3 system (3.62) to possess an elliptic solution, let us establish these solutions.

There exist two kinds of methods to possibly find these elliptic solutions. The first kind [320] has the advantage of requiring no additional assumption and of being able to find all the elliptic solutions, but it sometimes requires the use of computer algebra. It will be presented in Sect. 3.2.5. The second kind, which we now present, has the advantage of being very simple and the drawback of being unable to find those elliptic solutions which are outside some class of expressions since this class is given as an input.

Like all other methods introduced in Sects. 3.2.3.2 and 3.2.4, this latter method *a priori* assumes a given class of expressions compatible with the structure of singularities. In the KS case, knowing the triple pole $u \sim 120\nu\chi^{-3}$ of the KS ODE (2.27) and the double pole of the Weierstrass function $\wp$, a compatible assumption is the polynomial expression [153, 260]

$$u = c_0\wp' + c_1\wp + c_2, \; c_0 \neq 0. \tag{3.77}$$

After substitution in the KS ODE (2.27) and elimination of the derivatives of $\wp$ with

$$\wp'' = 6\wp^2 - \frac{g_2}{2}, \tag{3.78}$$

$$\wp'^2 = 4\wp^3 - g_2\wp - g_3, \tag{3.79}$$

the l.h.s. of (2.27) becomes an expression similar to (3.77), i.e. a polynomial in $\wp, \wp'$ of degree one in $\wp'$,

$$E(u) = E_{3,0}\wp^3 + E_{1,1}\wp\wp' + E_{2,0}\wp^2 + E_{0,1}\wp' + E_{1,0}\wp + E_{0,0} = 0, \quad (3.80)$$

and one has to solve the six *determining equations* $E_{j,k} = 0$

$$\begin{cases} E_{3,0} \equiv c_0(120\nu + 2c_0) = 0, \\ E_{1,1} \equiv 12bc_0 + c_0c_1 + 12\nu c_1 = 0, \\ E_{2,0} \equiv 6\mu c_0 + 6bc_1 + \dfrac{c_1^2}{2} = 0, \\ E_{0,1} \equiv c_0c_2 + \mu c_1 = 0, \\ E_{1,0} \equiv c_1c_2 - \dfrac{1}{2}g_2c_0^2 - 18\nu g_2c_0, \\ E_{0,0} \equiv A - 12\nu g_3c_0 + \dfrac{1}{2}\left(c_2^2 - bg_2c_1 - \mu g_2c_0 - g_3c_0^2\right), \end{cases} \quad (3.81)$$

for the five unknowns c_0, c_1, c_2, g_2, g_3. As a general rule, such determining equations *must* be solved by decreasing order of the singularity degree, here $-6, -5, -4, -3, -2, 0$, like for the computation of the Laurent series. The result is [153, 260]

$$\begin{cases} u = -60\nu\wp' - 15b\wp - \dfrac{b\mu}{4\nu}, \\ b^2 = 16\mu\nu,\ g_2 = \dfrac{\mu^2}{12\nu^2},\ g_3 = \dfrac{13\mu^3 + \nu A}{1080\nu^3}, \end{cases} \quad (3.82)$$

therefore the necessary condition (3.70) is sufficient.

Consider now the CGL3 system (3.62) with the necessary conditions (3.76). In the CGL3 case properly defined as $d_i \neq 0$, the two Laurent series for M and φ' terminate,

$$d_i \neq 0,\ cs_i = g_r = g_i = 0,\ M = \frac{9d_r \pm 3\Delta}{2d_i^2(x - x_0)^2},\ \varphi' = \text{constant}(x - x_0)^{-1}, \quad (3.83)$$

therefore CGL3 admits no elliptic solution [224]. In the real Ginzburg–Landau case $d_i = 0$ with the necessary conditions $cs_i = 0, g_r = 0$ (3.76), the notation (2.52) shows the identity of the system (3.62) to that of the NLS case, therefore the solution is elliptic and mathematically identical to (3.52). This elliptic traveling wave has been beautifully extrapolated [54, Eqns. (2.11), (3.7), (3.8ab)] to a solution representing the collision of two fronts, with a dependence on both x and t, given by

$$\begin{cases} d_i = 0,\ s_i = 0:\ |A|^2 = -\dfrac{2}{d_r}\dfrac{\wp(\Phi(x,t))}{\Phi_x^2}, \\ \Phi = \displaystyle\sum_{j=1}^{3} C_j e^{k_jx + \omega_jt},\ (k_j, \omega_j) = \text{constant}. \end{cases} \quad (3.84)$$

3.2.3 *Trigonometric Traveling Waves of KS*

Let us determine particular traveling waves of the KS ODE (2.27) which are rational in one exponential function e^{kx} (in the present section the independent variable ξ is renamed x for coherence with other sections).

3.2.3.1 Gauge Transformation

There exists a trivial case when the general analytic solution is known, this is

$$b = 0,\ \mu = 0,\ A = 0,\ U = 120\nu(x - x_0)^{-3}. \tag{3.85}$$

Despite the lack of interest of this solution, the natural question of whether this case is gauge-equivalent to a nontrivial case must be investigated. This question is: under a $T(\alpha,\beta;g)$ transformation

$$(U,x) \mapsto (V,X)\ :\ U(x) = \alpha(x)V(X) + \beta(x),\ X = g(x), \tag{3.86}$$

depending on three adjustable functions (α,β,g), can the ODE (2.27) be converted to the ODE

$$\nu\frac{\mathrm{d}^3V}{\mathrm{d}X^3} + \frac{V^2}{2} = 0 \tag{3.87}$$

by some choice of the three adjustable functions? With the notation

$$\Lambda = \frac{\alpha'}{\alpha} + \frac{b}{3\nu}, \tag{3.88}$$

the transformed ODE reads

$$\begin{aligned}
&\nu\frac{\mathrm{d}^3V}{\mathrm{d}X^3} + 3\nu(g')^{-1}\left(\Lambda + \frac{g''}{g'}\right)\frac{\mathrm{d}^2V}{\mathrm{d}X^2}\\
&+(g')^{-2}\left[\mu - \frac{b^2}{3\nu} + 3\nu\left(\Lambda' + \Lambda^2 + \Lambda\frac{g''}{g'} + \frac{g'''}{3g'}\right)\right]\frac{\mathrm{d}V}{\mathrm{d}X} + \alpha(g')^{-3}\frac{V^2}{2}\\
&+(g')^{-3}\left[\beta + \frac{2b^3}{27\nu^2} - \frac{b\mu}{3\nu} + \left(\mu - \frac{b^2}{3\nu}\right)\Lambda + \nu\Lambda'' + 3\nu\Lambda\Lambda' + \nu\Lambda^3\right]V\\
&+\alpha^{-2}\left(A + \nu\beta''' + b\beta'' + \mu\beta' + \frac{\beta^2}{2}\right) = 0.
\end{aligned} \tag{3.89}$$

The identification requires five conditions (constant nonzero value for the coefficient of V^2, zero value for the coefficients of V'',V',V and for the term independent of V''',V'',V',V). The first condition yields $g' = \alpha^{1/3}$, the second one $\Lambda = b/(12\nu)$, the third one $b^2/(\mu\nu) = 144/47$, the fourth one $\beta = 5b\mu/(47\nu)$, the fifth one $A = -1800/47^3$. The gauges α and g result from the two successive quadratures

$$\frac{\alpha'}{\alpha} = -\frac{b}{4\nu}, \ \alpha = c_1^3 e^{-bx/(4\nu)},$$
$$g' = c_1 e^{-bx/(12\nu)}, \ g = c_1 \left(-\frac{12\nu}{b} e^{-bx/(12\nu)} - g_0\right), \quad (3.90)$$

with (c_1, g_0) arbitrary constants of integration. In the original variable $U(x)$, the solution is independent of c_1,

$$U = e^{-bx/(4\nu)} 120\nu \left(-\frac{12\nu}{b} e^{-bx/(12\nu)} - g_0\right)^{-3} + \frac{5b\mu}{47\nu}. \quad (3.91)$$

Using elementary trigonometry, this is a third degree polynomial in $\tanh(k/2)(x - x_0)$, with $\nu k^2/\mu = 1/47$ and x_0 arbitrary, a result first obtained by Kudryashov [259] by a different method.

More solutions expressible as such third degree polynomials can be found by the method explained in the next section.

3.2.3.2 Polynomials in tanh

The assumption (3.77) made in Sect. 3.2.2.2 expressed the consistency between the triple pole of the KS ODE (2.27) and the double pole of the elementary elliptic function $\wp$.

If one chooses for the elementary function one with a simple pole instead of a double pole, one should be able to find more solutions. This is one of the two principles of the method introduced in this section. The second principle is that this elementary function with a simple pole, let us call it τ, should be *defined* as the general solution of a first order nonlinear ODE possessing the Painlevé property, so as to be sure that this general solution is single valued. As a direct consequence of the residue condition recalled in Sect. 3.2.2.1, this desired function cannot be elliptic.

All the first order nonlinear ODEs with the PP have been characterized by the classical authors (Briot and Bouquet, Lazarus Fuchs, Poincaré, Painlevé) and there exists only one such first order nonlinear ODE with the two additional properties: (i) to be autonomous, (ii) to have only one family of movable simple poles. This privileged equation is the *Riccati equation* with constant coefficients, normalized so as to have residue one,

$$\tau' + \tau^2 + \frac{S}{2} = 0, \ S = -\frac{k^2}{2} = \text{constant} \in \mathcal{C}. \quad (3.92)$$

Because of the elementary transformations between tanh, cotanh, tan, cotan, these four functions are not considered as distinct in the present volume and the general solution of (3.92) is written as

$$\tau = \frac{k}{2} \tanh \frac{k}{2}(x - x_0), \quad (3.93)$$

with k and x_0 complex. The limit $1/(x-x_0)$ of τ when $k \to 0$ is recovered by first transforming tanh to cotanh by a shift of x_0.

An *a priori* assumption for a closed form solution expressing the consistency between the structure of singularities of the KS ODE (2.27) (one family of movable poles of order $-p=3$) and the structure of singularities of the Riccati ODE (3.92) (one family of movable simple poles) is therefore that u is a polynomial in τ [69, 84]

$$u = \sum_{j=0}^{-p} c_j \tau^{-j-p},\ c_0 \neq 0, \tag{3.94}$$

with c_j constants to be determined. By elimination of the derivatives of τ from (3.92), the l.h.s. of (2.27) is a similar polynomial ($-q=6$ for this ODE)

$$E(u) = \sum_{j=0}^{-q} E_j \tau^{-j-q} = 0, \tag{3.95}$$

and this polynomial must identically vanish,

$$\forall j \in [0,-q]:\ E_j = 0. \tag{3.96}$$

Before writing down and solving these *determining equations*, let us make this assumption even simpler, by taking advantage of the singular part operator $\mathscr{D}$ (2.44). Introducing an entire function ψ as the general solution of the second order linear ODE with constant coefficients

$$\psi'' + \frac{S}{2}\psi = 0, \tag{3.97}$$

one has

$$\tau = \frac{\psi'}{\psi} = \frac{\mathrm{d}}{\mathrm{d}\xi}\log\psi, \tag{3.98}$$

therefore $\mathscr{D}\log\psi$ is a polynomial similar to (3.94). Moreover, since the Laurent series (2.36) near ξ_0 is unique, the difference between these two polynomials of τ is regular near ξ_0, therefore this is a constant.

As a consequence, the previous assumption

$$\begin{cases} u = \sum_{j=0}^{-p} c_j \tau^{-j-p},\ c_0 \neq 0, \\ \tau' + \tau^2 + \frac{S}{2} = 0,\ S = -\frac{k^2}{2} = \text{constant}, \\ E(u) = \sum_{j=0}^{-q} E_j \tau^{-j-q} = 0, \\ \forall j \in [0,-q]:\ E_j = 0. \end{cases} \tag{3.99}$$

is identical to and can be replaced by the simpler one

$$\begin{cases} u = \mathscr{D} \log \psi + U, \\ \psi'' + \dfrac{S}{2}\psi = 0, \ S = -\dfrac{k^2}{2} = \text{ constant}, \\ E(u) = \sum\limits_{j=-p}^{-q} E_j \chi^{j+q} = 0, \ \chi = \dfrac{\psi}{\psi'} = \tau^{-1}, \\ \forall j \in [-p, -q] : \ E_j = 0. \end{cases} \tag{3.100}$$

in which $\mathscr{D}$ is the singular part operator (2.44), and U and k^2 are constants to be determined. The computation of E_j only involves the elimination of ψ'' and higher derivatives of ψ.

The advantages of (3.100) over (3.99) are numerous:

1. fewer unknowns (U, k^2 instead of c_j, k^2),
2. fewer equations since the first equations $E_j = 0, j = 0, \dots, -p-1$ are identically zero by definition of $\mathscr{D}$,
3. and, most important, the possibility to replace the linear ODE defining ψ by any other linear ODE, since this respects the singularity structure.

This assumption (3.100) is nothing other than the reduction to ODEs of the famous *truncation method* initially designed for PDEs by Weiss et al. (WTC) [431]. It amounts to looking for solutions of the type

$$u = \mathscr{D} \log \psi + U, \tag{3.101}$$

in which $\mathscr{D}$ is the *singular part operator* $\mathscr{D}$, (2.44), ψ a given entire function (i.e. without any singularity at a finite distance, such as $e^x, \cosh x$), and U an adjustable function which has no singularities at the zeros of ψ. Naturally, ψ is defined by a differential equation, for instance a linear one with constant coefficients. Assuming a first order linear ODE for ψ yields nothing, for $\mathscr{D} \log(c_1 e^{kx})$ has no movable singularities. The assumption of a second order linear ODE for ψ as in (3.100) generates the system [84]

$$\begin{cases} E_j = 0, \ j = 0, 1, 2, \\ E_3 = 120\nu U - 300 b\nu S + \dfrac{195 b^3}{76\nu} - \dfrac{210}{19} b\mu, \\ E_4 = 60\left(-4\nu^2 S^2 - \dfrac{20}{19}\mu\nu S + \dfrac{11}{19^2}\mu^2\right) + \dfrac{15^2 b^4}{32 \times 19^2 \nu^2} \\ \quad - \dfrac{615 b^2 \mu}{4 \times 19^2 \nu} - 15 bU + \dfrac{1575}{38} b^2 S, \\ E_5 = \dfrac{S}{2} E_3 - \dfrac{15(b^2 - 16\mu\nu)}{76\nu} U, \\ E_6 = E(U) - 30\left(\dfrac{\mu^2}{19} S + \dfrac{20}{19}\mu\nu S^2 - 4\nu^2 S^3\right) \\ \quad - \dfrac{15}{2} bUS + \dfrac{15 b^2 \mu}{152\nu} S + \dfrac{1575}{76} b^2 S^2. \end{cases} \tag{3.102}$$

This system of four equations in two unknowns (U,S) is easy to solve because it is overdetermined. The linear equation $j=-p$ is first solved for U,

$$U=\frac{5}{2}bS+\frac{7b\mu}{4\times 19\nu}-\frac{13b^3}{32\times 19\nu^2} \tag{3.103}$$

Then the system $j=4,5,6$ admits six solutions, listed in Table 3.2, all represented by

$$u=120\nu\tau^3-15b\tau^2+\left(\frac{60}{19}\mu-30\nu k^2-\frac{15b^2}{4\times 19\nu}\right)\tau$$
$$+\frac{5}{2}bk^2-\frac{13b^3}{32\times 19\nu^2}+\frac{7\mu b}{4\times 19\nu},\ \tau=\frac{k}{2}\tanh\frac{k}{2}(\xi-\xi_0). \tag{3.104}$$

Table 3.2 The six trigonometric solutions of KS, (2.27), with the notation $k^2=-2S$. They all have the form (3.104). The last line is a degeneracy of the elliptic solution (3.82).

$\|b^2/(\mu\nu)\|$	$\nu A/\mu^3$	$\nu k^2/\mu$
0	$-4950/19^3,\ 450/19^3$	$11/19,\ -1/19$
$144/47$	$-1800/47^3$	$1/47$
$256/73$	$-4050/73^3$	$1/73$
16	$-18,\ -8$	$1,\ -1$

The solitary waves $b=0$ (see Fig. 3.2) were found by Kuramoto and Tsuzuki [263], and the three other values of $b^2/(\mu\nu)$ were added by Kudryashov [259].

The distance (as defined in Sect. 3.2.1.2) of these six trigonometric solutions to the general analytic solution is two, while the elliptic solution (3.82) has distance one. Therefore the problem remains open to find the general analytic solution, this will be investigated again in Sect. 3.2.5.

3.2.4 Trigonometric Traveling Waves of CGL3

The local singularity analysis has been performed in Sect. 2.1.3, and the presence of two irrational Fuchs indices forbids the existence of the three-parameter general solution of the third order system (3.62) and only allows a one-parameter particular solution, see the counting in Sect. 3.2.1.2. The problem is then, for any value of (p,q,γ,c,ω), to find this general analytic solution in closed form.

In order to apply a truncation method such as the one defined in Sect. 3.2.3.2, one must first determine a variable whose dominant behavior is single valued. This is the case of neither $(A,\bar{A})$, nor $(|A|,\arg A)$, and the limiting case NLS (Sect. 3.1.3)

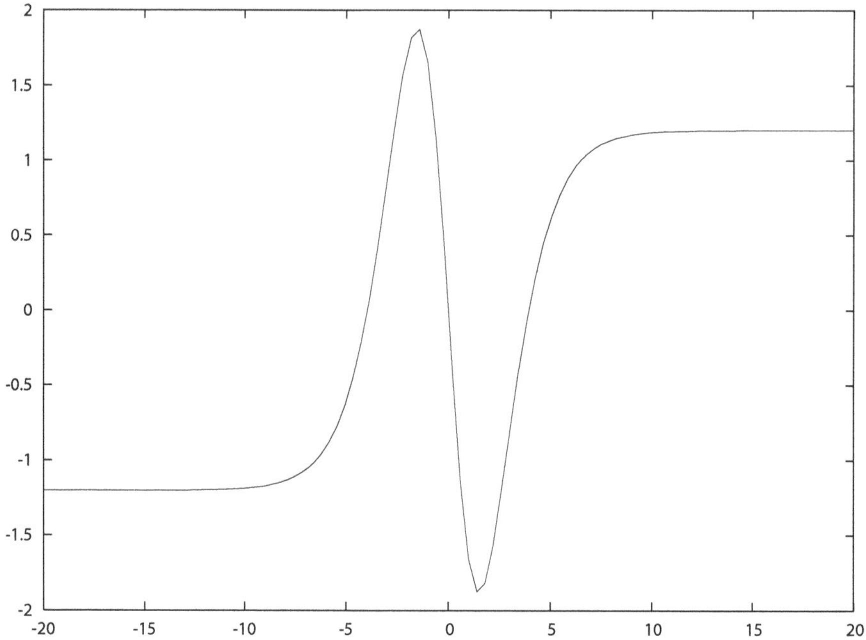

Fig. 3.2 KS, heteroclinic traveling wave of Kuramoto and Tsuzuki, (3.104) with $\nu=1,\mu=1,b=0,k^2=11/19$.

suggests considering either $M=|A|^2$ or φ', variables which obey the system (3.62) of two real equations.

Making two independent assumptions, one for M and one for φ', might create some undesirable constraints and prevent finding some solutions. Let us therefore first eliminate φ between the system (3.62), which results in

$$\varphi'=\frac{cs_r}{2}+\frac{G'-2cs_iG}{2M^2(g_r-d_iM)},\ \left(\varphi'-\frac{cs_r}{2}\right)^2=\frac{G}{M^2}, \tag{3.105}$$

$$(G'-2cs_iG)^2-4GM^2(d_iM-g_r)^2=0, \tag{3.106}$$

$$G=\frac{1}{2}MM''-\frac{1}{4}M'^2-\frac{cs_i}{2}MM'+d_rM^3+g_iM^2, \tag{3.107}$$

and let us concentrate on the single third order second degree equation (3.106) for $M=|A|^2$.

The variable M admits two families of movable double pole-like singularities (3.71) with singular part operators

$$\mathscr{D}_\pm=\frac{9d_r\pm 3\Delta}{2d_i^2}\left(-\frac{\mathrm{d}^2}{\mathrm{d}\xi^2}+\frac{cs_i}{3}\frac{\mathrm{d}}{\mathrm{d}\xi}\right). \tag{3.108}$$

The ODE (3.106) depends on five fixed parameters d_r, d_i, g_r, g_i, cs_i, out of which only three are essential (g_r, g_i, c, equivalent to γ, ω, c). Indeed, just like for NLS, p and q (i.e. $d_r + id_i$ and $s_r - is_i$) can be rescaled to convenient numerical values making rational the Laurent series coefficients, such as

$$\begin{aligned} & p = -1 - 3i,\ q = 4 - 3i, \\ & d_r = \frac{1}{2},\ d_i = \frac{3}{2},\ s_r = -\frac{1}{10},\ s_i = -\frac{3}{10},\ \Delta = \frac{9}{2}. \end{aligned} \tag{3.109}$$

Since the field M presents two families of movable double pole-like singularities, the one-family truncation introduced in Sect. 3.2.3.2 cannot yield the general analytic solution and the method must be extended. In Sect. 3.2.4.1 one performs this one-family truncation (polynomials in τ), and one finds all the solutions known to date except one.

In Sect. 3.2.4.2, a two-family truncation method is introduced, but, in the CGL3 case, it fails to provide any solution.

3.2.4.1 Polynomials in tanh

The solutions in which M is a polynomial in τ are searched for by the one-family truncation (3.99) or equivalently (3.100). In the present case, the assumption

$$\begin{cases} M = \mathscr{D}_{\pm} \log \psi + m, \\ \psi'' + \dfrac{S}{2}\psi = 0,\ S = -\dfrac{k^2}{2} = \text{constant}, \end{cases} \tag{3.110}$$

with $\mathscr{D}_{\pm}$ defined in (3.108), transforms (3.106) into a truncated Laurent series

$$\sum_{j=2}^{14} E_j \chi^{j-14} = 0,\ E_0 \equiv 0,\ E_1 \equiv 0, \tag{3.111}$$

and one must solve the 13 determining equations $E_j = 0$ for two unknowns (S, m) and, essentially, three parameters (g_r, g_i, c). As seen from the first few determining equations written for the numerical values (3.109),

$$(\mathscr{D}_{+}\ \text{case}) : \begin{cases} M = 4\left(-\dfrac{\mathrm{d}^2}{\mathrm{d}\xi^2} - \dfrac{c}{10}\dfrac{\mathrm{d}}{\mathrm{d}\xi} + m\right)\log\psi, \\ E_2 \equiv \dfrac{57}{100}c^2 - 156m + 13k^2 + 4g_i + 16g_r = 0, \\ E_3 \equiv \left(\dfrac{39}{25}c^2 + 432m - 28k^2 - 16g_i - 48g_r\right)c = 0, \end{cases} \tag{3.112}$$

this presents no difficulty and, in the CGL3 case properly defined as $d_i \neq 0$, one obtains three solutions for each sign (physically, only three solutions are admissible), whose respective distance to the general analytic solution is one, two, two. These are

$$\begin{cases} M = -2\left[\left(\tau - \frac{c}{20}\right)^2 + \left(\frac{c}{10}\right)^2\right],\ \varphi' - \frac{cs_r}{2} = \tau + \frac{c}{20} + \frac{c}{5M}\left(\tau^2 - \frac{k^2}{4}\right), \\ k^2 = -7\left(\frac{c}{10}\right)^2 - \frac{4}{3}g_r,\ 3g_i + 2g_r + \frac{3c^2}{50} = 0, \end{cases} \tag{3.113}$$

$$\begin{cases} M = -2\left(\tau^2 - \frac{k^2}{4}\right),\ \varphi' - \frac{cs_r}{2} = \tau, \\ k^2 = 2g_r,\ c = 0,\ g_i = 0, \end{cases} \tag{3.114}$$

$$\begin{cases} M = -2\left(\tau \pm \frac{k}{2}\right)^2,\ \varphi' - \frac{cs_r}{2} = \tau - \frac{c}{20}, \\ k^2 = \left(\frac{c}{10}\right)^2,\ g_r = 0,\ g_i - \frac{c^2}{50} = 0, \end{cases} \tag{3.115}$$

and

$$\begin{cases} M = 4\left[\left(\tau - \frac{c}{20}\right)^2 + \left(\frac{c}{20}\right)^2\right],\ \varphi' - \frac{cs_r}{2} = -2\tau + \frac{c}{20} + \frac{c}{5M}\left(\tau^2 - \frac{k^2}{4}\right), \\ k^2 = -\left(\frac{c}{10}\right)^2 + \frac{2}{3}g_r,\ 3g_i - g_r + \frac{3c^2}{80} = 0, \end{cases} \tag{3.116}$$

$$\begin{cases} M = 4\tau^2,\ \varphi' - \frac{cs_r}{2} = -2\tau, \\ k^2 = \frac{2}{3}g_r,\ c = 0,\ 3g_i - g_r = 0, \end{cases} \tag{3.117}$$

$$\begin{cases} M = 4\left(\tau \pm \frac{k}{2}\right)^2,\ \varphi' - \frac{cs_r}{2} = -2\tau - \frac{c}{10}, \\ k^2 = \left(\frac{c}{10}\right)^2,\ g_r = 0,\ g_i - \frac{c^2}{50} = 0. \end{cases} \tag{3.118}$$

In the original variable A, these solutions are easier to express in complex notation [86]. Denoting (A_0^2, α) one of the two real roots of the complex equation

$$(-1 + i\alpha)(-2 + i\alpha)p + A_0^2 q = 0, \tag{3.119}$$

these six complex solutions are the following.

1. A heteroclinic source or propagating hole [25] (see Fig. 3.3)

$$\begin{cases} A = A_0\left[\frac{k}{2}\tanh\frac{k}{2}\xi - \frac{iqp_i}{2(1 - i\alpha)p|p|^2 d_i}c\right] e^{i[-\omega t + \Phi]}, \\ \Phi = \alpha \log\cosh\frac{k}{2}\xi + \frac{q_i}{2|p|^2 d_i}c\xi, \\ \frac{i\gamma - \omega}{p} = \left(\frac{c}{2p}\right)^2 - (2 - 3i\alpha)\frac{k^2}{4}, \end{cases} \tag{3.120}$$

in which the velocity c is arbitrary. Indeed, the real and imaginary parts of the last equation define the value of ω and a linear relation between c^2 and k^2, see [86, (79)].

2. A homoclinic pulse or solitary wave [352]

$$\begin{cases} A = A_0(-ik\,\mathrm{sech}\,kx)e^{i[\alpha \log\cosh kx - \omega t]}, \\ \dfrac{i\gamma - \omega}{p} = (1-i\alpha)^2 k^2,\ c = 0. \end{cases} \tag{3.121}$$

3. A heteroclinic front or shock [334]

$$\begin{cases} A = A_0 \dfrac{k}{2}\left[\tanh\dfrac{k}{2}\xi + \varepsilon\right] e^{i\left[\alpha\log\cosh\frac{k}{2}\xi + \frac{3p_r+\alpha p_i}{6|p|^2}c\xi - \omega t\right]}, \\ \dfrac{i\gamma-\omega}{p} = \left(\dfrac{c}{2p}\right)^2 + \dfrac{k^2}{4},\ \dfrac{k}{2} = \varepsilon\dfrac{p_i c}{6|p|^2},\ \varepsilon^2 = 1. \end{cases} \tag{3.122}$$

None of these six solutions requires any constraint on p, q, γ.

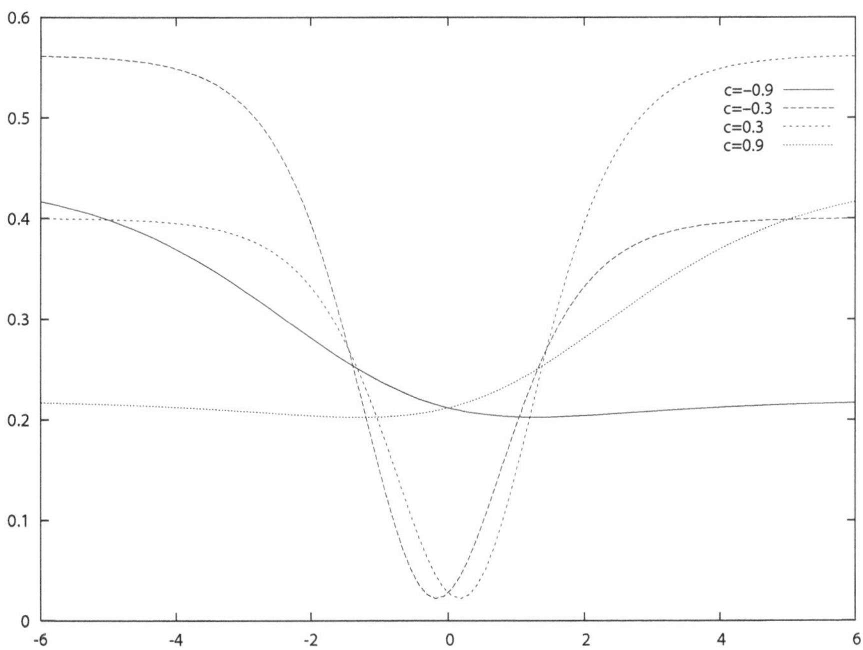

Fig. 3.3 CGL3, heteroclinic source or propagating hole (3.120). The plotted curves $x \to ((k/2)\tanh k\xi/2 + ac)^2 + (bc)^2$, with $a = 0.2, b = 0.5, k^2 + 2c^2 = 2, \xi = x - ct, t = 1$, display the qualitative dependence of the solution on the arbitrary velocity c.

3.2.4.2 Polynomials in tanh and sech

The existence of a fourth physically interesting solution has been predicted [196], which is a homoclinic hole with an arbitrary velocity c, which should exist for one constraint among the fixed coefficients of the system (3.62). This is another motivation to investigate it by a method, which we now describe, which fully respects the singularity structure (two families for M).

Since the elementary variable τ defined in (3.92) has only one family, one must introduce another elementary variable, let us call it σ, with the following properties: (i) to possess two families of movable simple poles, (ii) to be *defined* by some autonomous first order ODE with the PP, (iii) not to be elliptic. From the classical results on first order nonlinear ODEs with the PP, the defining ODE must have a genus equal to zero and a degree at least equal to two. Such an ODE has been encountered at the very beginning of this volume, see (1.2). In order to couple its singularities (two simple poles with opposite residues) with those of (3.92) (one simple pole), the variables (τ,σ) will be *defined* as the general solution of the following coupled system [236, 85] (the reason for the replacement $k \to 2k$ will soon be apparent),

$$\begin{cases} \tau' + \tau^2 - k^2 + k\mu\sigma = 0, \\ \sigma' + \sigma\tau = 0. \end{cases} \tag{3.123}$$

This system is part of a larger class of linearizable systems called *projective Riccati systems* [12]. It admits a first integral which we conveniently denote μ since the system only depends on (μ,σ) by the product $\mu\sigma$,

$$\frac{k^2 - \tau^2 - 2k\mu\sigma}{(\mu\sigma)^2} = \text{constant} = -\mu^{-2}, \tag{3.124}$$

and its general solution is

$$\begin{cases} \tau = \dfrac{k \sinh k(x-x_0)}{\cosh k(x-x_0) + \cosh ka}, \quad \mu = \coth ka, \\ \sigma = \dfrac{k \sinh ka}{\cosh k(x-x_0) + \cosh ka}, \end{cases} \tag{3.125}$$

in which the two constants of integration are x_0 and a.

In place of (3.99), the assumption for solutions polynomial in (τ,σ) is (taking account of the first integral)

$$\begin{cases} u = \sum_{l=0}^{1} \sum_{j=0}^{-p-l} c_{j,l} \sigma^l \tau^j, \ (c_{-p,0}, c_{-p-1,1}) \neq (0,0), \\ \tau' + \tau^2 - k^2 + k\mu\sigma = 0, \ \sigma' + \sigma\tau = 0, \\ k^2 - \tau^2 - 2\mu k\sigma + \sigma^2 = 0, \\ E(u) = \sum_{l=0}^{1} \sum_{j=0}^{-q-l} E_{j,l} \sigma^l \tau^j, \\ \forall j \in [0, -q], \forall l : \ E_{j,l} = 0. \end{cases} \tag{3.126}$$

The practical computation [85] presents no difficulty, except that one must discard the cases of linear dependence between τ and σ, i.e. the two values $\mu = \pm 1$. The value $\mu = 0$ represents the particular case

$$\mu = 0: \ \tau = k \tanh k(x - x_0), \ \sigma = ik \operatorname{sech} k(x - x_0). \tag{3.127}$$

The truncation (3.126) has the disadvantage of breaking the symmetry between τ and σ, since a degree one must be assumed in one of the two elementary variables to take the first integral into account. A more elegant formulation consists in replacing the above assumption (3.126) by the equivalent of (3.100). The projective Riccati system (3.123) is mapped by the transformation (note the symmetry between τ and σ) [86]

$$\begin{cases} \tau = \frac{\mathrm{d}}{\mathrm{d}x} \log \psi_1 + \frac{\mathrm{d}}{\mathrm{d}x} \log \psi_2, \\ \sigma = \frac{\mathrm{d}}{\mathrm{d}x} \log \psi_1 - \frac{\mathrm{d}}{\mathrm{d}x} \log \psi_2, \end{cases} \tag{3.128}$$

into the linear system

$$\begin{pmatrix} \psi_1 \\ \psi_2 \end{pmatrix}_x = \frac{k}{2 \sinh ka} \begin{pmatrix} \cosh ka & -1 \\ 1 & -\cosh ka \end{pmatrix} \begin{pmatrix} \psi_1 \\ \psi_2 \end{pmatrix}, \tag{3.129}$$

whose general solution is

$$\psi_1 = \cosh \frac{k}{2}(x - x_0 + a), \ \psi_2 = \cosh \frac{k}{2}(x - x_0 - a), \tag{3.130}$$

i.e. ψ_1 and ψ_2 are two different solutions of the same linear ODE

$$\psi_j'' - \frac{k^2}{4} \psi_j = 0, \ j = 1, 2. \tag{3.131}$$

Given the two singular part operators $\mathscr{D}\pm$, (3.108), the sum $M = \mathscr{D}_+ \log \psi_1 + \mathscr{D}_- \log \psi_2$, when evaluated *modulo* (3.131), is a polynomial in the two variables $(\log \psi_j)', j = 1, 2$ linked by the first integral μ. The advantage over (3.124) is the characterization of this first integral, the ratio of two constant Wronskians of (3.131), by the relation

$$\frac{\psi_{1,x}}{\psi_1}\frac{\psi_{2,x}}{\psi_2} = \frac{k^2}{4} - \mu\frac{k}{2}\left[\frac{\psi_{1,x}}{\psi_1} - \frac{\psi_{2,x}}{\psi_2}\right], \tag{3.132}$$

which splits the polynomial of two variables into the sum of two polynomials of one variable.

To conclude, there is much advantage in replacing the dissymetric assumption (3.126) with the symmetric one (two-family version of (3.100)) [86, App. A]

$$\begin{cases} u = \mathscr{D}_+ \log\psi_1 + \mathscr{D}_- \log\psi_2 + U, \\ \psi_1'' + \frac{S}{2}\psi_1 = 0,\ \psi_2'' + \frac{S}{2}\psi_2 = 0,\ S = -\frac{k^2}{2} = \text{constant}, \\ \frac{\psi_1'}{\psi_1}\frac{\psi_2'}{\psi_2} = \frac{k^2}{4} - \mu\frac{k}{2}\left[\frac{\psi_1'}{\psi_1} - \frac{\psi_2'}{\psi_2}\right], \\ E = \left(\sum_{j=1}^{-q} E_{-j}\left(\frac{\psi_1'}{\psi_1}\right)^j\right) + E_0 + \left(\sum_{j=1}^{-q} E_j\left(\frac{\psi_2'}{\psi_2}\right)^j\right), \\ \forall j \in [q,-q] :\ E_j = 0. \end{cases} \tag{3.133}$$

In the particular case $\mu = 0$, this truncation reduces to the one presented in [356].

When applied to the variable $u = M$ in the CGL3 case $d_i \neq 0$, the determining equations $E_j = 0$ in the movable parameters k^2, μ, m admit no solution, and this again reflects the difficulty of CGL3. Would such a solution exist, its expression would be

$$M = \left(\frac{3\sqrt{9d_r^2 + 8d_i^2}}{2d_i^2}\tau + c_1\right)\sigma + \frac{9d_r}{2d_i^2}\tau^2 + c_3\tau + c_4, \tag{3.134}$$

and the constraint $c_3 = 0$ would define a homoclinic hole solution, just like the (yet analytically unknown) one of van Hecke.

Remark. As can be guessed from the writings (3.120) and (3.121), the homoclinic pulse, which presents a sech two-family behavior, is quite easy to obtain by (3.133) if one represents the two multivalued original fields $(A, \bar{A})$ by three fields $(Z, \bar{Z}, \Theta)$ defined by [86]

$$A = Ze^{i\Theta},\ \bar{A} = \bar{Z}e^{-i\Theta},\ Z \in \mathscr{C},\ \Theta \in \mathscr{R}. \tag{3.135}$$

The three fields $(Z, \bar{Z}, \operatorname{grad}\Theta)$ are then locally single valued near a movable singularity, and they have two families of movable simple poles (in the notation of (2.53), the sign ε_1 is fixed),

$$\begin{cases} \frac{Z}{A_0} = \chi^{-1} + \frac{p_i + 3ip_r + ip_i\alpha}{6|p|^2}c - i\Theta_{0,x} + \mathscr{O}(\chi), \\ \Theta = \alpha\log\psi + \Theta_0,\ \frac{\psi_x}{\psi} = \chi^{-1}, \\ A_0^2 = \frac{3(3d_r + \varepsilon_1\Delta)}{2d_i^2},\ \alpha = \frac{3d_r + \varepsilon_1\Delta}{2d_i},\ \Delta = \sqrt{9d_r^2 + 8d_i^2},\ \varepsilon_1^2 = 1. \end{cases} \tag{3.136}$$

The truncation (3.133) of this set of three fields

$$\begin{cases} Z = A_0 \partial_x \log \psi_1 - A_0 \partial_x \log \psi_2 + X + iY, \\ \bar{Z} = A_0 \partial_x \log \psi_1 - A_0 \partial_x \log \psi_2 + X - iY, \\ \Theta = \alpha \log \psi_1 + \alpha \log \psi_2 + \Theta_0, \end{cases} \tag{3.137}$$

is the most efficient way [86, App. A] to obtain the pulse solution (3.121).

Remark. One important advantage of (3.133) over (3.126) is the possibility to change the definition of the linear system for ψ_j. In particular, taking a third order linear equation with constant coefficients instead of a second order one provides immediately the solution "collision of two shocks" [333, 334], which only exists for $p_r = 0$, and whose degeneracy $q_r = 0$ is identical to the similar solution (5.261) of KPP.

Let us terminate this section by an important remark. Given a one-family equation, performing its two-family truncation is useless and cannot yield results additional to those of the one-family truncation. This is a direct consequence of the elementary identities

$$\tanh z - \frac{1}{\tanh z} = -2i \operatorname{sech}\left[2z + i\frac{\pi}{2}\right], \ \tanh z + \frac{1}{\tanh z} = 2\tanh\left[2z + i\frac{\pi}{2}\right]. \tag{3.138}$$

The sum of two logarithmic derivatives cannot generate a sech.

3.2.5 General Method to Find Elliptic Traveling Waves

The methods described in Sects. 3.2.2.2, 3.2.3.2 and 3.2.4.2 share a built-in restriction, which prevents them from finding the desired result even in some elementary cases. For instance, if one considers the rational trigonometric solution

$$u = \frac{\tanh(\xi - \xi_0)}{2 + \tanh^2(\xi - \xi_0)}, \tag{3.139}$$

and builds the first order ODE which it obeys (this is a common way to construct examples),

$$u'^2 + \left(12u^2 - \frac{3}{2}\right)u' + 36u^4 - \frac{17}{2}u^2 + \frac{1}{2} = 0, \tag{3.140}$$

then none of the above truncation methods ((3.77), (3.100), (3.133)) succeeds in finding its solution.

Similarly, in our two main examples of the traveling waves of KS and CGL3, the general analytic solution could not be obtained, the best achievement being solutions with a distance unity to the general analytic solution, and the problem remains open,

1. in the KS equation (2.27), to extrapolate the traveling wave (3.104) by removing the constraints between the fixed parameters listed in Table 3.2,
2. in the CGL3 equation (3.106), to extrapolate the propagating hole solution (3.120) of Bekki and Nozaki to one more arbitrary constant by removing the constraint between the fixed parameters.

The natural generalization of such a situation is an N-th order autonomous algebraic ODE (2.1), for which any solution is necessarily

$$u = f(\xi - \xi_0), \tag{3.141}$$

in which ξ_0 is movable. Provided the elimination of ξ_0 between (3.141) and its derivative is possible, one obtains the first order nonlinear autonomous ODE

$$F(u, u') = 0, \tag{3.142}$$

in which F is as unknown as f.

However, $f(\xi - \xi_0)$ is now the *general solution* of (3.142) (while it is only a particular solution of (2.1)), and there exist classical results on first order autonomous ODEs which are in addition *algebraic*. Let us therefore assume from now on that the dependence of f on ξ_0 is algebraic (this is a sufficient condition for F to be algebraic).

To summarize: given the N-th order ODE (2.1) and its particular solution f (3.141), and assuming the dependence of f on ξ_0 to be algebraic, one is able to derive a first order ODE (3.142) which is algebraic.

Conversely, given an algebraic first order ODE $F = 0$ (3.142), is it possible to go back to f? This question has been answered positively by Briot and Bouquet, L. Fuchs, Poincaré and put in final form by Painlevé [346, pp. 58–59].

Theorem 3.1. *Given the algebraic first order autonomous ODE $F = 0$ (3.142), if its general solution is singlevalued, then*

1. *Its general solution is an elliptic function, possibly degenerate, and its expression is known in closed form.*
2. *The genus of the algebraic curve (3.142) is one or zero.*
3. *There exists a positive integer m and $(m+1)^2$ complex constants $a_{j,k}$, with $a_{0,m} \neq 0$, such that the polynomial F has the form*

$$F(u, u') \equiv \sum_{k=0}^{m} \sum_{j=0}^{2m-2k} a_{j,k} u^j u'^k = 0, \; a_{0,m} \neq 0. \tag{3.143}$$

Then, assuming f to be singlevalued with an algebraic dependence on ξ_0,

1. It is equivalent to search for the solution f or for the first order equation $F = 0$.
2. The solution f can only be elliptic (i.e. rational in $\wp$ and $\wp'$), or a rational function of e^{ax} with a constant, or a rational function of x.

Since the most singular term u'^m must balance another term in (3.143), this condition sets m to have the lower bound $-p$. For instance, with $p=-3$ as in the KS ODE (2.27), for $m=1$ or $m=2$ no other term can match the power $a_{0,m}(\xi-\xi_0)^{(p-1)m}$, therefore one must set $m \geq 3$. There also exists an upper bound for m, whose value depends on the structure of singularities of (2.1). Indeed, if the local analysis of (2.1) displays n_k distinct families of movable poles of order d_k, $k=1,\dots,K$, assuming the solution to be elliptic, this upper bound is [43, p. 277] [193, part II, chap. IX, p. 329], [207, p. 424]

$$\text{upper bound } m = \sum_k n_k d_k = \text{ number of poles, counting multiplicity.} \quad (3.144)$$

The explicit form (3.143) of F makes it much easier to look for F than for f, and an algorithm has been devised [320] which yields explicitly all possible subequations $F=0$, (3.143), then, using an algorithm [222] due to Poincaré, all solutions f in the canonical form

$$u = R(\wp',\wp) = R_1(\wp) + \wp' R_2(\wp), \quad (3.145)$$

in which R_1, R_2 are two rational functions, with the possible degeneracies

$$R(\wp',\wp) \longrightarrow R(e^{a\xi}) \longrightarrow R(\xi), \quad (3.146)$$

in which R denotes rational functions.

The input data and assumptions of this algorithm are:

1. an N-th order autonomous algebraic ODE (2.1), $N \geq 2$,
2. a Laurent series representing its general analytic solution,
3. a first order autonomous algebraic ODE (3.142) sharing its general solution with (2.1).

The algorithm is [320, Sect. 5]:

1. Compute finitely many terms of the Laurent series,

$$u = \chi^p \left(\sum_{j=0}^{J} u_j \chi^j + \mathcal{O}(\chi^{J+1}) \right), \ \chi = \xi - \xi_0. \quad (3.147)$$

   where p is -3 for the KS equation (2.27), and -2 for the CGL3 equation (3.106). This series excludes the contribution of the irrational Fuchs indices such as $(13 \pm i\sqrt{71})/2$ for KS or $(7 \pm \sqrt{1-24\alpha^2})/2$ for CGL3. It depends on at least one movable constant ξ_0.
2. Choose a positive integer m in the range $-p \leq m \leq$ total number of poles and define the first order ODE (3.148),

$$F(u,u') \equiv \sum_{k=0}^{m} \sum_{j=0}^{[(m-k)(p-1)/p]} a_{j,k} u^j u'^k = 0, \ a_{0,m} \neq 0, \quad (3.148)$$

in which $[z]$ denotes the integer part function. The upper bound on j implements the selection rule

$$m(p-1) \leq pj+(p-1)k, \tag{3.149}$$

identically satisfied if $p=-1$, that no term can be more singular than u'^m. The polynomial F contains at most $(m+1)^2$ unknown constants $a_{j,k}$.

3. Require the Laurent series to satisfy the Briot and Bouquet ODE (3.148), i.e. require the identical vanishing of the Laurent series for the l.h.s. $F(u,u')$ up to the order J

$$F \equiv \chi^{m(p-1)}\left(\sum_{j=0}^{J} F_j \chi^j + \mathcal{O}(\chi^{J+1})\right), \ \forall j \ : \ F_j = 0. \tag{3.150}$$

If it has no solution for $a_{j,k}$, increase m and return to first step.

4. For every solution, integrate the first order autonomous ODE (3.148).

The third step generates a *linear*, infinitely overdetermined, system of equations $F_j=0$ for the unknown coefficients $a_{j,k}$. This is quite an easy task to solve such a system. Practically, using the algorithm given in App. F, one computes just enough terms of the Laurent series to define a slightly overdetermined system.

If the ODE (2.1) is autonomous, the present method delivers *all* its solutions which are either elliptic or degenerate elliptic, i.e. rational in one exponential e^{ax} or rational in x.

Before proceeding with examples, let us examine how the results of this method compare with those of the one- and two-family truncation methods. The class of solutions u captured by these truncation procedures is typically [86] the polynomials in $\tanh(k/2)(x-x_0)$ of degree $-p$ for the one-family truncation, and the polynomials in tanh and sech (or more generally in τ and σ, (3.125)) of global degree $-p$ for the two-family truncation. By elimination of x_0 between u and u', these two classes satisfy a first order algebraic differential equation (3.143), and it is easy to check that its degree m is respectively $-p$ for the polynomials in tanh, and $-2p$ for the polynomials in tanh and sech. Indeed, for instance in the case of second degree polynomials in tanh, this amounts to eliminating tanh between the two polynomial equations

$$\begin{cases} \tanh^2 + 2a\tanh + b - u = 0, \\ 2(\tanh + a)(1-\tanh^2) - u' = 0, \end{cases} \tag{3.151}$$

which results in[5]

[5] This formula, due to Sylvester, expresses the resultant of two polynomials of degrees m and n as a determinant of order $m+n$.

$$\begin{vmatrix} 0 & -2 & -2a & 2 & 2a-u' \\ -2 & -2a & 2 & 2a-u' & 0 \\ 0 & 0 & 1 & 2a & b-u \\ 0 & 1 & 2a & b-u & 0 \\ 1 & 2a & b-u & 0 & 0 \end{vmatrix} \tag{3.152}$$

$$= (u' - 4a(u-b+a^2))^2 - 4(u-b+2a^2-1)^2(u-b+a^2) = 0, \tag{3.153}$$

a subequation with a degree $m = 2 = -p$, having genus zero.

Therefore, with the choice $m = -p$, one finds at least all the results of the one-family truncation, and with the choice $m = -2p$, at least all the results of the two-family truncation. Moreover, since (3.143) is form invariant under an homography on u, one also finds all the rational functions respectively in $\tanh(k/2)(x-x_0)$ of degree $-p$ (like (3.140)) and in tanh and sech of global degree $-p$.

Before applying this method to the two previously studied chaotic equations, let us give a tutorial example.

3.2.5.1 Application to KdV

Consider the third order autonomous ODE (3.34), which is the traveling wave reduction of KdV, with the aim of finding the first order elliptic subequation (3.36).

Since $p = -2$, the bounds on m are $2 \le m \le 2$, and one defines F with the selection rule (3.149) and $a_{0,2} = 1$,

$$F \equiv U'^2 + a_{0,1}U' + a_{1,1}UU' + a_{0,0} + a_{1,0}U + a_{2,0}U^2 + a_{3,0}U^3. \tag{3.154}$$

One first computes, with the algorithm given in Appendix F, slightly more than six Laurent coefficients,

$$U = 2a\chi^{-2} + U_4\chi^2 + U_6\chi^4 + \frac{U_4^2}{6a}\chi^5 + \dots, \tag{3.155}$$

then one inserts this series in (3.154) to generate the linear system (3.150),

$$\begin{cases} F_0 \equiv 16a^2a_{0,2} + 8a^3a_{3,0} = 0, \\ F_1 \equiv -8a^2a_{1,1} = 0, \\ F_2 \equiv 4a^2a_{2,0} = 0, \\ F_3 \equiv -4aa_{0,1} = 0, \\ F_4 \equiv 2aa_{1,0} - 16aa_{0,2}U_4 + 12a^2a_{3,0}U_4 = 0, \\ F_5 \equiv 0, \\ F_6 \equiv a_{0,0} + 4aa_{2,0}U_4 - 32aa_{0,2}U_6 + 12a^2a_{3,0}U_6 = 0, \\ \dots \end{cases} \tag{3.156}$$

Therefore the subequation is

$$U'^2 - (2/a)U^3 + 20U_4U + 56aU_6 = 0, \tag{3.157}$$

in which U_4 and U_6 are the two arbitrary constants arising from the Fuchs indices 4 and 6. This equation is of course identical to (3.36), but the way to obtain it is systematic and does not require any ability to find first integrals.

3.2.5.2 Application to KS

The Laurent series of (2.27) is (2.36).

In the second step, the bounds on m are $3 \le m \le 3$. With the normalization $a_{0,3} = 1$, the subequation contains ten coefficients, which are first determined by the Cramer system of ten equations $F_j = 0, j = 0 : 6, 8, 9, 12$. The remaining infinitely overdetermined nonlinear system for (ν, b, μ, A) contains as greatest common divisor (gcd) $b^2 - 16\mu\nu$ (see (3.70)), which defines a first solution

$$\frac{b^2}{\mu\nu} = 16,\ u_s = u + \frac{3b^3}{32\nu^2},$$
$$\left(u' + \frac{b}{2\nu}u_s\right)^2 \left(u' - \frac{b}{4\nu}u_s\right) + \frac{9}{40\nu}\left(u_s^2 + \frac{15b^6}{1024\nu^4} + \frac{10A}{3}\right)^2 = 0. \quad (3.158)$$

After division by this factor, the remaining system for (ν, b, μ, A) with $b^2 - 16\mu\nu \neq 0$ admits exactly four solutions (stopping the series at $j = 16$ is enough to obtain the result), identical to those listed in Table 3.2 (Sect. 3.2.3.2), each solution defining a first order third degree subequation,

$$b = 0,$$
$$\left(u' + \frac{180\mu^2}{19^2\nu}\right)^2 \left(u' - \frac{360\mu^2}{19^2\nu}\right) + \frac{9}{40\nu}\left(u^2 + \frac{30\mu}{19}u' - \frac{30^2\mu^3}{19^2\nu}\right)^2 = 0, \quad (3.159)$$

$$b = 0,\ u'^3 + \frac{9}{40\nu}\left(u^2 + \frac{30\mu}{19}u' + \frac{30^2\mu^3}{19^3\nu}\right)^2 = 0, \quad (3.160)$$

$$\frac{b^2}{\mu\nu} = \frac{144}{47},\ u_s = u - \frac{5b^3}{144\nu^2},\ \left(u' + \frac{b}{4\nu}u_s\right)^3 + \frac{9}{40\nu}u_s^4 = 0, \quad (3.161)$$

$$\frac{b^2}{\mu\nu} = \frac{256}{73},\ u_s = u - \frac{45b^3}{2048\nu^2},$$
$$\left(u' + \frac{b}{8\nu}u_s\right)^2 \left(u' + \frac{b}{2\nu}u_s\right) + \frac{9}{40\nu}\left(u_s^2 + \frac{5b^3}{1024\nu^2}u_s + \frac{5b^2}{128\nu}u'\right)^2 = 0. \quad (3.162)$$

In order to integrate the two sets of subequations (3.158), (3.159)–(3.162), one must first compute their genus[6], which is one for (3.158), and zero for (3.159)–(3.162). Therefore (3.158) has the elliptic general solution,

---

[6] For instance with the Maple command *genus* of the package *algcurves* [222], which implements an algorithm of Poincaré.

$$\begin{cases} u = -60\nu\wp' - 15b\wp - \dfrac{b\mu}{4\nu}, \\ b^2 = 16\mu\nu,\ g_2 = \dfrac{\mu^2}{12\nu^2},\ g_3 = \dfrac{13\mu^3 + \nu A}{1080\nu^3}, \end{cases} \tag{3.163}$$

initially found [153, 260] by assuming u polynomial in $\wp$ and $\wp'$. This solution, which succeeds in extrapolating the trigonometric functions of the solitary wave of Kuramoto and Tsuzuki, is best expressed with the singular part operator $\mathscr{D}$ (2.44), as

$$u = \mathscr{D} \log \sigma - \frac{b\mu}{4\nu}, \tag{3.164}$$

in which σ is the *entire* function defined by Weierstrass as

$$(\log \sigma)' = \zeta,\ \zeta' = -\wp. \tag{3.165}$$

This result is equivalent to having extrapolated the entire function cosh to the entire function σ of Weierstrass, see (C.21) in the Appendix.

As to the general solution of the four others (3.159)–(3.162), this is the third degree polynomial (3.104) in $\tanh k(\xi - \xi_0)/2$ already found by the one-family truncation method.

These four solutions constitute all the analytic results currently known on (2.27).

3.2.5.3 Application to CGL3

The first order subequation of (3.106) is defined as

$$F \equiv \sum_{k=0}^{m} \sum_{j=0}^{[3(m-k)/2]} a_{j,k} M^j M'^k = 0, \tag{3.166}$$

and the two Laurent series as (3.71).

To avoid carrying unpleasant square roots, we take the numerical values (3.109), which imply $\Delta = 9/2$, and the two Laurent series are

$$M_- = \chi^{-2} \left(-2 + \frac{c}{5}\chi + \left(\frac{g_r}{3} - \frac{g_i}{6} - \frac{c^2}{200} \right) \chi^2 + \mathscr{O}(\chi^3) \right), \tag{3.167}$$

$$M_+ = \chi^{-2} \left(4 - \frac{2c}{5}\chi + \left(\frac{16g_r}{39} + \frac{4g_i}{39} + \frac{19c^2}{1300} \right) \chi^2 + \mathscr{O}(\chi^3) \right). \tag{3.168}$$

The existence of two Laurent series, rather than only one, is a feature which the subequation must also possess, and this has the effect of setting the lower bound to $m = 4$ instead of 2. Indeed, the lowest degree subequations

$$F_2 \equiv M'^2 + M'(a_{1,1}M + a_{0,1}) + a_{3,0}M^3 + a_{2,0}M^2 + a_{1,0}M + a_{0,0} = 0, \tag{3.169}$$

$$F_3 \equiv M'^3 + M'^2(a_{1,2}M + a_{0,2}) + M'(a_{3,1}M^3 + a_{2,1}M^2 + a_{1,1}M + a_{0,1}) \\ + a_{4,0}M^4 + a_{3,0}M^3 + a_{2,0}M^2 + a_{1,0}M + a_{0,0} \quad = 0, \tag{3.170}$$

have the respective dominant terms $M'^2 + a_{3,0}M^3$ and $M'^3 + a_{3,1}M'M^3$, which define only one family of movable double poles.

If one nevertheless sets $m = 2$, the subequation (3.169) can only be satisfied by one series, thus preventing the full desired result to be obtained. However, the computation is much simpler and already provides results, so let us perform it first.

Let us consider the series (3.167). The six coefficients $a_{j,k}$ of (3.169) are first computed as the unique solution of the linear system of six equations $F_j = 0, j = 0,1,2,3,4,6$. Then the $J+1-6$ remaining equations $F_j = 0, j = 5,7:J$, which only depend on the fixed parameters (g_r, g_i, c), have the greatest common divisor (gcd) $3g_i + 2g_r + 3c^2/50$, and this factor defines the first solution

$$3g_i + 2g_r + \frac{3c^2}{50} = 0, \\ \left(M' + \frac{c}{5}M + \frac{c^3}{250}\right)^2 + 2\left(M + \frac{c^2}{50}\right)\left(M - \frac{c^2}{50} - \frac{2}{3}g_r\right)^2 = 0. \tag{3.171}$$

After division par this gcd, the system of three equations $F_j = 0, j = 5,7,8$, provides two and only two other solutions, which are

$$M'^2 + 2(M - g_r)M^2 = 0,\ c = 0,\ g_i = 0, \tag{3.172}$$

$$\left(M' + \frac{2c}{5}M\right)^2 + 2M^3 = 0,\ g_r = 0,\ g_i - \frac{c^2}{50} = 0. \tag{3.173}$$

and all the remaining equations $F_j = 0, j \geq 9$, are identically satisfied.

With the other series (3.168), the results are similarly

$$3g_i - g_r + \frac{3c^2}{80} = 0, \\ \left(M' + \frac{c}{5}M - \frac{c^3}{500}\right)^2 - \left(M - \frac{c^2}{100}\right)\left(M + \frac{c^2}{100} - \frac{2}{3}g_r\right)^2 = 0, \tag{3.174}$$

$$M'^2 - M\left(M - \frac{2}{3}g_r\right)^2 = 0,\ c = 0,\ g_i = \frac{1}{3}g_r, \tag{3.175}$$

$$\left(M' + \frac{c}{5}M\right)^2 - M^3 = 0,\ g_r = 0,\ g_i - \frac{c^2}{50} = 0. \tag{3.176}$$

Finally, for each of these two sets of three subequations, the fourth step finds a zero value for the genus and returns the general solution of the first order subequation as a rational function of $e^{a(\xi-\xi_0)}$, which basic trigonometric identities allow us to convert to the second degree polynomials in $(k/2)\tanh k(\xi - \xi_0)/2$ listed in (3.113)–(3.115) and (3.116)–(3.118). These solutions are therefore identical to (3.120), (3.121), and (3.122).

Therefore, with this lower bound $m = 2$, one already recovers all the presently known first order subequations. With the correct two-family lower (and upper) bound $m = 4$, which corresponds to 18 unknowns $a_{j,k}$ and at least 24 terms in the series, there is no solution other than the six above (three for each Laurent series). This situation is quite similar to the absence of solution in the class (3.134), and it reflects the difficulty of the CGL3 equation.

For the CGL5 equation (3.65), one new elliptic solution has been obtained [417] by this method.

3.2.6 First Integral of the Duffing–van der Pol Oscillator

Since it passes the weak Painlevé test, the Duffing–van der Pol oscillator ((2.63) with $d = 0$) may have a general solution with a finite amount of movable algebraic branching [346, Leçons 5–10,13,19], but finding it is still an open problem. No first integral is known, except in one particular case isolated by the method of infinitesimal symmetries [57]

$$3ab\beta + a^2c - 9\beta^2 = 0: \; K = \left(3a\frac{\mathrm{d}u}{\mathrm{d}x} + (3ab - 9\beta)u + a^2u^3\right) e^{3\beta t/a}, \tag{3.177}$$

and this first integral has the only singularity degree which is allowed by the Fuchs indices, $3/2$. This first order equation, which still has the algebraic branching $u \sim u_0(x-x_0)^{-1/2}$, can be mapped in at least two cases [57] to an ODE with the Painlevé property. The transformation involved is in each case a *hodograph transformation*, defined in the more general PDE case in 4.12. In the first case $K = 0$, the hodograph transformation $(u,x) \to (U,X)$

$$3ab\beta + a^2c - 9\beta^2 = 0, \; K = 0: \; \mathrm{d}x = u^{-1}\mathrm{d}X, \; u = U \tag{3.178}$$

maps the Abel equation (3.177) to the Riccati equation

$$\frac{\mathrm{d}U}{\mathrm{d}X} + \frac{a}{3}U^2 + b - \frac{3\beta}{a} = 0. \tag{3.179}$$

In the second case $K \neq 0$, after a preliminary transformation which conserves the PP,

$$u = e^{-\beta x/a}U, \; X = k_3 e^{-2\beta x/a}, \tag{3.180}$$

(3.177) becomes

$$k_3\frac{\mathrm{d}U}{\mathrm{d}X} - \frac{a^2}{6\beta}U^3 + \left(\frac{1}{2} + \frac{a^2c}{6\beta^2}\right)\frac{k_3}{X}U + \frac{K}{6\beta} = 0. \tag{3.181}$$

In the particular case when the coefficient of U vanishes this ODE for $U(X)$ is again an Abel equation, which a similar hodograph transformation maps to a Riccati equation

$$\begin{cases} b=4\frac{\beta}{a},\ c=-3\left(\frac{\beta}{a}\right)^2,\ K=-8a^2K_1^3,\ k_3=-\frac{a^2}{6\beta}, \\ U=V+K_1,\ \mathrm{d}X=\frac{\mathrm{d}Y}{V+3K_1},\ \frac{\mathrm{d}V}{\mathrm{d}Y}+V^2+3K_1^2=0. \end{cases} \tag{3.182}$$

3.2.7 *Singlevalued Solutions of the Bianchi IX Cosmological Model*

Sometimes, the no-log conditions generated by the test provide some global information, which can then be used to integrate.

The Bianchi IX cosmological model in vacuum can be defined by the metric [270]

$$\mathrm{d}s^2=\sigma^2\mathrm{d}t^2-\gamma_{\alpha\beta}\mathrm{d}x^\alpha\mathrm{d}x^\beta, \tag{3.183}$$

$$\gamma_{\alpha\beta}=\eta_{ab}e_\alpha^a e_\beta^b,\ \eta=\mathrm{diag}(A,B,C), \tag{3.184}$$

in which e_α^a are the components of the three frame vectors, and $\sigma^2=\pm 1$ according to whether the metric is Minkovskian or Euclidean. Introducing the so-called logarithmic time τ by the hodograph transformation

$$\mathrm{d}\tau=\frac{\mathrm{d}t}{\sqrt{ABC}}, \tag{3.185}$$

this gives rise to the six-dimensional system of three second order ODEs

$$\sigma^2(\log A)''=A^2-(B-C)^2 \text{ and cyclically},\ {}'=\mathrm{d}/\mathrm{d}\tau, \tag{3.186}$$

or equivalently

$$\sigma^2(\log\omega_1)''=\omega_2^2+\omega_3^2-\omega_2^2\omega_3^2/\omega_1^2 \text{ and cyclically}, \tag{3.187}$$

under the change of variables

$$A=\omega_2\omega_3/\omega_1,\ \omega_1^2=BC \text{ and cyclically}. \tag{3.188}$$

The sign σ^2 is irrelevant for the singularity structure, so we choose $\sigma=1$.

One of the families [98, 273]

$$\begin{aligned} &A=\chi^{-1}+a_2\chi+O(\chi^3),\ \chi=\tau-\tau_2,\\ &B=\chi^{-1}+b_2\chi+O(\chi^3), \end{aligned} \tag{3.189}$$

$$C = \chi^{-1} + c_2\chi + O(\chi^3),$$

has the Fuchs indices $-1, -1, -1, 2, 2, 2$, and the Kowalevski–Gambier test detects no logarithms at the triple index 2. The Fuchsian perturbative method (see Sect. 2.2)

$$A = \chi^{-1} \sum_{n=0}^{N} \varepsilon^n \sum_{j=-n}^{2+N-n} a_j^{(n)} \chi^j, \ \chi = \tau - \tau_2, \ \text{and cyclically,} \tag{3.190}$$

then detects movable logarithms at $(n, j) = (3, -1)$ and $(5, -1)$ [273], and the no-log conditions depend on the six movable coefficients $a_2^{(0)}, b_2^{(0)}, c_2^{(0)}, a_{-1}^{(1)}, b_{-1}^{(1)}, c_{-1}^{(1)}$. This first proves the nonintegrable nature of Bianchi IX, a question which had remained open for a long time [119]. In addition, the enforcement of these no-log conditions generates the three solutions:

$$(b_2^{(0)} = c_2^{(0)} \text{ and } b_{-1}^{(1)} = c_{-1}^{(1)}) \text{ or cyclically,} \tag{3.191}$$

$$a_2^{(0)} = b_2^{(0)} = c_2^{(0)} = 0, \tag{3.192}$$

$$a_{-1}^{(1)} = b_{-1}^{(1)} = c_{-1}^{(1)}. \tag{3.193}$$

These are constraints which reduce the number of arbitrary coefficients to, respectively, four, three and four, thus defining particular solutions which may have no movable critical points.

The first constraint (3.191) implies the equality of two of the components (A, B, C), and thus defines the 4-dimensional subsystem $B = C$ [401], whose general solution is single valued,

$$A = \frac{k_1}{\sinh k_1(\tau - \tau_1)}, \ B = C = \frac{k_2^2 \sinh k_1(\tau - \tau_1)}{k_1 \sinh^2 k_2(\tau - \tau_2)}. \tag{3.194}$$

The second constraint (3.192) amounts to suppressing the triple Fuchs index 2, thus defining a 3-dimensional subsystem with a triple Fuchs index -1. One can indeed check that the perturbed Laurent series (3.190) is identical to that of the *Darboux–Halphen system* [109, 192]

$$\omega_1' = \omega_2\omega_3 - \omega_1\omega_2 - \omega_1\omega_3, \ \text{and cyclically,} \tag{3.195}$$

whose general solution is single valued[7].

The third and last constraint (3.193) amounts to suppressing two of the three Fuchs indices -1, thus defining a 4-dimensional subsystem whose explicit writing is yet unknown. With the additional constraint

$$a_2^{(0)} + b_2^{(0)} + c_2^{(0)} = 0, \tag{3.196}$$

[7] This is a particular case of the class studied by Hoyer [227], however explicitly discarded by this author.

the Laurent series (3.189) is identical to that of the 3-dimensional *Euler system* (1750) [26], describing the motion of a rigid body around its center of mass

$$\omega_1' = \omega_2 \omega_3, \text{ and cyclically,} \tag{3.197}$$

whose general solution is elliptic,

$$\omega_j = -\sqrt{\wp(\tau - \tau_0, g_2, g_3) - e_j},\ j = 1,2,3,\ (\tau_0, g_2, g_3) \text{ arbitrary}, \tag{3.198}$$

$$\wp'^2 = 4(\wp - e_1)(\wp - e_2)(\wp - e_3) = 4\wp^3 - g_2\wp - g_3. \tag{3.199}$$

The 4-dimensional subsystem (the one without (3.196)) defines an extrapolation to four parameters of this elliptic solution, quite probably single valued, whose closed form is still unknown (see [81] for further investigations).

One thus retrieves by analysis all the results of the geometric assumption of self-duality [171], and even slightly more.

A similar analysis performed on the dynamical system defined in the proper time t [395] does not provide new closed form solutions.

3.2.8 Results of the Nevanlinna Theory on KS and CGL3

By studying the growth of solutions $u(x)$ near the complex point $x = \infty$, one can sometimes deduce important global properties, such as a rational dependence of u on x. This theory, known as Nevanlinna theory, is briefly introduced in Appendix D. There are however two main differences with the approach of Painlevé:

1. the solution $u(x)$ is assumed to be meromorphic, as opposed to being without movable critical singularities,
2. the solution $u(x)$ is a particular solution of the considered ODE, as opposed to the general solution.

By using only the two inequalities (D.15) and (D.16) of the Appendix, the application of this theory to the KS example (2.27) yields the following quite remarkable results [132]. If $u(x)$ is assumed meromorphic with finitely many poles, then it can only be $u = 120\nu(x - x_0)^{-3}$, with $b = \mu = A = 0$. If $u(x)$ is assumed meromorphic with infinitely many poles, then it can only be the elliptic solution (3.82) or one of the six trigonometric solutions of Table 3.2. Let us recall that Nevanlinna theory can say nothing when the considered solution is assumed nonmeromorphic. Similar results can probably be obtained for the CLG3 traveling wave.

This proves that the unknown general analytic solution of (2.27) is surely not meromorphic. In order to get some hint on the possible analytic form of this general analytic solution, one can experimentally investigate the singularities of this solution in the complex plane, by computing the *Padé approximants* [42] of the Laurent series (2.36). Padé approximants are a powerful tool to study the singularities of the unknown sum of a given Taylor series, and more generally to perform the summation of divergent series.

Given the first $N+1$ terms of a Taylor series near $x=0$,

$$S_N = \sum_{j=0}^{N} c_j x^j, \tag{3.200}$$

the Padé approximant $[L,M]$ of the series is the unique rational function

$$[L,M] = \frac{\sum_{l=0}^{L} a_l x^l}{\sum_{m=0}^{M} b_m x^m}, \; b_0 = 1, \tag{3.201}$$

obeying the condition

$$S_N - [L,M] = \mathscr{O}(x^{N+1}), \qquad L+M=N. \tag{3.202}$$

The extension to Laurent series presents no difficulty. In particular, for L and M large enough, Padé approximants are exact on rational functions.

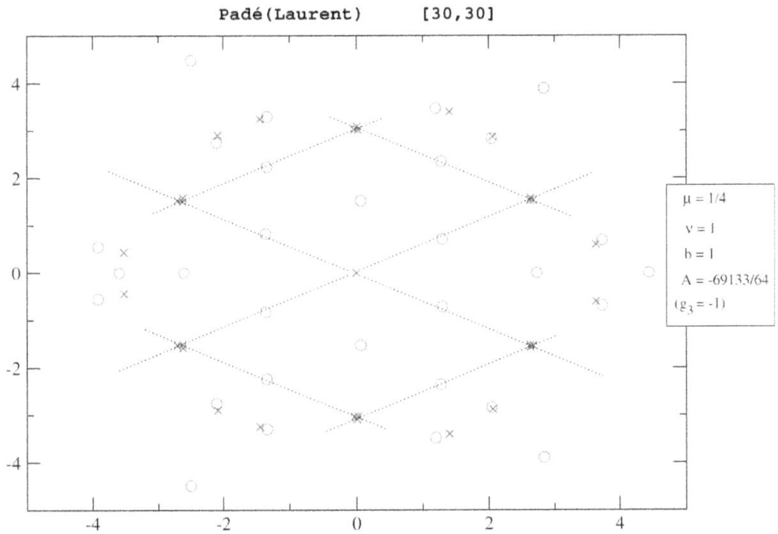

Fig. 3.4 Singularities of the Padé approximant [L/M] of the unknown general analytic solution of KS ODE (2.27) (courtesy of Tony Yee Tat-leung). The numerical values are $L=30, M=30, \nu=1, b=1, \mu=1/4, A=-69133/64$.

The advantage of $[L,M]$ over S_N (which has no poles) is to display the *global structure* of singularities of the series.

From a thorough investigation [439] of the singularities of the sum of the Laurent series (2.36) one concludes (this is not a proof): for generic values of (ν, b, μ, A), no multivaluedness is detected, no cuts are detected, and the singularities seem to be arranged in a nearly doubly periodic pattern, the elementary cell being made of one triple pole and three simple zeroes (Fig. 3.4). This suggests that the unknown general analytic solution, which is quite probably single valued but surely not meromorphic, could be expressible as a deformation of an elliptic function, something like

$$\forall \nu, b, \mu, A: \; u = f_0(\xi - \xi_0) + \mathscr{D} \log \sigma(\xi - \xi_0, G_2(\xi - \xi_0), G_3(\xi - \xi_0)), \qquad (3.203)$$

in which σ would be a deformation of the entire function of Weierstrass, G_2, G_3 deformations of the usual constant arguments of the σ function. As shown by a direct computation, this assumption is however insufficient to yield any new result, and other directions are currently under investigation [446].

Chapter 4
Partial Differential Equations: Painlevé Test

Abstract This chapter deals with the extension to nonlinear *partial differential equations* (PDEs) of the Painlevé property and Painlevé test previously introduced for ODEs. After mentioning reductions, we introduce the quite important class of *soliton equations*, together with their main properties: existence of a N-soliton solution and of a remarkable transformation called the Bäcklund transformation (BT). We then extrapolate to PDEs the notion of integrability and the definition of Painlevé property. After defining the expansion variable χ which minimizes the computation of the Laurent series representing the local solution, we present the successive steps of the Painlevé test, on the example of the KdV equation in order to establish necessary conditions for the Painlevé property. Finally, we apply the test to the equation of Kolmogorov–Petrovski–Piskunov (KPP) to generate necessary conditions for the existence of closed form particular solutions.

Let us first recall that our ultimate goal is to present methods, based only on singularity considerations, in order to build explicit *solutions* to the PDEs under consideration.

Consider a nonlinear PDE in several independent variables $x, t, \ldots$,

$$E(u, x, t, \ldots) = 0. \tag{4.1}$$

Its singularities are not isolated in the space of the independent variables $(x, t, \ldots)$, that is to say they are not located in the neighborhood of a point (x_0, t_0), but they lie on a codimension one manifold

$$\varphi(x, t, \ldots) - \varphi_0 = 0, \tag{4.2}$$

in which the *singular manifold variable* φ is an arbitrary function of the independent variables and φ_0 an arbitrary movable constant.

A similar manifold, also represented by (4.2) but this time regular, is involved in the famous theorem of Cauchy–Kowalevski which states the local existence, unicity, analyticity, etc of a solution in its neighborhood. This theorem, whose precise formulation can be found in classical textbooks [178], introduces a notion which is also relevant to the point of view of singularities.

Definition 4.1. A movable manifold is called **characteristic** if it makes inapplicable the existence theorem of Cauchy–Kowalevski.

Practically, a characteristic manifold (4.2) is one for which the Cauchy series involves no contribution from the highest derivatives of the PDE (4.1).

For instance, the movable manifold $\varphi - \varphi_0 \equiv t - t_0 = 0$ is characteristic for the KdV equation (3.32) since the relation $\varphi_x = 0$ forbids the series of Cauchy to contain any contribution from the highest derivative u_{xxx}.

Even in the ODE case, the movable singularity can be defined as $\varphi(x) - \varphi_0 = 0$, since the implicit functions theorem allows this to be locally inverted to $x - x_0 = 0$ provided the condition $\varphi'(x_0) \neq 0$ to be noncharacteristic is satisfied; this freedom in choosing the arbitrary function φ may then be used profitably to construct exact solutions which would be impossible to find with the restriction $\varphi(x) = x$ [426, 325].

The main difficulty with PDEs as compared to ODEs is that the notion of *general solution* is not well defined [178, Vol. III Chap. XXIV], so the question of integration is addressed differently. The principal method is to build a link to some linear system, and, for "integrable" (in a sense to be defined below) PDEs there exist two classes of such links.

The first class is made of the explicitly linearizable PDEs, such as the Burgers equation,

$$u_t + \frac{2}{a} u u_x + u_{xx} = 0, \tag{4.3}$$

linearizable into the heat equation by the so-called Hopf–Cole transformation [152, Part IV pp. 101, 106]

$$u = a \frac{\psi_x}{\psi}, \ \psi_{xx} + \psi_t = 0. \tag{4.4}$$

The second class is made of PDEs which admit a nonlinear version of the Fourier transform, called the *inverse spectral transform* (IST). This IST technique, discovered for the KdV equation by Gardner, Greene, Kruskal and Miura [165] and generalized to other equations independently by Zakharov and Shabat [442] and Ablowitz, Kaup, Newell and Segur [3], allows one, given some class of initial data, to perform a *global* resolution of the Cauchy problem. Presenting this powerful technique is outside the scope of this volume, and the interested reader can refer to e.g. [271] or [1].

4.1 On Reductions

There exist certain classes of solutions of a PDE which are also solutions of some ODE called *reductions*. We have already seen in the previous chapter the traveling wave reduction of various PDEs. For instance, the KdV and KS equations admit another reduction $u(x,t) \mapsto U(X)$ [5, 399]

$$u = 2a(U - t/b),\ X = x - 6(t/b)^2,\ U'' - 6U^2 - X + K = 0, \tag{4.5}$$

$$u = U - 2at,\ X = x + at^2,\ \nu U''' + bU'' + \mu U' + \frac{U^2}{2} - 2aX + K = 0, \tag{4.6}$$

in which K denotes an integration constant. For KdV, $U(X)$ obeys the first Painlevé equation, while for KS no analytic solution is known to this ODE. In such a case, the question of finding solutions of the PDE *via* reductions splits into the search for reductions, followed by the search for solutions of an ODE.

Searching for reductions is too large a domain to be covered by this volume, and we refer to textbooks [345, 343] or to reviews adapted to the present context such as [67]. As to the search for solutions of an ODE, it has been dealt with in Chap. 3 on the integration of ODEs.

From the point of view of the Painlevé property, one will also have to distinguish two types of reductions, the characteristic ones and the noncharacteristic ones.

Definition 4.2. A reduction of a PDE in N independent variables to a PDE in $N-1$ independent variables is called **noncharacteristic** if it preserves the differential order.

A conjecture has been proposed by Ablowitz, Ramani and Segur [6] according which, given a PDE integrable in the sense of the inverse spectral transform method (IST), any noncharacteristic reduction to an ODE implies the Painlevé property for that ODE. This is verified using many examples, of which we present a few here.

1. The self-dual Yang–Mills equations admit reductions to all six Painlevé equations [300].
2. The KdV equation (3.32) admits the reduction [143]

$$u = a(3t/b)^{-2/3}\left(U(\xi) - \frac{\xi}{2}\right),\ \xi = (3t/b)^{-1/3}x,$$
$$U''' - 6UU' + 2\xi U' + U = 0, \tag{4.7}$$

the reduced ODE has the PP [59, pp. 339, 343] and its general solution is an algebraic transform of P2,

$$U = W' + W^2 + \frac{\xi}{2} = 0,\ W'' - 2W^3 - \xi W - A = 0,\ A = \text{arbitrary}. \tag{4.8}$$

Another method for integrating (4.7) will be given in Sect. 5.8.
3. The NLS equation (2.47) admits the reduction [35]

$$A = t^{ia}\sqrt{Y'(\xi)}e^{i(\xi^2 + \int (Y/Y')\mathrm{d}\xi)},\ \xi = xt^{-1/2},\ a \text{ real}, \tag{4.9}$$

and the ODE for $Y(\xi)$ is an algebraic transform of the generic fourth Painlevé equation.

4. The three-wave resonant interaction system [441],

$$\begin{cases} u_{j,t} + c_j u_{j,x} - i\bar{u}_k\bar{u}_l = 0, \\ \bar{u}_{j,t} + c_j\bar{u}_{j,x} + iu_k u_l = 0,\ i^2 = -1, \end{cases} \tag{4.10}$$

in which (j,k,l) denotes any permutation of $(1,2,3)$, c_j are the constant values of the group velocities, with $(c_2-c_3)(c_3-c_1)(c_1-c_2) \neq 0$, admits a reduction $\xi = x/t$ [252] whose general solution [298, 83] is an algebraic transform of the generic sixth Painlevé equation.

We will come back to this conjecture at the end of Sect. 4.3.

4.2 Soliton Equations

There exists a quite important class of PDEs which display many physically interesting solutions; these are the so-called *soliton equations*. A number of them have been encountered in previous chapters: the *Korteweg–de Vries equation*, the *nonlinear Schrödinger equation*, the three PDEs (2.82) associated to the cubic Hénon–Heiles Hamiltonian.

A presentation of these equations can be found in e.g. [271, 49], [1, p. 19], or [125, p. 15]. In a few words, they are characterized by their possessing a class of solutions describing the nonlinear interaction of an arbitrary number N of elementary waves (often called "solitons" by physicists), with the nice property that these elementary waves may retain their shape after the nonlinear interaction.

The existence of this class of solutions with N arbitrary is one criterium of integrability, among many others [440]. Indeed, as opposed to the case of ODEs, there is no unique definition for the *integrability* of a nonlinear PDE.

In order to provide a consistent exposition, we first need a precise definition for these various items.

Definition 4.3. (Already given on p. xi) A **traveling wave** of a given PDE $E(u,x,t) = 0$ is any solution of the reduction $\xi = x - ct$ if it exists.

Definition 4.4. A **solitary wave** of a given PDE $E(u,x,t) = 0$ is a traveling wave such that the solution itself or its derivative obeys some decreasing conditions when the real variable $\xi = x - ct$ goes to $\pm\infty$.

For instance, the traveling wave (3.38) of KdV is not a solitary wave, unless it reduces to (3.45), a solution which decreases exponentially fast to the *background B*.

Other examples of solitary waves are the various traveling waves encountered in previous chapters, such as[1] (1.11), (1.14), (3.45), (3.59), (3.60), (3.104), (3.120), (3.121), (3.122).

[1] All these waves decrease exponentially fast at infinity, but this is not a rule. An example of algebraic decreasing is the algebraic solitary wave of the Benjamin–Ono PDE [27, 344, 1].

Before a correct definition can be given for the above mentioned solution representing the nonlinear interaction of N elementary waves, a quite important notion is required.

Definition 4.5. ([20], [114, Vol. III Chap. XII], [374]) A **Bäcklund transformation** (BT) between two given PDEs

$$E_1(u,x,t)=0,\ E_2(U,X,T)=0 \tag{4.11}$$

is a pair of relations

$$F_j(u,x,t,U,X,T)=0,\ j=1,2 \tag{4.12}$$

with some transformation between (x,t) and (X,T), in which F_j depends on the derivatives of $u(x,t)$ and $U(X,T)$, such that the elimination of u (resp. U) between (F_1,F_2) implies $E_2(U,X,T)=0$ (resp. $E_1(u,x,t)=0$). In case the two PDEs are the same, the BT is also called the *auto-BT*.

Example 4.1. The Liouville equation

$$\text{Liouville: } E(u)\equiv u_{xt}+\alpha e^u=0 \tag{4.13}$$

admits two BTs, one not so interesting auto-BT and the BT

$$(u-v)_x=\alpha\lambda e^{(u+v)/2}, \tag{4.14}$$

$$(u+v)_t=-2\lambda^{-1}e^{(u-v)/2}, \tag{4.15}$$

to a linear equation called the d'Alembert equation

$$\text{d'Alembert: } E(v)\equiv v_{xt}=0. \tag{4.16}$$

Since the general solution of the latter equation is $v=f(x)+g(t)$, with f and g arbitrary functions, this last BT allows one to obtain the general solution of the Liouville equation, an easy exercise with the result

$$e^u=-\frac{2}{\alpha}\frac{F'(x)G'(t)}{(F(x)+G(t))^2}. \tag{4.17}$$

Needless to say that such a situation is nongeneric.

The practical use of an auto-BT is to build a new solution from an old one, even if this seed solution is trivial.

Definition 4.6. One calls any constant solution of a given PDE the **vacuum solution**.

Such trivial solutions are $u=$ arbitrary constant for KdV, mKdV, $u=0$ for sine-Gordon, NLS, etc. It is precisely by iterating such trivial solutions with the BT that physically interesting solutions can be built (this will be done in Sect. 5.6).

Remark. A *plane wave solution* generically does not exist for a nonlinear PDE, with the notable exception of phase invariant equations such as the NLS (2.47), the CGL3 (2.45), or the CGL5 (3.65), therefore it cannot contribute to the definitions involving the soliton equations.

Definition 4.7. A **one-soliton solution** is the first iterate of the vacuum solution *via* the Bäcklund transformation.

Many authors also require the one-soliton solution to be a solitary wave. The name *soliton* commonly used in place of either "one-soliton solution" (the *bright one-soliton* and *dark one-soliton* of the NLS have been seen in Sect. 3.1.3) or "solitary wave" is not used in this volume in order to avoid such a confusion.

Definition 4.8. An N**-soliton solution**, with N an arbitrary positive integer, is the N-th iterate of the vacuum solution *via* the Bäcklund transformation.

Definition 4.9. A **soliton equation** is a partial differential equation admitting an N-soliton solution with N arbitrary.

Examples of soliton PDEs are: the KdV equation, the NLS equation, the Benjamin–Ono equation [27, 344, 1]. These PDEs possess many remarkable properties [146], such as a Hamiltonian or a bi-Hamiltonian structure, which implies an infinite number of *conservation laws*[2]. In fact, these soliton PDEs come into hierarchies, and for instance KdV is the base member of the so-called KdV hierarchy. There exist only a finite number of such hierarchies, which can be beautifully described in the framework of the theory of Sato [381], see tutorial introductions in [337, 384].

Examples of nonsoliton PDEs (we will say, equivalently, *partially integrable equations*) are: the CGL3 equation, the KS equation, the KPP equation. They only admit a finite number of conservations laws and N-soliton solutions with N small ($N = 1$ for CGL3 and KS, $N = 2$ for KPP).

The quite important difference between soliton and nonsoliton PDEs best appears on the solitary wave solution. In the "dark one-soliton solution" (3.60) of the NLS (a soliton PDE), the parameters k and c are arbitrary and independent (movable in the language of singularities). On the contrary, in the "propagating hole solution" (3.120) of CGL3 (a nonsoliton PDE), these two parameters are constrained by the relation $a_1k^2 + a_2c^2 = 1$, in which a_1 and a_2 are real constants only depending on the coefficients (p, q, γ) of CGL3 (fixed in the language of singularities).

4.3 Painlevé Property for PDEs

In order to find these physically interesting solutions (one-soliton, N-soliton) by singularity-based methods, a prerequisite is to extrapolate to PDEs the powerful

[2] A conservation law of a PDE like (4.1) is any relation of the form $\partial_x X(u, u_x, \ldots) + \partial_t T(u, u_x, \ldots) = 0$, e.g. the definition itself (3.32) of the KdV equation.

methods of Chap. 3 (truncations, etc), a task which in turn requires the extrapolation to PDEs of the notions up to now only defined for ODEs: Painlevé property (Definition 1.1 Sect. 1.2), Painlevé test (Chap. 2). Skipping this task would just turn the methods presented in Chap. 5 into recipes. We insist that the definitions for PDEs should be an extrapolation of those for ODEs.

Since the Painlevé test investigates whether the Painlevé property may be satisfied, the property must be defined before the test, and such a definition must involve a *global* property. The concept of general solution, which is central in the ODE case, is insufficient since it is ill-defined (see the very beginning of this chapter). Indeed, this is only in nongeneric cases like the Liouville equation (4.13) that the general solution of a PDE can be built explicitly. The Bäcklund transformation is also insufficient because, under a reduction PDE→ODE, the BT reduces to a birational transformation (note the same initials BT; an example will be given in Sect. 5.8), and a birational transformation is *not* involved in the definition of the PP for ODEs. However, an adequate combination of these two items (general solution, Bäcklund transformation) does provide definitions [94] which are indeed extrapolations of the ODE case.

Definition 4.10. A PDE in N independent variables is said to be **integrable** if at least one of the following properties holds.

1. Its general solution can be obtained, and it is an explicit closed form expression, possibly presenting movable critical singularities.
2. It is linearizable.
3. For $N > 1$, it possesses an auto-BT which, if $N = 2$, depends on an arbitrary complex constant, the Bäcklund parameter.
4. It possesses a BT to another integrable PDE.

Examples of these various situations are, respectively:

1. the Liouville PDE $u_{xt} + e^u = 0$ with its general solution (4.17); also the PDE $u_x u_t + u u_{xt} = 0$ with the general solution $u = \sqrt{f(x) + g(t)}$ which presents movable critical singularities;
2. the Burgers PDE $u_t + u_{xx} + 2uu_x = 0$, linearizable into the heat equation $\psi_t + \psi_{xx} = 0$;
3. the KdV PDE $u_t + u_{xxx} - 6uu_x = 0$, which possesses an auto-BT depending on an arbitrary complex cosntant;
4. the modified KdV PDE $u_t + u_{xxx} - 6u^2 u_x = 0$, which possesses a BT to the KdV equation.

We now have enough elements to give a definition of the Painlevé property for PDEs which is indeed an extrapolation of the one for ODEs. Such a definition must evidently refer to the movable singularities, defined by the movable singular manifold (4.2).

Definition 4.11. The **Painlevé property** (PP) of a PDE is the absence of movable critical singularities near any noncharacteristic manifold, and its integrability (Definition 4.10).

Practically, checking the absence of movable critical singularities can only be done locally, this is precisely the purpose of the Painlevé test, which is why the construction of a Bäcklund transformation is the second part of the definition.

A *weak Painlevé property* can be defined similarly to the ODE case (see Sect. 2.1.4). In particular one can include in this category all equations which can be mapped to an equation having the (full) PP by the following transformation.

Definition 4.12. Given a PDE $E(u,x,t)$, one calls a **hodograph transformation** a transformation in which one of the new independent variables depends on the old dependent variable.

Such transformations have been studied e.g. in [374, 66]. Their interest is to possibly map PDEs having only the weak PP to PDEs having the (full) PP. Consider for instance the quasilinear equation

$$U_T + U^3 U_{XXX} = 0, \tag{4.18}$$

introduced by Harry Dym [258] in relation to the classical problem of vibrating strings. Under the hodograph transformation

$$(U,X,T) \to (u,x,t): \ \mathrm{d}X = U\mathrm{d}x,\ U = u, T = t, \tag{4.19}$$

it is mapped to the intermediate equation [248] (see details in [200])

$$u_t + u_{xxx} - 3u^{-1}u_x u_{xx} + \frac{3}{2}u^{-2}u_x^3 = 0, \tag{4.20}$$

an equation which is then mapped to the modified KdV equation, a semilinear equation

$$v = \frac{u_x}{u}: \ v_t + v_{xxx} - \frac{3}{2}v^2 v_x = 0. \tag{4.21}$$

Since the modified KdV has the PP, one concludes that the Harry Dym equation (4.18) possesses the weak PP.

Another example is the quasilinear equation[226]

$$m_T + U m_X + b U_X m = 0,\ U_{XX} - U - m = 0, \tag{4.22}$$

which has the weak PP in only two cases [173], $b = 2$ and $b = 3$, defining respectively the *Camassa–Holm equation* [159, 50] for $b = 2$, and the *Degasperis–Procesi equation* [151, 118] for $b = 3$. Under the hodograph transformation [158, 226]

$$(U,X,T) \to (u,x,t): \ m = -p^b,\ \mathrm{d}x = p(\mathrm{d}X - u\mathrm{d}T),\ t = T, \tag{4.23}$$

it is mapped to the semilinear PDE

$$\left(p^{-1}\right)_t + \left(p(\log p)_{xt} + p^b\right)_x = 0, \tag{4.24}$$

and this latter equation can be shown [159, 226] to have the (full) PP only for $b=2$ and $b=3$.

The PP for PDEs is invariant under the natural extension of the homographic group (1.43), and *classifications* similar to those of ODEs have also been performed for PDEs, see Appendix A.3.7. Classifications based on other criteria, such as the existence of an infinite number of conservation laws [304], isolate more PDEs, and it would be interesting to check that, under the group of transformations generated by Bäcklund transformations and hodograph transformations, each of them is equivalent to a PDE with the PP.

From the above definition of the PP for PDEs, it follows that the conjecture stated in Sect. 4.1 is true at least for those IST-integrable PDEs which in addition have the PP.

4.4 Painlevé Test

After introducing the expansion variable which minimizes the size of the computations, we present the PDE Painlevé test on two examples, one integrable to display the advantage of the optimal variable, one partially integrable to build some constructive information.

4.4.1 Optimal Expansion Variable

According to Definition 4.11, one must check the existence of a Laurent series near every noncharacteristic movable singular manifold, therefore one must first define an *expansion variable* χ for the Laurent series. There is indeed no reason to confuse the roles of the singular manifold variable φ and the expansion variable χ. Two requirements must be respected: firstly, χ must vanish as $\varphi-\varphi_0$ when $\varphi\to\varphi_0$; secondly, the structure of singularities in the φ complex plane must be in one-to-one correspondence with that in the χ complex plane, so χ must be a homographic transform of $\varphi-\varphi_0$ (with coefficients depending on the derivatives of φ).

The Laurent series for u and E involved in the Kowalevski–Gambier part of the test are defined as

$$u=\sum_{j=0}^{+\infty} u_j\chi^{j+p},\ -p\in\mathscr{N},\ E=\sum_{j=0}^{+\infty} E_j\chi^{j+q},\ -q\in\mathscr{N}^* \tag{4.25}$$

with coefficients u_j, E_j independent of χ and only depending on the derivatives of φ.

In order to minimize the size of the computation, which quickly becomes very large, one must carefully choose the expansion variable χ.

There exists an optimal choice of χ for which the coefficients exhibit the highest invariance and therefore are the shortest possible, this is [70]

$$\chi = \frac{\varphi - \varphi_0}{\varphi_x - \dfrac{\varphi_{xx}}{2\varphi_x}(\varphi - \varphi_0)} = \left[\frac{\varphi_x}{\varphi - \varphi_0} - \frac{\varphi_{xx}}{2\varphi_x}\right]^{-1}, \ \varphi_x \neq 0, \tag{4.26}$$

in which x denotes any independent variable whose component of grad φ does not vanish.

Indeed, the gradient of χ evaluates to

$$\chi_x = 1 + \frac{S}{2}\chi^2, \tag{4.27}$$

$$\chi_t = -C + C_x\chi - \frac{1}{2}(CS + C_{xx})\chi^2. \tag{4.28}$$

a Riccati system whose coefficients only depend on (S, C) defined as

$$S = \{\varphi; x\} = \frac{\varphi_{xxx}}{\varphi_x} - \frac{3}{2}\left(\frac{\varphi_{xx}}{\varphi_x}\right)^2, \tag{4.29}$$

$$C = -\varphi_t/\varphi_x. \tag{4.30}$$

The quantity S is named the *Schwarzian* (or the Schwarzian derivative) of φ. These two quantities are invariant under the group of homographic transformations

$$\varphi \mapsto \frac{a\varphi + b}{c\varphi + d}, \ ad - bc \neq 0, \tag{4.31}$$

in which a, b, c, d are arbitrary complex constants. Therefore, the expansion coefficients u_j, E_j, which by definition only depend on the coefficients of the second degree polynomials of χ in the r.h.s. of (4.27)–(4.28), are invariant under this homographic group. These two invariants are linked by the cross-derivative condition

$$X \equiv ((\varphi_{xxx})_t - (\varphi_t)_{xxx})/\varphi_x = S_t + C_{xxx} + 2C_xS + CS_x = 0, \tag{4.32}$$

identically satisfied in terms of φ.

For the practical computation of (u_j, E_j) as functions of (S, C) only, i.e. what is called the *invariant Painlevé analysis*, the above explicit expressions of (S, C, χ) in terms of φ are *not* required, the variable φ disappears, and the only necessary information is the gradient of the expansion variable χ defined by (4.26).

Note that the Riccati system (4.27)–(4.28), under the transformation

$$\chi = \frac{\psi}{\psi_x}, \tag{4.33}$$

is linearized into the second order scalar system,

$$L_1\psi \equiv \psi_{xx} + \frac{S}{2}\psi = 0, \tag{4.34}$$

$$L_2\psi \equiv \psi_t + C\psi_x - \frac{C_x}{2}\psi = 0, \tag{4.35}$$

$$2[L_1, L_2] \equiv X = S_t + C_{xxx} + CS_x + 2C_xS = 0. \tag{4.36}$$

The above choice (4.26) of χ which generates homographically invariant coefficients is the simplest one, but it is not the most general one. The most general choice is the variable Y which must satisfy the same two requirements as above: firstly, Y must vanish as $\varphi - \varphi_0$ (i.e. as χ) when $\varphi \to \varphi_0$; secondly, the structure of singularities in the φ complex plane must be in one-to-one correspondence with that in the Y complex plane, so Y must be a homographic transform of $\varphi - \varphi_0$ (with coefficients depending on the derivatives of φ). The most general such variable is defined by [315, 357]

$$Y^{-1} = B(\chi^{-1} + A),\ B \neq 0. \tag{4.37}$$

Since a homography conserves the Riccati nature of an ODE, the variable Y satisfies a Riccati system, easily deduced from the canonical one (4.27)–(4.28) satisfied by χ. One will take advantage of this freedom on A and B in Sect. 5.6.2.

This replacement of $\varphi - \varphi_0$ by either χ or Y looks very much like a resummation of the Laurent series, just like the geometric series

$$\sum_{j=0}^{+\infty} x^j,\ x \to 0, \tag{4.38}$$

becomes a finite sum in the resummation variable $X = x/(1-x)$

$$\sum_{J=0}^{1} X^J,\ X \to 0. \tag{4.39}$$

4.4.2 Integrable Situation, Example of KdV

To illustrate the test, let us take as an example the Korteweg-de Vries equation (3.32)

$$E \equiv bu_t + u_{xxx} - \frac{6}{a}uu_x = 0. \tag{4.40}$$

The successive steps are the same as for an ODE, see Sect. 2.1.1, so we will only briefly recall them.

In the *first step*, the dominant terms are found to be $\hat{E}(u) \equiv u_{xxx} - (6/a)uu_x$, and the leading powers p and q do not depend on the choice of χ and result from $p-3 = 2p-1 = q$, i.e. $p = -2, q = -5$.

With the choice of expansion variable $\chi = \varphi - \varphi_0$ originally made by Weiss et al. [431], the leading coefficient is given by

$$(-2)(-3)(-4)\varphi_x^3 u_0 - (6/a)u_0(-2)\varphi_x u_0 = 0,\ u_0\varphi_x \neq 0, \tag{4.41}$$

resulting in the single family $u_0 = 2a\varphi_x^2$. With the optimal choice (4.26), this is just the constant $u_0 = 2a$.

In the *second step*, the Fuchs indices are evidently independent of the choice of χ. With the choice (4.26), their computation is

$$\lim_{\varepsilon\to 0} \frac{\hat{E}(u+\varepsilon w) - \hat{E}(u)}{\varepsilon} = (\partial_x^3 - (6/a)u_0\chi^p\partial_x - (6/a)pu_0\chi^{p-1})w = 0, \tag{4.42}$$

$$\lim_{\chi\to 0} \chi^{-j-q}(\partial_x^3 - (6/a)u_0\chi^p\partial_x - (6/a)pu_0\chi^{p-1})\chi^{j+p} \tag{4.43}$$

$$= (j-2)(j-3)(j-4) - 12(j-2) + 24 = (j+1)(j-4)(j-6) = 0, \tag{4.44}$$

therefore the test will end after checking the highest positive index 6 for possible movable logarithms.

As to the Laurent series to be computed in the *third step*, its terms strongly depend on the choice of χ. With the choice $\chi = \varphi - \varphi_0$, the series is

$$u = 2a\varphi_x^2\chi^{-2} - 2a\varphi_{xx}\chi^{-1} + ab\frac{\varphi_t}{6\varphi_x} + \frac{2a}{3}\frac{\varphi_{xxx}}{\varphi_x} - \frac{a}{2}\left[\frac{\varphi_{xx}}{\varphi_x}\right]^2 + O(\chi),\ \chi = \varphi - \varphi_0. \tag{4.45}$$

and its coefficients u_j depend on all the derivatives of φ, making tedious the computation up to $j = 6$. With the choice (4.26), it is much easier to check that no logarithms enter the series at $j = 4$ and $j = 6$, leading to the much shorter series

$$\begin{aligned} u = {} & 2a\chi^{-2} - ab\frac{C}{6} + \frac{2aS}{3} - 2a(bC - S)_x\chi + u_4\chi^2 \\ & + \left(\frac{a}{3}(SS_{xx} + 2S_{xxx}) + \frac{ab}{3}(4SC_x - 5C_{xxx}) - \frac{ab^2}{6}(C_t + CC_x) + u_{4,x}\right)\chi^3 \\ & + u_6\chi^4 + O(\chi^5), \end{aligned} \tag{4.46}$$

in which the coefficients u_4 and u_6 are arbitrary functions of (x,t). This ends the test, which therefore passes for this particular equation.

A frequent worry when performing the test is: is there any restriction (or advantage, or inconvenient) to perform the test with χ (4.26) or Y (4.37) rather than with $\varphi - \varphi_0$? The precise answer is : the three Laurent series are equivalent (their set of coefficients are in one-to-one correspondence, only their radii of convergence are different). As a consequence, the Painlevé test, which involves the *infinite* series, is insensitive to the choice, and the costless choice (the one which minimizes the computations) is undoubtedly χ defined by its gradient (4.27)–(4.28) (to perform the test, one can even set, following Kruskal [238], $S = 0, C_x = 0$). If the same question were asked not about the Painlevé test but about proving the Painlevé property by the singular manifold method formulated in Sect. 5.5.1, the answer would be quite different, and it is given in Sect. 5.5.2.

When processing a quasilinear equation such as (4.18) or (4.22), the typical output is rational values (instead of integer values) for the leading powers and the Fuchs indices, like in the ODE case (2.68). Detailed examples can be found in e.g. [425, 173].

Remark. One might wonder whether the KdV equation (4.40) can be "generalized" by replacing the constant coefficients by functions of (x,t) and adding subdominant terms like u_{xx}, etc, also with variable coefficients. If one requires this variable coefficient KdV to pass the Painlevé test, then the answer [44] is negative. Indeed, after the variable coefficients have been constrained by the Painlevé test, the resulting equation KdV(U,X,T) can always be mapped to the constant coefficient KdV(u,x,t) by a homographic transformation

$$u = \frac{\alpha(x,t)U + \beta(x,t)}{\gamma(x,t)U + \delta(x,t)},\ X = \xi(x,t),\ T = \tau(x,t). \tag{4.47}$$

The same can be proven for the NLS equation [161, 65].

4.4.3 Partially Integrable Situation, Example of KPP

The Kolmogorov–Petrovskii–Piskunov (KPP) equation [255, 322]

$$E(u) \equiv bu_t - u_{xx} + \gamma uu_x + 2d^{-2}(u-e_1)(u-e_2)(u-e_3) = 0, \tag{4.48}$$

with (b,γ,d^2) real and e_j real and distinct, is encountered in reaction-diffusion systems and prey-predator models (the optional convection term uu_x [383] is quite important in physical applications to prey-predator models). Keeping symbolic values for the three fixed points $u = e_j$ instead of, e.g., $e_j = (0,1,a)$, has the advantage of displaying the identity of solutions often presented as different.

The *first step*, to search for the families $u \sim u_0\chi^p, E \sim E_0\chi^q, u_0 \neq 0$, results in the dominant terms

$$\hat{E}(u) \equiv -u_{xx} + \gamma uu_x + 2d^{-2}u^3, \tag{4.49}$$

which provide two families (p,u_0)

$$p = -1, q = -3, -2 - \gamma u_0 + 2d^{-2}u_0^2 = 0. \tag{4.50}$$

The necessary condition that all values of p be integer is satisfied.

In the *second step*, the linearized equation is

$$(\hat{E}'(u))w \equiv (-\partial_x^2 + \gamma u\partial_x + \gamma u_x + 6d^{-2}u^2)w = 0, \tag{4.51}$$

then its Fuchs indices i near $\chi = 0$ are the roots of the indicial equation

$$P(i) = -(i+1)(i-4-\gamma u_0) = 0, \tag{4.52}$$

and one must enforce the necessary condition that, for each family, these two indices be distinct integers [149, 82]. Considering each family separately would produce a countable number of solutions, which is incorrect. The two families must be considered simultaneously

$$\gamma u_{0,j} = i_j - 4,\ (i_j-4)^2 - \frac{\gamma^2 d^2}{2}(i_j-4) - \gamma^2 d^2 = 0,\ j = 1,2, \tag{4.53}$$

with the aim of solving the *diophantine condition* that the two values i_1, i_2 of the Fuchs index $4+\gamma u_0$ be integer. The elimination of the adimensional parameter $\gamma^2 d^2$ between the sum and the product of the two roots

$$(i_1-4)+(i_2-4) = \frac{\gamma^2 d^2}{2},\ (i_1-4)(i_2-4) = -\gamma^2 d^2, \tag{4.54}$$

yields

$$\frac{1}{i_1-4} + \frac{1}{i_2-4} = -\frac{1}{2}. \tag{4.55}$$

This is typically the kind of diophantine equation encountered when systematically looking for ODEs possessing the Painlevé property (see (2.95) and Appendix A), its advantage is to have a finite number of solutions, which are [45, (9.2)]

$$\gamma^2 d^2 = 0,\ (i_1,i_2) = (4,4),\ u_0 = (-d,d), \tag{4.56}$$
$$\gamma^2 d^2 = 2,\ (i_1,i_2) = (3,6),\ \gamma u_0 = (-1,2), \tag{4.57}$$
$$\gamma^2 d^2 = -16,\ (i_1,i_2) = (0,0),\ \gamma u_0 = (-4,-4), \tag{4.58}$$
$$\gamma^2 d^2 = -18,\ (i_1,i_2) = (-2,1),\ \gamma u_0 = (-6,-3). \tag{4.59}$$

It would be wrong at this stage to discard negative integer indices. Indeed, in linear ODEs such as (4.51), the single valuedness required by the Painlevé test restricts the Fuchs indices to integers, whatever their sign.

The recurrence relation for the next coefficients u_j,

$$\forall j \ge 1:\ E_j \equiv P(j)u_j + Q_j(\{u_l \mid l < j\}) = 0 \tag{4.60}$$

depends linearly on u_j and nonlinearly on the previous coefficients u_l. Let us proceed with the first case only, $\gamma = 0$ (the usual KPP equation).

The *third and last step* is then to require, for any admissible family and any Fuchs index i, that the *no-logarithm condition*

$$\forall i \in \mathscr{Z},\ P(i) = 0:\ Q_i = 0 \tag{4.61}$$

holds true. One thus finds the Laurent series

$$\frac{u}{d} = \chi^{-1} + \frac{s_1}{3d} - \frac{b}{6}C + \left(a_2 - \frac{b}{36}C^2 + \frac{S + bC_x}{6}\right)\chi \tag{4.62}$$
$$+ \left(\frac{a_1}{2} + \frac{a_2 bC}{2} - \frac{S_x}{24} - \frac{bC_{xx}}{12} + b^2\frac{7CC_x + 3C_t}{72} - \frac{b^3C^3}{54}\right)\chi^2 + \mathcal{O}(\chi^3),$$

and, at index $i = 4$, the two conditions, one for each sign of d [69],

$$Q_4 \equiv C[(bdC + s_1 - 3e_1)(bdC + s_1 - 3e_2)(bdC + s_1 - 3e_3)$$
$$- 3b^2d^3(C_t + CC_x)] = 0,\ s_1 = e_1 + e_2 + e_3, \tag{4.63}$$

are not identically satisfied, so the PDE fails the test. This ends the test.

The system of two no-log conditions (4.63) implies

$$(2e_1 - e_2 - e_3)(2e_2 - e_3 - e_1)(2e_3 - e_1 - e_1)C = 0, \tag{4.64}$$

therefore a necessary condition for the Painlevé property is that either $C = 0$ or one of the three fixed points be equidistant from the two others. In Sect. 5.7.2, it will be seen that the first possibility $C = 0$ is sufficient for the Painlevé property, while the second one is not sufficient.

If instead of the PDE (4.48) one considers its reduction $u(x,t) = U(\xi), \xi = x - ct$ to an ODE,

$$-\frac{d^2U}{d\xi^2} + (\gamma U - bc)\frac{dU}{d\xi} + 2d^{-2}(U - e_1)(U - e_2)(U - e_3) = 0,\ \gamma = 0, \tag{4.65}$$

then $C =$ constant $= c$, and the two conditions $Q_4 = 0$ select the seven values $c = 0$ and $c^2 = (s_1 - 3e_k)^2(bd)^{-2}, k = 1,2,3$. For all these values, the necessary conditions are then sufficient since the general solution $U(\xi)$ is singlevalued (equation numbered 8 in the list of Gambier [163] reproduced in [232]).

This KPP equation is partially integrable in the sense that, despite its failing the Painlevé test, it admits particular solutions which have no movable critical singularities, as will be shown in Sect. 5.7.2.

Chapter 5
From the Test to Explicit Solutions of PDEs

Abstract In this chapter we exploit the information provided by the Painlevé test in order to obtain, firstly the Bäcklund transformation so as to prove the Painlevé property, secondly solutions in closed form. Although partially integrable and nonintegrable equations, i.e. the majority of physical equations, admit no Bäcklund transformation, they retain part of the properties of (fully) integrable PDEs, and this is why the methods presented in this chapter apply to both cases as well. Such partially integrable examples are handled in Sect. 5.7.

5.1 Global Information from the Test

Despite its local nature, the Painlevé test yields a lot of global information, of which the most important pieces are the following.

– The number of families. This information may exclude some types of solutions. For instance, if an equation admits only one family, like the Fisher equation or the KdV equation, it cannot admit solitary waves of the type $u = \text{sech}$. A necessary condition for solutions $u = \text{sech}$ to exist is the existence of at least two families (u_0, p) and (u_0', p) with the same singularity degree p, like mKdV or CGL3.
– The singular part operator of each family, introduced in Sect. 2.1.2. This is of direct use to make the correct assumption for searching the Lax pair (case of an integrable equation) or particular solutions (integrable or nonintegrable equation).
– The subset of positive integer values of the Fuchs indices of a given family. In case the equation is an ODE, these integers are the only possible values for the singularity degree of a first integral (see example of Lorenz model Sect. 3.1.1).
– The conditions for the absence of movable logarithmic branching (no-log conditions).

5.2 Building N-Soliton Solutions

Most of the explicit solutions which will be found are in fact N-soliton solutions (see Definition 4.8) or, in the partially integrable case, their degeneracy. There exist two main approaches to build this N-soliton solution.

The first approach uses the following tools, which will be defined precisely in Sect. 5.3:

– the *Lax pair*, a set of two linear differential operators, the commutativity condition of which is identical to the nonlinear PDE $E(u) = 0$. Its existence is a criterium of integrability, however insufficient in itself to build solutions.
– the *Darboux transformation*, a gauge transformation which preserves the covariance of the Lax pair. This is the elementary tool to build a new solution from an existing solution, even if this existing solution is trivial, such as the vacuum solution $u = 0$.
– the *Crum transformation*, which is the Nth iterate of the Darboux transformation. Starting from any known solution, this transformation allows one to build a solution depending on $2N$ more arbitrary parameters, and the only computation is that of a determinant.

The second approach is more easily implemented in the framework of singularity analysis, and the tools involved are:

– the *Lax pair*,
– the *singular part transformation*, expressing the difference $u - U$ of two solutions as a logarithmic derivative defined with the singular part operator $\mathscr{D}$. In the nonintegrable case, this transformation still exists but it only allows one to build particular solutions [84], and it would be an error to conclude the existence of a BT.
– the *Bäcklund transformation* [20] (BT), already defined in Def. 4.5. Historically defined before the Darboux transformation, this transformation, which is made of two nonlinear PDEs, has been later shown [60] to be nonelementary since it results from the Lax pair and the singular part transformation by the elimination of the wave vector ψ.
– the *nonlinear superposition formula* (NLSF), which, given three solutions involved in two copies of the BT, allows one to obtain a fourth solution by a purely algebraic process, i.e. without integration.

This is this second approach which will be developed in the remaining of this chapter, but let us first give more precise definitions.

5.3 Tools of Integrability

In order to illustrate the definitions given in this section, we will mainly take two examples, the Korteweg-de Vries equation already presented as (4.40),

$$\mathrm{KdV}(u) \equiv bu_t + u_{xxx} - \frac{6}{a} u u_x = 0,$$

and the Boussinesq equation

$$\mathrm{Bq}(u) \equiv \varepsilon^2 \left(\frac{\beta^2}{3} u_{xx} + u^2 \right)_{xx} + u_{tt} = 0,\ \varepsilon^2 \text{ and } \beta^2 \text{ real.} \tag{5.1}$$

Both PDEs are two different reductions of the same $2+1$-dimensional PDE, the Kadomtsev–Petviashvili (KP) equation [243],

$$\mathrm{KP}(u) \equiv \varepsilon^2 \left(\frac{\beta^2}{3} \left(u_{x_1 x_1 x_1} + u_{x_3} \right) + \left(u^2 \right)_{x_1} \right)_{x_1} + u_{x_2 x_2} = 0, \tag{5.2}$$

Physically, x_1 and x_2 are space coordinates, x_3 is the time, u is real, and the constants ε^2 and β^2 are real.

The behavior of the solutions strongly depends on the sign of $\varepsilon^2\beta^2$, and two different names are given to KP: KP-I for $\varepsilon^2\beta^2 = -1$, KP-II for $\varepsilon^2\beta^2 = 1$. The reduction of KP to KdV is

$$\partial_{x_2} u = 0,\ x_1 = x,\ x_3 = t/b,\ \beta^2 = -a, \tag{5.3}$$

and the reduction of KP to Boussinesq is

$$\partial_{x_3} u = 0,\ x_1 = x,\ x_2 = t. \tag{5.4}$$

5.3.1 Lax Pair

Definition 5.1. Given a nonlinear PDE $E(U) = 0$, one calls the **Lax pair** a set of two linear operators L_1, L_2 depending on a solution U of the PDE, whose commutativity condition is equivalent to the condition $E(U) = 0$,

$$([L_1, L_2] = 0) \Longleftrightarrow E(U) = 0. \tag{5.5}$$

If u depends on two independent variables, the Lax pair must in addition depend on an arbitrary complex parameter called the *spectral parameter.*

These two operators can be scalar or matrix operators. Scalar representations of the Lax pairs of our examples are,
for the KdV equation (4.40) [274]

$$\begin{cases} L_1 \equiv \partial_x^2 - U/a - \lambda, \\ L_2 \equiv b\partial_t + 4\partial_x^3 - 6(U/a)\partial_x - 3U_x/a,\ [L_1, L_2] = (1/a)E(U), \end{cases} \tag{5.6}$$

for the Boussinesq equation (5.1) [443],

$$\begin{cases} L_1 \equiv \partial_x^3 + (3/(2\beta^2))(U\partial_x + (1/2)U_x) + (3/(4\varepsilon\beta^3))R - \lambda, \\ L_2 \equiv \partial_t - \beta\varepsilon\partial_x^2 - (\varepsilon/\beta)U, \\ [L_1, L_2] = \{R_t + \varepsilon^2(U^2 + (\beta^2/3)U_{xx})_x)\} + (U_t - R_x)\partial_x, \end{cases} \tag{5.7}$$

and for the KP equation (5.2) [129, 445],

$$\begin{cases} L_1 \equiv -\varepsilon^{-1}\beta^{-1}\partial_{x_2} + \partial_{x_1}^2 + \beta^{-2}U, \\ L_2 \equiv \partial_{x_1}^3 + (3/(2\beta^2))(U\partial_{x_1} + (1/2)U_{x_1}) + (3/(4\varepsilon\beta^3))R - (1/4)\partial_{x_3}, \\ [L_1, L_2] = \left\{R_{x_2} + \varepsilon^2\left(\frac{\beta^2}{3}\left(U_{x_1x_1x_1} + U_{x_3}\right) + \left(U^2\right)_{x_1}\right)\right\} + (U_{x_2} - R_{x_1})\partial_{x_1}. \end{cases} \tag{5.8}$$

In (5.7) or (5.8), by the elimination of the auxiliary field R defined by either $U_t - R_x = 0$ or $U_{x_2} - R_{x_1} = 0$, one checks that the vanishing of the commutator $[L_1, L_2]$ is indeed equivalent to the Boussinesq or the KP equation for U.

In the two 1+1-dimensional cases, the Lax pair depends on the spectral parameter λ, and the operator L_1 is called *scattering operator*. In the 2+1 case of KP, the two equations defining the Lax pair are linear partial differential equations, therefore the notion of spectral parameter is meaningless.

The reductions (5.3) and (5.4) which act on KP to yield KdV and Boussinesq also act on the Lax pair of KP to yield those of the reduced PDEs,

$$L_{j,\mathrm{KdV}} = P^{-1}L_{j,\mathrm{KP}}P,\ j = 1, 2,\ P = e^{\varepsilon\lambda x_2}, \tag{5.9}$$

$$L_{j,\mathrm{Bq}} = P^{-1}L_{3-j,\mathrm{KP}}P,\ j = 1, 2,\ P = e^{-\lambda x_3}, \tag{5.10}$$

and these are the reductions which introduce the spectral parameter λ.

It is possible to obtain these Lax pairs from singularity considerations only, this will be done in Sects. 5.6.1.1, 5.6.1.2, 5.6.2, and 5.6.3.

Besides the above scalar representation for a Lax pair, there exist several other equivalent representations.

The *Lax representation* [274] is a pair of linear operators (L, P) (scalar or matrix) defined by

$$L_1 = L - \lambda,\ L_2 = \partial_t - P,\ L_1\psi = 0,\ L_2\psi = 0,\ \lambda_t = 0, \tag{5.11}$$

in which the elimination of the scalar λ yields

$$L_t = [P, L], \tag{5.12}$$

i.e. , thanks to the *isospectral* condition $\lambda_t = 0$, a time evolution analogous to the one in Hamiltonian dynamics.

The *zero-curvature representation* is a pair (L, M) of linear operators independent of (∂_x, ∂_t)

$$\begin{aligned} &L_1 = \partial_x - L,\ L_2 = \partial_t - M,\ L_1\psi = 0,\ L_2\psi = 0, \\ &[\partial_x - L, \partial_t - M] = L_t - M_x + LM - ML = 0. \end{aligned} \tag{5.13}$$

The common order N of the matrices is called the *order* of the Lax pair.

The *projective Riccati representation* is a first order system of $2N-2$ Riccati equations in the unknowns $\psi_j/\psi_1, j=2,\dots,N$, equivalent to the zero-curvature representation (5.13).

The *string representation* or *Sato representation* [241]

$$[P,Q]=1. \tag{5.14}$$

This very elegant representation, reminiscent of Hamiltonian formalism, uses the Sato definition of a *microdifferential operator* (a differential operator with positive and negative powers of the differential operator ∂) and of its *differential part* denoted $()_+$ (the subset of its nonnegative powers), e.g.

$$Q=\partial_x^2-u, \tag{5.15}$$
$$L=Q^{1/2}=\partial_x-(1/4)\{u,\partial_x^{-1}\}, \tag{5.16}$$
$$(L)_+=\partial_x, \tag{5.17}$$
$$\left(L^3\right)_+=\partial_x^3-(3/4)\{u,\partial_x\}=P, \tag{5.18}$$
$$\left(L^5\right)_+=\partial_x^5-(5/4)\{u,\partial_x^3\}+(5/16)\{3u^2+u_{xx},\partial_x\}, \tag{5.19}$$

in which $\{a,b\}$ denotes the anticommutator $ab+ba$. See [120] for a tutorial presentation.

Examples of these various representations will be encountered later in the text.

5.3.2 Darboux Transformation

Definition 5.2. Given a nonlinear PDE $E(u)=0$ and its Lax pair, one calls a **Darboux transformation** (DT) [110, 301] a gauge transformation which at the same time preserves the covariance of the Lax pair and increases the number of arbitrary constants in the solution U appearing in the Lax pair.

To write explicitly this transformation, it is now necessary to introduce the *wave vector* ψ and to write the scalar Lax pairs as $L_1(U,\lambda)\psi=0, L_2(U,\lambda)\psi=0$, because the DT acts on the triplet (ψ,U,λ).

In the case of KdV, starting from the scalar Lax pair (5.6) for the triplet (Ψ,U,Λ) (we use upper case notation for the input, lower case for the iterate),

$$L_1(U,\Lambda)\Psi=0,\ L_2(U,\Lambda)\Psi=0, \tag{5.20}$$

following Darboux [110], one first introduces an intermediate triplet (θ,U,μ) also satisfying the Lax pair,

$$L_1(U,\mu)\theta=0,\ L_2(U,\mu)\theta=0, \tag{5.21}$$

in which U is the same as in (5.20), μ is different from Λ and the wave vector θ is nonzero. Then, under the transformation

$$(\Psi,U,\Lambda) \to (\psi,u,\lambda) : \begin{cases} \psi = G\Psi = \theta\partial_x(\theta^{-1}\Psi), \\ u = U + 2a\partial_x^2 \log\theta, \\ \lambda = \Lambda, \end{cases} \tag{5.22}$$

the new triplet (ψ,u,λ) is such that

$$G(\theta)L_j(U,\Lambda) = L_j(u,\lambda)G(\theta),\ j = 1,2, \tag{5.23}$$

therefore the commutativity condition of the transformed operators $L_j(u,\lambda)$ is equivalent to $E(u) = 0$. The iterated field u in (5.22) depends on two more arbitrary constants than the initial field U in (5.20).

Remark 1. The simpler covariant transformation

$$(\Psi,U,\Lambda) \to (\psi,u,\lambda) : \begin{cases} \psi = \Psi^{-1}, \\ u = U + 2a\partial_x^2 \log\Psi, \\ \lambda = \Lambda, \end{cases} \tag{5.24}$$

does not increase the number of arbitrary constants and is therefore not a DT. This involution, which shares with the true DT the transformation on U, is however quite easy to derive knowing the singular part operator $\mathscr{D} = 2a\partial_x^2$ of KdV.

Remark 2. The formulae (5.22) for the DT do not involve t-derivatives, and in fact the original work of Darboux [110] only dealt with the Sturm–Liouville ordinary differential equation

$$\psi_{xx} + (-U/a - \lambda)\psi = 0. \tag{5.25}$$

Remark 3. At the KP level, depending on the sign of $\varepsilon^2\beta^2$, the Lax pair [129] is (KP-II case) or is not (KP-I case) self-adjoint, therefore the intermediate wave vector θ is generically complex for KP-I, and so is the solution u appearing in the iterated Lax pair. If one wants the iterated u to be real, which may be required by physical considerations, one must also consider the adjoint Lax pair and introduce [278, 329] two gauge operators such as the G in (5.22). The resulting transformation, which will not be described here, is called the *binary Darboux transformation.*

Remark 4. When the scalar scattering problem (5.25) is written in matrix form

$$\partial_x \begin{pmatrix} \psi \\ \psi_x \end{pmatrix} = \begin{pmatrix} 0 & 1 \\ \lambda - u & 0 \end{pmatrix} \begin{pmatrix} \psi \\ \psi_x \end{pmatrix}, \tag{5.26}$$

it is convenient to also write the Darboux transformation in matrix form [223],

$$\begin{pmatrix} \psi \\ \psi_x \end{pmatrix} = (\mu - \Lambda)^{-1/2} \begin{pmatrix} -y & 1 \\ y^2 + \Lambda - \mu & -y \end{pmatrix} \begin{pmatrix} \Psi \\ \Psi_x \end{pmatrix},\ y = \frac{\theta_x}{\theta}, \tag{5.27}$$

in which $(\mu - \Lambda)^{-1/2}$ is an irrelevant normalization factor intended at making unity the determinant of the matrix. This will be used in Sect. 7.6.1.

5.3.3 Crum Transformation

Definition 5.3. Given a nonlinear PDE $E(u) = 0$ and its Darboux transformation, one calls the **Crum transformation** [108, 301] the Nth iterate of the Darboux transformation.

With the Darboux transformation, one can devise an iteration on (ψ, u, λ) [170] to build solutions depending on the set of successive values $\lambda_1, \dots, \lambda_n, \dots$ The total cost is the integration of just *one* linear system, since the potential denoted U in (5.21) does not change throughout the iteration. The N-th iterate $(\psi, u, \lambda)_N$ can then be expressed in a closed form involving N different copies of the starting point $(\psi, u, \lambda)_0 = (\theta_0, U, \lambda)$, each copy corresponding to a different λ_k.

In the KdV case for instance, starting from the *vacuum* $u_0 = 0$, the N wave functions $\theta_j, j = 1, \dots, N$, are chosen as particular solutions of the linear system (5.6) with a zero potential,

$$\begin{cases} \theta_{xx} - \lambda_j \theta = 0, \\ \theta_t + 4\theta_{xxx} = 0, \end{cases} \tag{5.28}$$

i.e.

$$\theta_j = A_j \cosh k_j (x - 4k_j^2 t + \delta_j),\ k_j^2 = \lambda_j,\ j = 1, \dots, N, \tag{5.29}$$

in which A_j and δ_j are $2N$ arbitrary constants of integration. Let us also denote θ_0 another solution of (5.28) for another value $\lambda_0 = \lambda$. The Nth iterate is then [422]

$$\begin{cases} u_N = -2a\partial_x^2 \log \mathrm{W}(\theta_1, \dots, \theta_N), \\ \psi_N = \dfrac{\mathrm{W}(\theta_1, \dots, \theta_N, \theta_0)}{\mathrm{W}(\theta_1, \dots, \theta_N)}, \end{cases} \tag{5.30}$$

in which the *Wronskian* is the determinant

$$\mathrm{W}(\theta_1, \dots, \theta_N) = \det(\partial_x^{i-1} \theta_j) = \begin{vmatrix} \theta_1 & \theta_1' & \dots & \theta_1^{(N-1)} \\ \theta_2 & \theta_2' & \dots & \theta_2^{(N-1)} \\ \dots & \dots & \dots & \dots \\ \theta_N & \theta_N' & \dots & \theta_N^{(N-1)} \end{vmatrix}. \tag{5.31}$$

The wave vectors ψ_N and θ_0 obey the Lax pair (5.6) for the same spectral parameter λ and for the respective potentials u_N and 0.

This solution, which depends on $2N$ arbitrary complex constants k_j, δ_j, is called the *N-soliton solution*. It enjoys nice properties, well documented in [125], among them elastic collisions (see Fig. 5.1).

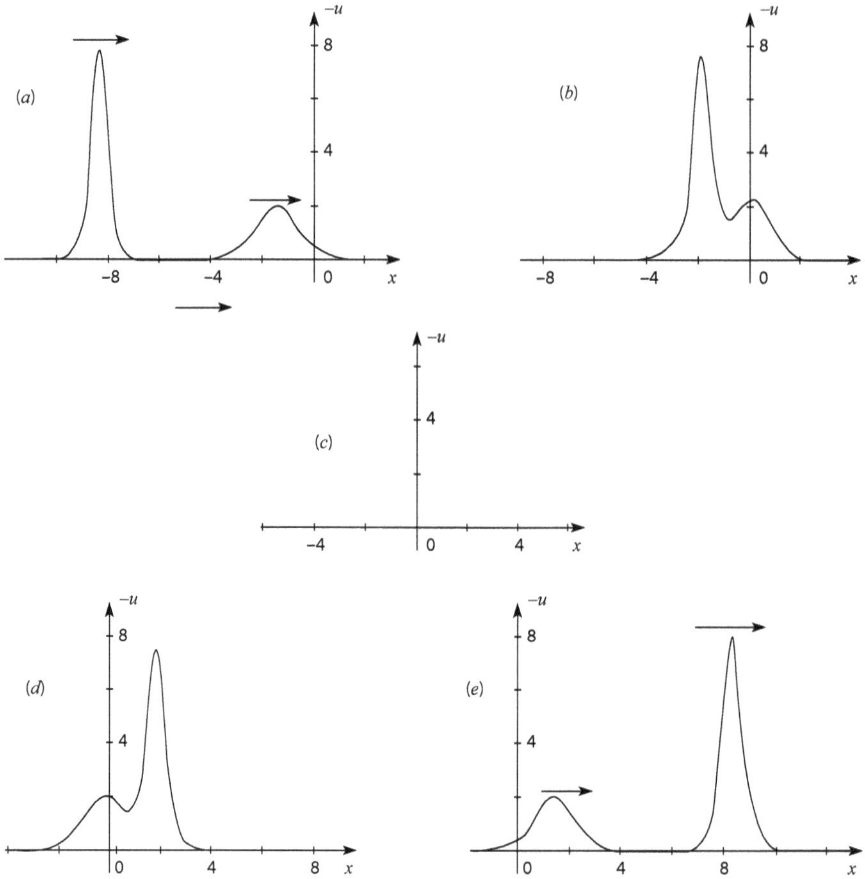

Fig. 5.1 KdV. Two-soliton solution displaying the elastic collision of two one-soliton solutions (reproduced from Figure 4.3 page 76 in [125]).

5.3.4 Singular Part Transformation

The Laurent series (4.25) is the sum of a singular part (usually called principal part, made of the first $-p$ terms $j = 0,\dots,-p-1$) and a regular part made of the remaining terms $j = -p,\dots,+\infty$. The singular part can be represented by a *linear* differential operator.

Definition 5.4. Given a family of movable singularities represented by the Laurent series (4.25), one calls the **singular part operator** of this family the linear differential operator $\mathscr{D}$ uniquely defined by the property

$$u - \mathscr{D}\log\psi = \mathscr{O}(1),\ \chi = \frac{\psi}{\psi_x}. \tag{5.32}$$

For the various families previously encountered ((4.41) for KdV, (4.50) for KPP, (2.31) for the KS PDE (2.25)), these operators are, respectively, $\mathscr{D} = -2a\partial_x^2$, $\mathscr{D} = u_0\partial_x$ and (2.44) in which ∂_ξ is replaced by ∂_x.

As to the regular part of the Laurent series, it is of no direct help because the Laurent series is not a closed-form solution. Since this is the only directly available piece of information and since a finite (closed form) expression is required for an exact solution, let us represent, following the idea of Weiss, Tabor, and Carnevale [431], an unknown exact solution u as the sum of a singular part, characterized by the singular part operator $\mathscr{D}$, and of a regular part denoted U.

Definition 5.5. Given a PDE with only one family of movable singularities, one calls the **singular part transformation** the representation of a solution u as the sum

$$u = \mathscr{D} \log \tau + U,\ E(u) = 0, \tag{5.33}$$

in which $\mathscr{D}$ is the singular part operator of the family, τ is a function ("tau-function"[1]) linked to the wave vector ψ of the Lax pair to be found, and U an unspecified field.

The aim of the representation (5.33) is to devise a choice of link between τ and ψ such that the field U in (5.33) be another solution of the same PDE, like it is in the Darboux transformation (second equation in (5.22)).

In case the PDE has two families, the representation (5.33) is replaced by [317]

$$u = \mathscr{D}_1 \log \tau_1 + \mathscr{D}_2 \log \tau_2 + U, \tag{5.34}$$

the operator $\mathscr{D}_f$ and the function τ_f being attached to each family f.

5.3.5 Nonlinear Superposition Formula

The Bäcklund transformation has been defined in Definition 4.5 Sect. 4.2. Before defining the nonlinear superposition formula, we need to state a theorem.

Theorem 5.1. *(Bianchi permutability theorem [31])*
Given a nonlinear PDE $E(u) = 0$ and its auto-Bäcklund transformation $BT(u, U, \lambda) = 0$, if one applies this BT to a given solution u_{n-1} with two different spectral parameters, $BT(u_{n-1}, u_n, \lambda_n) = 0$, $BT(u_{n-1}, \tilde{u}_n, \lambda_{n+1}) = 0$, then there exists a fourth solution u_{n+1} which can be obtained by either $BT(u_n, u_{n+1}, \lambda_{n+1}) = 0$ or $BT(\tilde{u}_n, u_{n+1}, \lambda_n) = 0$, i.e. by permuting the two spectral parameters.

The content of this theorem is schematically represented by the Bianchi diagram, Fig. 5.2.

[1] More precisely the ratio of two tau-functions in the sense of Sato [241].

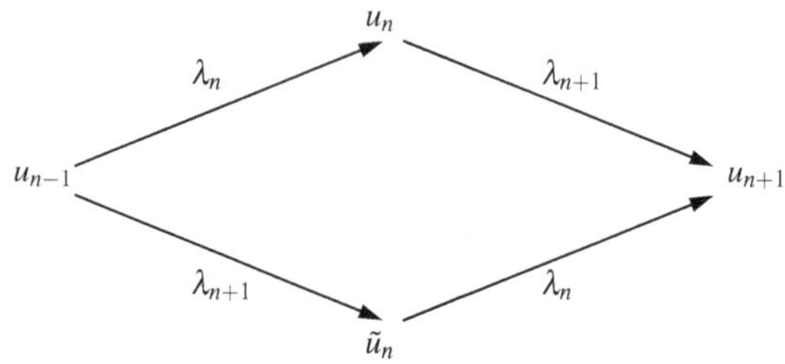

Fig. 5.2 Bianchi diagram. Its four branches are described in theorem 5.1.

No general demonstration of this theorem seems to be known [374], and one has to prove it for every PDE admitting an auto-Bäcklund transformation.

Definition 5.6. Given a nonlinear PDE $E(u) = 0$ and any four different solutions $u_{n-1}, u_n, \tilde{u}_n, u_{n+1}$ involved in the Bianchi permutability theorem, one calls the **nonlinear superposition formula** (NLSF) [113, 31] a relation yielding explicitly (i.e. without any integration) the fourth solution u_{n+1}[2] in terms of the three others.

If such an NLSF exists, this proves *ipso facto* the Bianchi permutability theorem.

Example. The most elementary example of an equation admitting a NLSF is the Riccati equation

$$\frac{\mathrm{d}u}{\mathrm{d}x} = a_2(x)u^2 + a_1(x)u + a_0(x), \tag{5.35}$$

already encountered in many instances, see (1.1), (3.92), (3.179) and (4.27). Given three particular solutions u_1, u_2, u_3 of (5.35), another solution is defined by the relation

$$(u, u_1, u_2, u_3) = c, \tag{5.36}$$

in which the l.h.s. is the crossratio

$$(x_1, x_2, x_3, x_4) = \frac{(x_3 - x_1)(x_4 - x_2)}{(x_3 - x_2)(x_4 - x_1)}, \tag{5.37}$$

of the four solutions and c is an arbitrary complex constant. Moreover, since c is arbitrary, this fourth solution is the general solution of (5.35).

The end of the present chapter is devoted to the derivation of the Bäcklund transformation, and therefore of explicit solutions, by methods only based on singularities, globally known as the *singular manifold method*. Since the BT results from the Lax pair and the singular part transformation by an elimination [60], one will first search for these two simpler elements.

Examples will be given in Sects. 5.6.1.1, 5.6.1.2, 5.6.2, and 5.6.3.

[2] Or some primitive of it.

5.4 Choosing the Order of Lax Pairs

In the algorithm to be presented in Sect. 5.5.1, it is necessary to provide as an input a Lax pair in scalar form and its order (two, three, etc), with undetermined coefficients. Such Lax pairs are defined in the present section, together with the constraints imposed on these coefficients by the commutativity condition.

It is sometimes appropriate to represent an n-th order Lax pair by the $2(n-1)$ equations satisfied by an equivalent $(n-1)$-component pseudopotential $\mathbf{Y}$ of Riccati type, the first component of which is chosen as

$$Y_1 = \psi_x/\psi, \tag{5.38}$$

in which ψ is a scalar component of the Lax pair.

5.4.1 Second-Order Lax Pairs and Their Privilege

The general second-order scalar Lax pair reads, in the case of two independent variables (x,t),

$$L_1\psi \equiv \psi_{xx} - d\psi_x - a\psi = 0, \tag{5.39}$$

$$L_2\psi \equiv \psi_t - b\psi_x - c\psi = 0, \tag{5.40}$$

$$[L_1, L_2] \equiv X_0 + X_1\partial_x, \tag{5.41}$$

$$X_0 \equiv -a_t + a_x b + 2ab_x + c_{xx} - c_x d = 0, \tag{5.42}$$

$$X_1 \equiv -d_t + (b_x + 2c - bd)_x = 0. \tag{5.43}$$

The coefficient d could be set to zero by the linear change $\psi \mapsto \exp[\int d\mathrm{d}x/2]\psi$, but this would modify X_0 and X_1 and thus the commutator could be no more equivalent to the nonlinear PDE under study.

The Lax pair (5.39)–(5.40) is identical to a linearized version of the Riccati system satisfied by the most general expansion variable Y defined by (4.37), under the correspondence

$$Y^{-1} = B\left(\frac{\psi_x}{\psi} + A\right),\ B \neq 0, \tag{5.44}$$

$$d = 2A,\ a = A_x - A^2 - S/2,\ b = -C,\ c = C_x/2 + AC + \int A_t \mathrm{d}x, \tag{5.45}$$

and the commutator of the Lax pair is (4.32).

In particular, when the coefficient d is zero, the correspondence with the Lax pair (4.34)–(4.35) is [315]

$$\chi = \frac{\psi}{\psi_x},\ B = 1,\ A = 0, \tag{5.46}$$

$$a = -S/2,\ b = -C,\ c = C_x/2,\ d = 0. \tag{5.47}$$

Therefore second order Lax pairs are privileged in the singularity approach, in the sense that their coefficients can be identified with the elementary homographic invariants S, C of the invariant Painlevé analysis and, if appropriate, A, B. Conversely, when the Lax order is higher than two, these homographic invariants S, C are useless and they should not be considered; this has historically been the reason of some errors described for instance in [78, section 8].

5.4.2 Third-Order Lax Pairs

The third-order scalar Lax pair will be defined as (the coefficient of ψ_{xx} in L_1 is assumed zero here for simplification)

$$\begin{aligned}
L_1\psi &\equiv \psi_{xxx} - a\psi_x - b\psi = 0, && (5.48)\\
L_2\psi &\equiv \psi_t - c\psi_{xx} - d\psi_x - e\psi = 0, && (5.49)\\
[L_1, L_2] &\equiv X_0 + X_1\partial_x + X_2\partial_x^2, && (5.50)\\
X_0 &\equiv -b_t - ae_x + e_{xxx} + b_{xx}c \\
&\quad +3bc_{xx} + 3b_xc_x + 3bd_x + b_xd = 0, && (5.51)\\
X_1 &\equiv -a_t + 3e_{xx} + 2b_xc + a_{xx}c + d_{xxx} + 3ac_{xx} + 2ad_x \\
&\quad +3a_xc_x + 3bc_x + a_xd = 0, && (5.52)\\
X_2 &\equiv (2ac + c_{xx} + 3d_x + 3e)_x = 0. && (5.53)
\end{aligned}$$

An equivalent two-component pseudopotential is the projective Riccati one $\mathbf{Y} = (Y_1, Y_2)$ [12, 315, 316],

$$\begin{aligned}
Y_1 &= \frac{\psi_x}{\psi},\ Y_2 = \frac{\psi_{xx}}{\psi}, && (5.54)\\
Y_{1,x} &= -Y_1^2 + Y_2, && (5.55)\\
Y_{2,x} &= -Y_1Y_2 + aY_1 + b, && (5.56)\\
Y_{1,t} &= -(dY_1 + cY_2 + ac + d_x)Y_1 + (c_x + d)Y_2 + e_x + bc && (5.57)\\
&= (cY_2 + dY_1 + e)_x, && (5.58)\\
Y_{2,t} &= -(dY_1 + cY_2)Y_2 + (2ac_x + a_xc + bc + d_{xx} + ad + 2e_x)Y_1 \\
&\quad +(c_{xx} + 2d_x + ac)Y_2 + 2bc_x + b_xc + bd + e_{xx}, && (5.59)\\
Y_{1,tx} - Y_{1,xt} &= X_1 + X_2Y_1, && (5.60)\\
Y_{2,tx} - Y_{2,xt} &= -X_0 + X_2Y_2. && (5.61)
\end{aligned}$$

5.5 Singular Manifold Method

We now have all the ingredients to give a general exposition of the method in the form of an algorithm. The present exposition largely follows the lines of [319], updated in the lecture notes [78].

This method is known as the *singular manifold method* or *truncation method* because it selects the beginning of a Laurent series and discards ("truncates") the remaining infinite part. Since its introduction by WTC [431], it has been improved in many directions [315, 134, 164, 317, 87, 357, 319], and we present here the current state of the method.

5.5.1 Algorithm

Consider a PDE (4.1) with either one family of movable singularities or two families of movable singularities with the same singularity degree p and different values of u_0, and denote $\mathscr{D}_1, \mathscr{D}_2$ the two singular part operators.

First step. Assume a singular part transformation defined either as (5.33) (one-family case) or as (5.34) (two-family case), with u a solution of the PDE under consideration, U an unspecified field, and τ_f one or two functions.

Second step. Choose the order two, then three, then ..., for the unknown Lax pair, and define one or two (as many as the number of families) *scalars* ψ_f from the component(s) of its wave vector (e.g. the scalar wave vector if the PDE has one family and the pair is defined in scalar form).

Third step. Choose an explicit link F

$$\forall f: \mathscr{D}_f \log \tau_f = F(\psi_f), \tag{5.62}$$

the same for each family f, between the functions τ_f and the scalars ψ_f defined from the Lax pair. As will be shown in Sect. 5.6.3.1, at each scattering order, there exists only a finite number of choices (5.62), among them the most frequent one

$$\forall f: \mathscr{D}_f \log \tau_f = \mathscr{D}_f \log \psi_f, \text{ i.e. } \tau_f = \psi_f. \tag{5.63}$$

Fourth step. Define the "truncation" and solve it, that is to say: with the assumptions (5.33) or (5.34) for a singular part transformation, (5.62) for a link between τ_f and ψ_f, (5.39)–(5.40) or (5.48)–(5.49) or other for the Lax pair in ψ, express $E(u)$ as a polynomial in the derivatives of ψ_f which is irreducible *modulo* the Lax pair. For the just mentioned Lax pairs and a one-family PDE, this amounts to eliminate any derivative of ψ of order in (x,t) higher than or equal to $(2,0)$ or $(0,1)$ (second order case) or to $(3,0)$ or $(0,1)$ (third order), thus resulting in a polynomial of one variable ψ_x/ψ (second order) or two variables $\psi_x/\psi, \psi_{xx}/\psi$ (third order).

The l.h.s. $E(u)$ of the equation thus becomes, for the second order Lax pair (5.39)–(5.40) in the particular "Gel'fand–Dikii" case $d=0$,

$$E(u) = \sum_{j=0}^{-q} E_j(S,C,U)(\psi/\psi_x)^{j+q} \tag{5.64}$$

and for the third order Lax pair (5.48)–(5.49),

$$E(u) = \sum_{k\geq 0}\sum_{l\geq 0} E_{k,l}(a,b,c,d,e,U)(\psi_x/\psi)^k(\psi_{xx}/\psi)^l. \tag{5.65}$$

Since one has no more information on this polynomial $E(u)$ except the fact that it must vanish, one requires that it identically vanishes, by solving the set of *determining equations*

$$\forall j \quad E_j(S,C,U) = 0 \text{ (one-family PDE, second order)} \tag{5.66}$$
$$\forall k\ \forall l \quad E_{k,l}(a,b,c,d,e,U) = 0 \text{ (one-family PDE, third order)} \tag{5.67}$$

for the unknown coefficients (S,C) or (a,b,c,d,e) as functions of U, and one establishes the constraint(s) on U by eliminating (S,C) or (a,b,c,d,e). The strategy of resolution relies on some basic principles (never increase the differential order, enforce all cross-derivative conditions, etc) which mathematically deal with the construction of a differential Groebner basis [297, 37], and a detailed example (the Kuramoto–Sivashinsky PDE (2.25)) is handled in [78].

The constraints on U reflect the integrability level of the PDE. If the only constraint on U is to satisfy some PDE, one is on the way to an auto-BT if the PDE for U is the same as the PDE for u, or to a BT between the two PDEs.

In the integrable case, the second, third and fourth steps must be repeated until a success occurs. The process is successful if and only if all the following conditions are met:

1. U comes out with one constraint exactly, namely to be a solution of some PDE,
2. (if an auto-BT is desired) the PDE satisfied by U is identical to (4.1),
3. the vanishing of the commutator $[L_1, L_2]$ is equivalent to the vanishing of the PDE satisfied by U,
4. in the $1+1$-dimensional case only, the PDE satisfied by U is identical to (4.1), the coefficients depend on an arbitrary constant λ, the spectral or Bäcklund parameter.

In the nonintegrable case, one is happy with any solution. For instance, solutions u polynomial in $\tanh k(x-ct)$ or in the two variables $\tanh k(x-ct), \operatorname{sech} k(x-ct)$ are described by constant values of (S,C,A,B) in (5.39)–(5.40), and the determining equations are no more differential but algebraic.

At this stage, one has obtained the singular part transformation and the Lax pair.

Fifth step. Obtain the two equations for the BT by eliminating τ_f and ψ_f [60] between the singular part transformation and the Lax pair. This sometimes difficult operation when the order n of the Lax pair is too high may become elementary by considering the equivalent Riccati representation of the Lax pair and eliminating the appropriate components of $\mathbf{Y}$ rather than ψ. Assume for instance that $\tau = \psi$, $\mathscr{D} = \partial_x$, and the PDE has only one family. Then (5.33) reads

$$Y_1 = u - U \tag{5.68}$$

with Y_1 defined in (5.38), and the BT is computed as follows: eliminate all the components of $\mathbf{Y}$ but Y_1 between the equations for the gradient of $\mathbf{Y}$, then in the resulting equations substitute Y_1 as defined in (5.68).

If the computation of the BT requires the elimination of Y_2 between (5.55)–(5.59), this BT is

$$Y_{1,xx} + 3Y_1Y_{1,x} + Y_1^3 - aY_1 - b = 0, \tag{5.69}$$

$$Y_{1,t} - (cY_{1,x} + cY_1^2 + dY_1 + e)_x = 0, \tag{5.70}$$

$$(Y_{1,xx})_t - (Y_{1,t})_{xx} = X_0 + X_1Y_1 + X_2Y_1^2 = 0, \tag{5.71}$$

in which Y_1 is replaced by an expression of $u - U$, e.g. (5.68).

Although, let us repeat, the method equally applies to integrable as well as non-integrable PDEs, examples are split according to that distinction, to help the reader to choose his/her field of interest.

5.5.2 Level of Truncation and Choice of Variable

This section is self-contained, and mainly aimed at those readers accustomed to performing the WTC truncation. Although some paragraphs might be redundant with Sect. 5.5.1, it may help the reader by presenting a complementary point of view.

Let us assume in this section that the unknown Lax pair is second order. Then the truncation defined in the fourth step of Sect. 5.5.1 is performed in the style of Weiss et al. [431], i.e. with a single variable. This WTC truncation consists in forcing the series (4.25) to terminate; let us denote p and q the singularity degrees of u and $E(u)$, $-p'$ the rank at which the series for u terminates, and $-q'$ the corresponding rank of the series for E

$$u = \sum_{j=0}^{-p'} u_j Z^{j+p},\ u_0 u_{-p'} \neq 0,\ E = \sum_{j=0}^{-q'} E_j Z^{j+q}, \tag{5.72}$$

in which the *truncation variable* Z chosen by WTC is $Z = \varphi - \varphi_0$. Since one has no more information on Z, the method of WTC is to require the separate vanishing of each of the *truncation equations*

$$\forall j = 0, \ldots, -q' : E_j = 0. \tag{5.73}$$

In earlier presentations of the method, one had to prove by recurrence that, assuming that enough consecutive coefficients u_j vanish beyond $j = -p'$, then all further coefficients u_j would vanish. This painful task is useless if one defines the process as done above.

The first question to be solved is: what are the admissible values of p', i.e. those which respect the condition $u_{-p'} \neq 0$?

The answer depends on the choice of the truncation variable Z. In Sect. 4.4.1 three choices were presented, $Z =$ either $\varphi - \varphi_0$, χ or Y, respectively defined by equations (4.2), (4.26), (4.37), with the property that any two of their inverse are linearly dependent.

The advantage of χ or Y over $\varphi - \varphi_0$ is the following. The gradient of χ (resp. Y) is a polynomial of degree two in χ (resp. Y), so each derivation of a monomial aZ^k increases the degree by one, while the gradient of $\varphi - \varphi_0$ is a polynomial of degree zero in $\varphi - \varphi_0$, so each derivation decreases the degree by one. Consequently, one finds two solutions and only two to the condition $u_{-p'} \neq 0$ [356]:

1. $p' = p, q' = q$, in which case the three truncations are identical, since the three sets of equations $E_j = 0$ are equivalent (the finite sum $\sum E_j Z^{j+q}$ is just the same polynomial of Z^{-1} written with three choices for its base variable),
2. for χ and Y only, $p' = 2p, q' = 2q$, in which case the two truncations are different since the two sets of equations $E_j = 0$ are inequivalent (they are equivalent only if $A = 0$).

To perform the first truncation $p' = p, q' = q$, one must then choose $Z = \chi$ since Y brings no more information and $\varphi - \varphi_0$ creates equivalent but lengthier expressions.

To perform the second truncation $p' = 2p, q' = 2q$, one must choose $Z = Y$, since χ would create the *a priori* constraint $A = 0$.

The second question to be solved is: given some PDE with such and such structure of singularities, and assuming that one of the above two truncations is relevant (which is a separate topic), which one should be selected?

The answer lies in the two elementary identities [86]

$$\tanh z - \frac{1}{\tanh z} = -2i \,\mathrm{sech}\left[2z + i\frac{\pi}{2}\right], \ \tanh z + \frac{1}{\tanh z} = 2\tanh\left[2z + i\frac{\pi}{2}\right]. \quad (5.74)$$

Let us explain why in the two examples already presented in Sect. 1.1, the ODEs whose general solution is $\tanh(x - x_0)$ and $\mathrm{sech}(x - x_0)$,

$$E \equiv u' + u^2 - 1 = 0, \ u = \tanh(x - x_0), \quad (5.75)$$

$$E \equiv v'^2 + a^{-2}v^4 - v^2 = 0, \ v = a\,\mathrm{sech}(x - x_0), \quad (5.76)$$

(it is for convenience that we do not set $a = 1$). Equation (5.75) has the single family

$$p = -1, q = -2, u_0 = 1, \text{ Fuchs indices} = (-1), \quad (5.77)$$

and (5.76) has the two opposite families

$$p = -1, q = -4, v_0 = ia, \text{ Fuchs indices} = (-1), \quad (5.78)$$

in which ia denotes any square root of $-a^2$. The first truncation

$$u = \sum_{j=0}^{-p} u_j \chi^{j+p},\ E = \sum_{j=0}^{-q} E_j \chi^{j+q},\ \forall j\ : E_j = 0, \tag{5.79}$$

generates the respective results

$$u = \chi^{-1},\ S = -2, \tag{5.80}$$
$$v = ia\chi^{-1},\ E_2 \equiv a^2(1-S) = 0,\ E_3 \equiv 0,\ E_4 \equiv -a^2S^2/4, \tag{5.81}$$

thus providing (after integration of the Riccati ODE (4.27)) the general solution of (5.75), and no solution at all for (5.76).

The second truncation

$$u = \sum_{j=0}^{-2p} u_j Y^{j+p},\ E = \sum_{j=0}^{-2q} E_j Y^{j+q},\ \forall j\ : E_j = 0, \tag{5.82}$$

generates the respective results

$$u = B^{-1}Y^{-1} + (1/4)BY,\ A = 0,\ S = -1/2,\ B \text{ arbitrary}, \tag{5.83}$$
$$v = iaB^{-1}Y^{-1} - (1/4)iaBY,\ A = 0,\ S = -1/2,\ B \text{ arbitrary}, \tag{5.84}$$

thus providing, thanks to the identities (5.74), the general solution for both equations.

The conclusions from this exercise which can be generalized are :

1. for PDEs with only one family, the second truncation brings no additional information as compared to the first one and is always useless;
2. for PDEs with two opposite families (two opposite values of u_0 for a same value of p), the first truncation can never provide the general solution and can only provide particular solutions, while the second one may provide the general solution.

This defines the guidelines to be followed in the next sections of this chapter. The question of the relevance of the parameter B, which seems useless in the above two examples, is addressed in Sect. 5.6.2.

5.6 Application to Integrable Equations

In the integrable case, the method must yield all the elements of integrability up to and including the nonlinear superposition formula. Let us apply the method successively to a one-family equation with a second order Lax pair (KdV, Sect. 5.6.1.1), a one-family equation with a third order Lax pair (Boussinesq, Sect. 5.6.1.2), two two-family equations with a second order Lax pair (sine-Gordon, modified KdV, Sect. 5.6.2), and two one-family equations with a third order Lax pair, Sect. 5.6.3.

5.6.1 One-Family Cases: KdV and Boussinesq

In this case the SMM can be viewed as an extrapolation to PDEs of the truncation defined for ODEs in Sect. 3.2.3.2.

5.6.1.1 The Case of KdV

As shown in Sect. 4.4.2, the Painlevé test is satisfied, and the question is now to prove the Painlevé property. The use of singularity analysis to derive the results of the present section is mostly due to WTC [431], and its homographically invariant presentation can be found in [312].

From the Laurent series (4.46), one deduces the singular part operator,

$$\mathcal{D} = -2a\partial_x^2. \tag{5.85}$$

If one assumes the singular part transformation,

$$u = U + \mathcal{D} \log \tau,\ E(u) = 0, \tag{5.86}$$

the choice of the second-order scalar Lax pair (4.34)–(4.35) for ψ, and the link (5.63) between τ and ψ, the solution (5.86) becomes a Laurent series in χ with finitely many terms,

$$u = 2a\chi^{-2} + U + aS = -2a\partial_x^2 \log \psi + U,\ \chi = \frac{\psi}{\psi_x}, \tag{5.87}$$

and the identification with the (infinite) Laurent series (4.46) first provides the value of U

$$U = \frac{4aS - abC}{6}. \tag{5.88}$$

Inserting the finite Laurent series (5.87) into the KdV equation, and eliminating any derivative of χ from (4.27)–(4.28), one generates a finite Laurent series which must be identically zero. This defines only two determining equations

$$E_3 \equiv -2a(bC - S)_x = 0, \tag{5.89}$$

$$E_5 \equiv \frac{a}{3}(SS_x + 2S_{xxx}) - \frac{ab}{6}(4SC_x + 5C_{xxx}) - \frac{ab^2}{6}(C_t + CC_x) = 0, \tag{5.90}$$

together with the cross-derivative condition (4.32). Indeed, E_0, E_1 are identically zero because of the expression (5.85) of $\mathcal{D}$, E_2 is zero because of the value (5.88) of U, and E_4 is zero because 4 is a Fuchs index. The elimination of C_{xxx} and S_x between E_3, E_5 and the cross-derivative condition (4.32) yields the relation

$$(bC - S)_t = 0, \tag{5.91}$$

which, together with $E_3 = 0$, integrates as

$$bC - S - 6\lambda = 0, \tag{5.92}$$

in which λ is an arbitrary complex constant of integration. This equation (5.92), which is called [431] the *singular manifold equation* (SME), is the constraint on φ for the truncation to exist. It has by definition the property that all the determining equations (here $E_3 = 0, E_5 = 0$) are differential consequences of the SME. The system (5.88), (5.92) is then solved for (S,C)

$$S = -2\lambda - (2/a)U,\ bC = 4\lambda - (2/a)U, \tag{5.93}$$

and the cross-derivative condition (4.32) expresses that U is another solution of KdV. Therefore one has obtained the singular part transformation (5.86) and the second order Lax pair

$$\begin{cases} \psi_{xx} - (U/a + \lambda)\psi = 0, \\ b\psi_t + (4\lambda - 2U/a)\psi_x + (U_x/a)\psi = 0. \end{cases} \tag{5.94}$$

This Lax pair can also be written, at the reader's taste, with a second equation independent of λ,

$$\begin{cases} \psi_{xx} - (U/a + \lambda)\psi = 0, \\ b\psi_t + 4\psi_{xxx} - 6(U/a)\psi_x - 3(U_x/a)\psi = 0, \end{cases} \tag{5.95}$$

or in the zero-curvature representation (5.13)

$$L = \begin{pmatrix} 0 & 1 \\ U/a + \lambda & 0 \end{pmatrix}, \tag{5.96}$$

$$M = b^{-1} \begin{pmatrix} -U_x/a & 2U/a - 4\lambda \\ -U_{xx}/a + 2(U/a + \lambda)(U/a - 2\lambda) & U_x/a \end{pmatrix}, \tag{5.97}$$

or in the Riccati representation for $\omega = \chi^{-1}$ (see (4.27)–(4.28) and (4.32))

$$\omega_x = -\frac{S}{2} - \omega^2 = \left(\frac{U}{a} + \lambda\right) - \omega^2, \tag{5.98}$$

$$\omega_t = (-C\omega + C_x/2)_x = b^{-1}((2U/a - 4\lambda)\omega - U_x/a)_x. \tag{5.99}$$

In order to derive the auto-Bäcklund transformation of KdV, one should eliminate τ and ψ between: the singular part transformation (5.86), the link (5.63) and one convenient form of the Lax pair. If one introduces the so-called potential field v of u defined by $u = v_x, U = V_x$, then the singular part transformation relation (5.86) can be integrated once to yield

$$\omega = (v - V)/(-2a), \tag{5.100}$$

so the elimination reduces to the substitution of the value of ω into the Riccati system (5.98)–(5.99). This auto-BT of KdV reads

$$\begin{aligned} a(v+V)_x &= -2a^2\lambda + (v-V)^2/2, && (5.101)\\ ab(v+V)_t &= -(v-V)(v-V)_{xx} + 2(V_x^2 + v_x V_x + v_x^2), && (5.102) \end{aligned}$$

after suitable differential consequences of the x-part have been added to the t-part in order to suppress λ and cubic terms in (5.102). One notices the invariance under a permutation of v and V.

The difference $u-U$ (or $v-V$) of the two solutions involved in the BT obeys a PDE called by definition the *modified equation* [60] of the considered PDE, here KdV. This modified equation is obtained by eliminating U between (5.98)–(5.99), and the result is the *modified Korteweg–de Vries equation* (mKdV),

$$\text{mKdV}(w) \equiv bw_t + \left(w_{xx} - 2w^3/\alpha^2 + 6\nu w\right)_x = 0, \tag{5.103}$$

with the identification

$$w = \alpha\omega, \nu = \lambda. \tag{5.104}$$

Remark. Two different names are sometimes given to the mKdV equation (5.103), depending on whether ν is zero (*modified KdV*) or nonzero (*extended modified KdV*), but there is no reason to distinguish them. Indeed, they can be identified by simply redefining the time variable: $\text{mKdV}(\nu \neq 0, x, t, u(x,t)) = \text{mKdV}(\nu = 0, x, T, u(x,T)), t = T - x/(6\nu)$. The reason why we do not set the constant ν to zero is the relation $\nu = \lambda$ with λ arbitrary, a relation which comes out of the above definition for *modified equation*.

This PDE (5.103) also has the Painlevé property, and the system (5.98)–(5.99) is the Bäcklund transformation between KdV and mKdV ([269, (5.16)], [424]). The first half (5.98) of this BT between KdV and mKdV is an algebraic transformation

$$U = a\left(\frac{w_x}{\alpha} + \frac{w^2}{\alpha^2} - \nu\right), \tag{5.105}$$

called by definition the *Miura transformation* [306] from mKdV to KdV.

Let us now derive the nonlinear superposition formula (NLSF). Among the two equations (5.101)–(5.102), the one with the lowest differential order is (5.101). Let us consider this latter equation written for the four branches of the Bianchi diagram represented in Fig. 5.2,

$$\begin{cases} a(v_{n-1}+v_n)_x - (v_{n-1}-v_n)^2/2 + 2a^2\lambda_n = 0,\\ a(v_{n-1}+\tilde{v}_n)_x - (v_{n-1}-\tilde{v}_n)^2/2 + 2a^2\lambda_{n+1} = 0,\\ a(v_n+v_{n+1})_x - (v_n-v_{n+1})^2/2 + 2a^2\lambda_{n+1} = 0,\\ a(\tilde{v}_n+v_{n+1})_x - (\tilde{v}_n-v_{n+1})^2/2 + 2a^2\lambda_n = 0. \end{cases} \tag{5.106}$$

There exist two linear combinations, $(1,-1,-1,1)$ and $(1,-1,1,-1)$, whose result yields v_{n+1} as an explicit algebraic transform of $(v_{n-1}, v_n, \tilde{v}_n)$. These two NLSF are

$$v_{n+1} = v_{n-1} + 4a^2 \frac{\lambda_{n+1} - \lambda_n}{\tilde{v}_n - v_n}, \tag{5.107}$$

$$v_{n+1} = -v_{n-1} + (v_n + \tilde{v}_n) - 2a\partial_x \log(v_n - \tilde{v}_n). \tag{5.108}$$

The first one [423] is simpler but the second one [218] is more general as will be seen at the end of Sect. 5.6.1.2. Both NLSF are able to yield the expression (5.30) for the N-soliton solution.

5.6.1.2 The Case of Boussinesq

Let us first briefly perform the Painlevé test on (5.1), as detailed in Sect. 4.4.2.

Since the Boussinesq equation is a conservation law, it is cheaper (in the sense of shorter expressions to compute) to process the one-component "potential" form

$$\text{pBq}(v) \equiv v_{tt} + \varepsilon^2 \left((v_x)^2 + (\beta^2/3) v_{xxx} \right)_x = 0,\ u = v_x, \tag{5.109}$$

or even the "second potential" equation

$$\text{ppBq}(w) \equiv w_{tt} + \varepsilon^2 \left((w_{xx})^2 + (\beta^2/3) w_{xxxx} \right) = 0,\ u = v_x = w_{xx}. \tag{5.110}$$

Whatever the considered equation among these three equivalent forms, there exists only one family of movable singularities

$$\begin{cases} \text{Bq}: \ p = -2,\ q = -6, \text{ indices } (-1,4,5,6),\ \mathscr{D} = 2\beta^2 \partial_x^2, \\ \text{pBq}: \ p = -1,\ q = -5, \text{ indices } (-1,1,4,6),\ \mathscr{D} = 2\beta^2 \partial_x, \\ \text{ppBq}: \ p = 0^-,\ q = -4, \text{ indices} (-1,0,1,6),\ \mathscr{D} = 2\beta^2, \end{cases} \tag{5.111}$$

(the notation $p = 0^-$ means a logarithmic behavior $w \sim 2\beta^2 \log \chi$), and one has to compute the coefficients of the Laurent series up to $j = 6$ in order to check that, indeed, the equation passes the Painlevé test [429].

Let us assume for the would-be singular part transformation the relation (in the "second potential" equation to shorten the computations),

$$w = 2\beta^2 \log \tau + W,\ \text{ppBq}(w) = 0, \tag{5.112}$$

and let us choose for the link between τ and ψ the identity (5.63).

Let us first assume that ψ satisfies the second-order scalar Lax pair (4.34)–(4.35). This is equivalent to the usual WTC truncation in the invariant formalism [70], as already performed in Sect. 5.6.1.1,

$$\text{ppBq}(w) \equiv \sum_{j=0}^{4} E_j \chi^{j-4} = 0, \tag{5.113}$$

and this generates the three determining equations

$$E_2 \equiv (4/3)\beta^2\varepsilon^2 S - 2C^2 - 4\varepsilon^2 W_{xx} = 0, \tag{5.114}$$
$$E_3 \equiv -2(C_t - CC_x - (\beta^2\varepsilon^2/3)S_x) = 0, \tag{5.115}$$
$$E_4 \equiv (SE_2 - E_{3,x})/2 + C_x^2 + \beta^2 \mathrm{ppBq}(W) = 0. \tag{5.116}$$

From the last equation $E_4 = 0$, the desired solution $\mathrm{ppBq}(W) = 0$ cannot be generic since this would imply $C_x = 0$, so this second-order assumption fails to provide the auto-BT. However, it does provide another information, namely a BT between the Boussinesq PDE and another PDE. Indeed, under the natural parametric representation of the conservation law $E_3 = 0$ (which, by the way, would be the singular manifold equation if the second order were the correct order),

$$S = 3z_t - 3(\beta\varepsilon)^2 z_x^2/2,\ C = (\beta\varepsilon)^2 z_x, \tag{5.117}$$

the field z, by the cross-derivative condition (4.32), satisfies the PDE [217]

$$z_{tt} + ((\beta\varepsilon)^2/3)z_{xxxx} + 2(\beta\varepsilon)^2 z_t z_{xx} - 2(\beta\varepsilon)^4 z_x^2 z_{xx} = 0. \tag{5.118}$$

Going to third order, the assumption (5.112) and (5.63), with ψ solution of the scalar Lax pair (5.48)–(5.49), generates the expression (5.65), i.e.

$$\mathrm{ppBq}(w) \equiv \sum_{k=0}^{2}\sum_{l=0}^{2} E_{k,l} Y_1^k Y_2^l,\ k + l \le 2. \tag{5.119}$$

These six determining equations $E_{k,l} = 0$, plus the three cross-derivative conditions $X_j = 0, j = 0,1,2$, are solved as follows in the Gel'fand–Dikii case $f = 0$: [316, 319]

$$\begin{array}{lll}
E_{02} &\equiv (\beta\varepsilon)^2 - c^2 = 0 & \Rightarrow c = \beta\varepsilon,\\
E_{11} &\equiv d = 0 & \Rightarrow d = 0,\\
E_{20} &\equiv 3V_x + 2\beta^2 a = 0 & \Rightarrow a = -3V_x/(2\beta^2),\\
E_{10} &\equiv \varepsilon V_{xx} - \beta e_x = 0 & \Rightarrow e_x = \beta^{-1}\varepsilon V_{xx},\\
X_1 &\equiv 3V_{xt} + 3\beta\varepsilon V_{xxx} + 4\beta^3\varepsilon b_x = 0 & \Rightarrow b = g(t) - 3(\beta^{-2}V_{xx} + \beta^{-3}\varepsilon^{-1}V_t)/4,\\
X_0 &\equiv (3/(4\varepsilon\beta^2))\mathrm{pBq}(V) = 0 & \Rightarrow V \text{ satisfies the PDE (5.109)},\\
E_{00} &\equiv 2\beta^2 g'(t) = 0 & \Rightarrow g(t) = \lambda,
\end{array} \tag{5.120}$$

in which λ is an arbitrary constant. Returning to the conservative field $U = V_x$, the coefficients a,b,c,d,e are

$$\begin{aligned}
&a = -(3/2)\beta^{-2}U,\ b = \lambda - (3/4)\beta^{-2}U_x - (3/4)\beta^{-3}\varepsilon^{-1}R,\\
&c = \beta\varepsilon,\ d = 0,\ e = \beta^{-1}\varepsilon U,\\
&X_0 = R_t + \varepsilon^2(U^2 + (\beta^2/3)U_{xx})_x,\ X_1 = U_t - R_x,\ X_2 = 0,
\end{aligned} \tag{5.121}$$

and they define a third-order Lax pair of the Boussinesq equation (5.1) [443, 445, 307], identical to (5.7).

The BT results from the substitution of $Y_1 = (v - V)/(2\beta^2)$ into (5.69)–(5.70),

$$(v-V)_{xx}+3\beta^{-1}\varepsilon^{-1}(v+V)_t+3\beta^{-2}(v-V)(v+V)_x$$
$$+\beta^{-4}(v-V)^3-8\beta^2\lambda=0, \qquad (5.122)$$
$$(v+V)_{xx}-\beta^{-1}\varepsilon^{-1}(v-V)_t+\beta^{-2}(v-V)(v-V)_x=0. \qquad (5.123)$$

The *modified Boussinesq equation* [217] which is by definition the PDE obeyed by the difference $v-V$, is obtained by elimination of $v+V$ and is found to be (5.118), with the correspondence $Y_1=\beta\varepsilon z_x$. Like for the KdV equation (Sect. 5.6.1.1), after a short computation left to the reader, this leads to the BT between the Boussinesq and the modified Boussinesq equations, whose x-part is

$$U=-\beta^3\varepsilon z_{xx}-\beta^4\varepsilon z_x^2+\beta^2 z_t. \qquad (5.124)$$

Let us now derive the nonlinear superposition formula (NLSF). The two equations (5.122)–(5.123) have the same differential order, but the second one has the advantage of being a conservation law, allowing us to lower the differential order with a quadrature after the elimination of the t-derivatives. In the notation of Sect. 5.6.1.1, there is only one combination of the four copies of (5.123) able to achieve that, $(1,-1,1,-1)$, the result being

$$v_{n+1}=-v_{n-1}+(v_n+\tilde{v}_n)+2\beta^2\partial_x\log(v_n-\tilde{v}_n). \qquad (5.125)$$

This ends the processing of the Boussinesq equation.

Various remarks are in order.

1. The NLSF (5.108) of KdV and (5.125) of Boussinesq are identical because they are in fact the NLSF [61] of their common parent KP, and because this NLSF of KP is independent of ∂_{x_2} and ∂_{x_3}. For the Painlevé analysis of KP, see [430].
2. According to Definition 4.7, the one-soliton solution of either KdV or Boussinesq or KP is given by the formula (5.30) for $N=1$, in which θ_1 is a particular solution of the linear system with constant coefficients defined by the Lax pair $L_1(U,\lambda)\theta_1=0, L_2(U,\lambda)\theta_1=0$ taken for $U=0$. This particular solution is chosen as the sum of two exponentials (even in the Boussinesq case where L_1 is a third order operator). The result for KdV is the particular solitary wave (3.45) in which the background is $B=0$,

$$u=2ak^2\,\mathrm{sech}^2(kx-\omega t),\ \omega=-\frac{2}{3}ak^3. \qquad (5.126)$$

The one-soliton of Boussinesq only differs from that of KdV by the dispersion relation,

$$u=-2\beta^2k^2\,\mathrm{sech}^2(kx-\omega t),\ \omega^2=-\frac{\varepsilon^2\beta^2}{3}k^4. \qquad (5.127)$$

3. For both KdV and Boussinesq, the N-soliton solution is expressed as the logarithmic derivative of the Wronskian (5.31), but the entries $\partial_x^{i-1}\theta_j$ are different

because their scattering operator L_1 (Eqs. (5.6) and (5.7) taken for $U=0$) has a different order (two for KdV, three for Boussinesq).

5.6.2 Two-Family Cases: Sine-Gordon and Modified KdV

The *sine-Gordon equation* (SG)

$$\mathrm{SG}(u) \equiv u_{xt} + \alpha e^{u} + a_1 e^{-u} = 0,\ \alpha a_1 \neq 0, \tag{5.128}$$

(we make no difference between sine- and sinh-Gordon since a change $u \to iu$ does not alter the structure of singularities in the complex plane) and the *modified Korteweg-de Vries equation* (mKdV) (5.103)

$$\mathrm{mKdV}(w) \equiv b w_t + \left(w_{xx} - 2w^3/\alpha^2 + 6\nu w\right)_x = 0, \tag{5.129}$$

share many features. Let us first notice that, in the field e^u, SG is algebraic. The main feature of interest to us is the structure of singularities. More precisely let us show that, for both equations, the field u (resp. w) possesses two families with opposite singular part operators.

The field e^u of SG possesses the two nonopposite families

$$\begin{cases} e^{u} \sim -(2/\alpha)\varphi_x\varphi_t(\varphi-\varphi_0)^{-2}, \text{ indices } (-1,2),\ \mathscr{D}_1 = (2/\alpha)\partial_x\partial_t, \\ e^{-u} \sim (2/a_1)\varphi_x\varphi_t(\varphi-\varphi_0)^{-2}, \text{ indices } (-1,2),\ \mathscr{D}_2 = -(2/a_1)\partial_x\partial_t, \end{cases} \tag{5.130}$$

therefore the same linear combination of e^u, e^{-u} as the one appearing in the definition of SG defines an assumption (5.34)

$$e^{u} + (a_1/\alpha)e^{-u} = (2/\alpha)\partial_x\partial_t(\log\tau_1 - \log\tau_2) + \tilde{W},\ E(u) = 0, \tag{5.131}$$

in which the two singular part operators $\pm(2/\alpha)\partial_x\partial_t$ are opposite.

In the case of mKdV, its definition as a conservation law makes its processing for the potential field r ($w = r_x$) cheaper, there exist only two families and they have opposite singular part operators (α denotes any square root of α^2),

$$\begin{cases} \mathrm{mKdV}:\ p=-1,\ q=-4,\ w_0=\alpha,\ \text{indices } (-1,3,4),\ \mathscr{D} = \alpha\partial_x, \\ \mathrm{pmKdV}:\ p=0^-,\ q=-3,\ r_0=\alpha,\ \text{indices}(-1,0,4),\ \mathscr{D} = \alpha. \end{cases} \tag{5.132}$$

To conclude this local analysis, both PDEs have the same assumption (5.34) (after taking two quadratures of (5.131)),

$$\mathrm{SG}:\ u = -2(\log\tau_1 - \log\tau_2) + W,\ \mathrm{SG}(u) = 0, \tag{5.133}$$

$$\mathrm{pmKdV}:\ r = \alpha(\log\tau_1 - \log\tau_2) + R,\ \mathrm{pmKdV}(r) = 0, \tag{5.134}$$

in which the sum of the two opposite singular parts

$$\mathscr{D} \log \tau_1 - \mathscr{D} \log \tau_2 \tag{5.135}$$

only depends on the variable

$$Y = \frac{\tau_1}{\tau_2}. \tag{5.136}$$

As compared to the one-family situation of Sects. 5.6.1.1 and 5.6.1.2, this new situation of two opposite families leads to an implementation of the SMM [317, 357] which used to be called the two-singular manifold method, and which is as follows.

Let us restrict ourselves here to identity links (5.63) between the two τ and the two ψ functions, so $Y = \psi_1/\psi_2$. If we restrict ourselves further to a second-order Lax pair, ψ_1 and ψ_2 are the two wave vector components of the most general second order Lax pair in matrix form and, since only the ratio ψ_1/ψ_2 contributes to (5.133) and (5.134), it is convenient to represent the scalar Lax pair (5.39)–(5.40) by its equivalent Riccati system for Y, defined by (4.37) and (4.27)–(4.28).

So the truncation to be solved is defined as introduced in (5.82) [357] (for second order Lax pairs only)

$$u = \mathscr{D} \log Y + U, \tag{5.137}$$

$$Y^{-1} = B(\chi^{-1} + A), \tag{5.138}$$

$$E(u) = \sum_{j=0}^{-2q} E_j(S,C,A,B,U) Y^{j+q}, \tag{5.139}$$

$$\forall j \; E_j(S,C,A,B,U) = 0, \tag{5.140}$$

in which nothing is imposed on U. Let us solve this truncation for SG and mKdV.

5.6.2.1 The Sine-Gordon Equation

Replacing (5.137) by (5.133), the system (5.140) is made of five determining equations in the unknowns (S,C,A,B,W) [357, 95]

$$E_0 \equiv \alpha B^2 e^W - 2C = 0, \tag{5.141}$$

$$E_1 \equiv 2(C_x + 2AC) = 0, \tag{5.142}$$

$$E_2 \equiv 0, \text{ (Fuchs index)} \tag{5.143}$$

$$E_3 \equiv -\sigma_t - \sigma(C_x + 2AC) = 0, \tag{5.144}$$

$$E_4 \equiv \sigma\,(C\sigma + (C_x + 2AC)_x)\,/2 + a_1 B^{-2} e^{-W} = 0, \tag{5.145}$$

with the abbreviation

$$\sigma = S + 2A^2 - 2A_x, \tag{5.146}$$

and, together with the cross-derivative condition (4.32), they are solved as follows,

$$E_0: \; e^W = \frac{2C}{\alpha B^2}, \tag{5.147}$$

$$E_1: \; A = -\frac{1}{2}(\log C)_x, \tag{5.148}$$

$$E_3: \; \sigma = -F(x), \; S = -F(x) + \frac{C_x^2}{2C^2} - \frac{C_{xx}}{C}, \tag{5.149}$$

$$E_4: \; CC_{xt} - C_x C_t + F(x)C^3 + a_1 \alpha F(x)^{-1} C = 0, \tag{5.150}$$

$$X: \; a_1 F'(x) = 0. \tag{5.151}$$

in which F is a function of integration. The last line introduces a spectral parameter by

$$F(x) = 2\lambda^2. \tag{5.152}$$

Then E_4 expresses that $\log C$ is linearly related to a second solution U of the PDE

$$C = \frac{\alpha}{2}\lambda^{-2} e^U, \; E(U) = 0, \tag{5.153}$$

and one has obtained the singular part transformation

$$u = -2\log y + U, \; y = \lambda BY, \tag{5.154}$$

in which y satisfies the Riccati system

$$y_x = \lambda + U_x y - \lambda y^2, \tag{5.155}$$

$$y_t = -\frac{\alpha}{2}\lambda^{-1}(e^U + (a_1/\alpha)e^{-U} y^2), \tag{5.156}$$

$$(\log y)_{xt} - (\log y)_{tx} = \mathrm{SG}(U). \tag{5.157}$$

The linearization

$$y = \psi_1/\psi_2 \tag{5.158}$$

yields the second-order matrix Lax pair

$$(\partial_x - L)\begin{pmatrix}\psi_1\\ \psi_2\end{pmatrix} = 0, \; L = \begin{pmatrix} U_x/2 & \lambda \\ \lambda & -U_x/2\end{pmatrix}, \tag{5.159}$$

$$(\partial_t - M)\begin{pmatrix}\psi_1\\ \psi_2\end{pmatrix} = 0, \; M = -(\alpha/2)\lambda^{-1}\begin{pmatrix} 0 & e^U \\ -(a_1/\alpha)e^{-U} & 0\end{pmatrix}. \tag{5.160}$$

The auto-BT [111] results from the substitution $y = e^{-(u-U)/2}$ into (5.155)–(5.156)

$$(u+U)_x = -4\lambda \sinh\frac{u-U}{2}, \tag{5.161}$$

$$(u-U)_t = \lambda^{-1}\left(\alpha e^{(u+U)/2} + a_1 e^{-(u+U)/2}\right). \tag{5.162}$$

The SME results from the elimination of U, this is [357]

$$S + C^{-1}C_{xx} - \frac{1}{2}C^{-2}C_x^2 + 2\lambda^2 = 0, \tag{5.163}$$

and it coincides, but this is not generic, with the one [428, 70] obtained from the one-family truncation in χ.

Remark. The reason for the presence of the apparently useless parameter B in Definition (4.37) is to allow the precise correspondence (5.63)

$$\tau_1 = \psi_1,\ \tau_2 = \psi_2 \tag{5.164}$$

for some choice of B, namely

$$B = \lambda^{-1},\ y = Y,\ W = U. \tag{5.165}$$

In order to derive the NLSF, one can choose either equation among (5.161)–(5.162) because of the Lorentz invariance $x \leftrightarrow t$. There exists a unique linear combination of the four copies of (5.161) which eliminates all first order derivatives, with a result [268]

$$\tanh\frac{u_{n+1} - u_{n-1}}{4} = \frac{\lambda_{n+1} + \lambda_n}{\lambda_{n+1} - \lambda_n} \tanh\frac{u_n - \tilde{u}_n}{4}, \tag{5.166}$$

a formula which can be solved for u_{n+1}, like in the KdV case (5.107).

In the pure sine-Gordon case $z_{xt} - \sin z = 0$, obtained with the choice $\alpha = -1/2, a_1 = 1/2, u = \varepsilon iz, \varepsilon^2 = 1$, let us write the one-soliton solution and the breather solution. Starting from the vacuum $U = 0$, one integrates the Riccati system (5.155)–(5.156) with constant coefficients as

$$y = \tanh\theta,\ \theta = \lambda x + \frac{t}{4\lambda} + \theta_0, \tag{5.167}$$

and the one-soliton solution results from the singular part transformation (5.154),

$$e^{-u/2} = \tanh\theta,\ \tanh(u/4) = e^{-2\theta},\ u_x = 4i\lambda\,\mathrm{sech}(2\theta - i\pi/2), \tag{5.168}$$

i.e. in the physical variable $z = -\varepsilon iu$,

$$\tan(z/4) = -\varepsilon e^{-2\theta + i\pi/2},\ z_x = 4\varepsilon\lambda\,\mathrm{sech}(2\theta - i\pi/2). \tag{5.169}$$

The two-soliton solution u_{2SS} is obtained without integration by setting $u_{n-1} = 0$ in the NLSF (5.166) and taking for u_n and $\tilde{u}_n$ two copies of this one-soliton,

$$\tanh\frac{u_{2SS}}{4} = \frac{\lambda_2 + \lambda_1}{\lambda_2 - \lambda_1}\frac{\sinh(\theta_1 - \theta_2)}{\sinh(\theta_1 + \theta_2)}. \tag{5.170}$$

In the physical variable $z = -\varepsilon iu$, the breather solution z_b [271] is defined as the particular two-soliton solution in which λ_1 and λ_2 are complex conjugate and the constant shifts θ_0 are chosen so that the denominator has no zeros on the real axis,

$$\tan\frac{z_{\rm b}}{4}=\varepsilon\frac{k}{q}\frac{\sin q(x-t/(k^2+q^2)+\delta_n)}{\cosh k(x+t/(k^2+q^2)+\delta_d)},\ \lambda_1=k+iq,\ \lambda_2=k-iq. \tag{5.171}$$

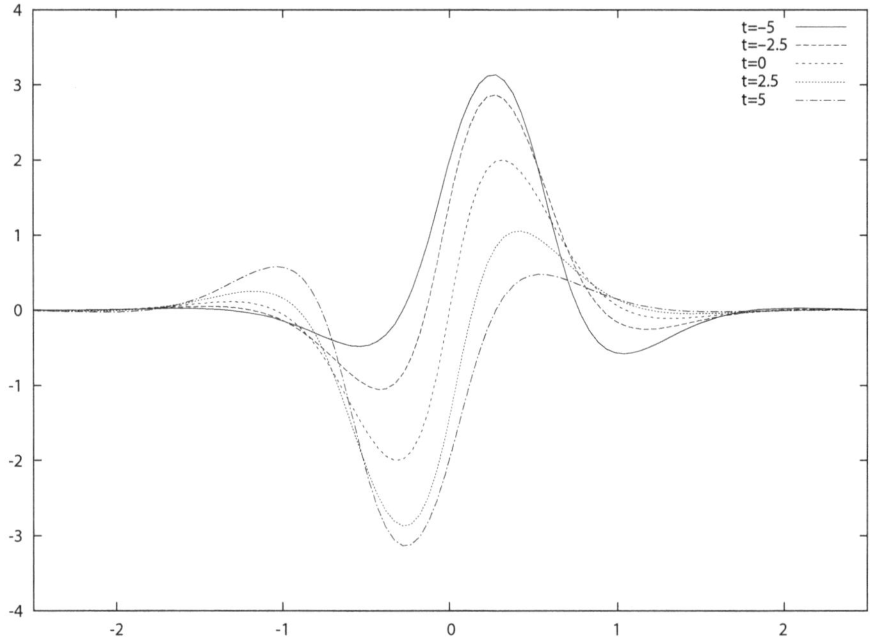

Fig. 5.3 The breather solution (5.171) of sine-Gordon as a function of x, with $k=3, q=3, \delta_n=0, \delta_d=0$, for the five equispaced values of $t=-5,-2.5,0,2.5,5$.

5.6.2.2 The Modified Korteweg–de Vries Equation

Replacing (5.137) by (5.134), the system (5.140) is also made of five determining equations in the unknowns (S,C,A,B,R) [357], with the notation (5.146) for σ

$$E_1\equiv 6\alpha A-6(R-\alpha\log B)_x=0, \tag{5.172}$$
$$E_2\equiv \alpha(2A_x+4A^2-bC-2\sigma+6\nu)-\alpha^{-1}(E_1+6\alpha A)^2/6=0, \tag{5.173}$$
$$E_3\equiv \text{p-mKdV}(R-\alpha\log B)-(3/2)\alpha^{-1}\sigma_x-2AE_2-E_{2,x}$$
$$+(\sigma-4A^2-(1/3)\alpha^{-1}E_{1,x}-2A_x)E_1-2AE_{1,x}, \tag{5.174}$$
$$E_4\equiv \text{expression vanishing with } E_1,E_2,E_3,E_5, \tag{5.175}$$
$$E_5\equiv (3/4)\alpha\sigma\sigma_x+(1/4)\sigma^2E_1=0, \tag{5.176}$$

$$X \equiv S_t + C_{xxx} + 2C_x S + C S_x = 0. \tag{5.177}$$

They depend on (R,B) only through the combination $R - \alpha \log B$. Equation $j = 4$ is a differential consequence of equations $j = 1,2,3,5$, because 4 is a Fuchs index, and the other equations have been written so as to display how they are solved:

$$E_1 : A = \alpha^{-1}(R - \alpha \log B)_x, \tag{5.178}$$
$$E_5 : \sigma = -2(\lambda^2(t) - \nu),\ \lambda \text{ arbitrary function}, \tag{5.179}$$
$$E_2 : bC = 2A_x - 2A^2 + 4\lambda^2(t) + 2\nu, \tag{5.180}$$
$$E_3 : \text{p-mKdV}(R - \alpha \log B) = 0, \tag{5.181}$$
$$X : \lambda'(t) = 0. \tag{5.182}$$

Thus, their general solution can be expressed in terms of a second solution W of the mKdV equation (5.103) and an arbitrary complex constant λ [357]

$$\begin{aligned} W &= (R - \alpha \log B)_x,\ A = W/\alpha, \\ bC &= 2W_x/\alpha - 2W^2/\alpha^2 + 2\nu + 4\lambda^2, \\ S &= 2W_x/\alpha - 2W^2/\alpha^2 + 2\nu - 2\lambda^2, \end{aligned} \tag{5.183}$$

and the cross-derivative condition $X_1 = 0$ ((5.43)) is equivalent to the mKdV equation (5.103) for W, which proves that one has obtained a singular part transformation and a Lax pair.

The singular part transformation is

$$w = \alpha \partial_x \log y + W,\ y = \lambda B Y, \tag{5.184}$$

and the Riccati representation of the Lax pair is

$$\frac{y_x}{y} = \lambda\Big(\frac{1}{y} - y\Big) - 2\frac{W}{\alpha} + \frac{\nu}{\lambda} y, \tag{5.185}$$
$$b\frac{y_t}{y} = \frac{1}{y}\left(-4\lambda \frac{W}{\alpha} + \left(2\frac{W_x}{\alpha} + 2\frac{W^2}{\alpha^2} - 2\nu - 4\lambda^2\right) y\right)_x. \tag{5.186}$$

In the same manner as in the KdV truncation, these two Riccati equations can also be interpreted as the BT between the mKdV equation and the PDE satisfied by the pseudopotential y, called the Chen–Calogero–Degasperis–Fokas PDE [60].

The auto-BT of mKdV is obtained by the substitution (we choose $B = \lambda^{-1}$)

$$y = \exp\left\{\alpha^{-1} \int (w - W)\mathrm{d}x\right\} = \exp\frac{r - R}{\alpha} \tag{5.187}$$

in the two equations for the gradient of y, resulting in [420, 269, 60]

$$(r + R)_x + 2\alpha\lambda \sinh\frac{r - R}{\alpha} = 0, \tag{5.188}$$

$$(r+R)_t - 2\lambda (r-R)_{xx} \cosh \frac{r-R}{\alpha} + 2\lambda \frac{R_x^2 + r_x^2}{\alpha} \sinh \frac{r-R}{\alpha} = 0. \quad (5.189)$$

The SME, obtained by the elimination of W between S and C,

$$bC - S - 6\lambda^2 = 0, \quad (5.190)$$

is identical to that of the KdV equation (5.92), and it does not coincide with the one [431] obtained from the one-family truncation in χ.

Since the x-parts (5.161) and (5.188) of the BT of SG and mKdV are identical, the nonlinear superposition formula of mKdV [420, 269, 60] is identical to that of SG,

$$\tanh \frac{r_{n+1} - r_{n-1}}{2\alpha} = \frac{\lambda_{n+1} + \lambda_n}{\lambda_{n+1} - \lambda_n} \tanh \frac{r_n - \tilde{r}_n}{2\alpha}. \quad (5.191)$$

Remark. The invariance of mKdV (5.103) under the involution $w \mapsto -w$ provides an elegant way [290] to derive the BT of the KdV equation and its hierarchy.

5.6.3 Third Order: Sawada–Kotera and Kaup–Kupershmidt

The fifth order nonlinear partial differential equation (2.82),

$$u_t + \left(u_{xxxx} + (8\alpha - 2\beta) u u_{xx} - 2(\alpha + \beta) u_x^2 - \frac{20}{3} \alpha \beta u^3 \right)_x = 0, \quad (5.192)$$

has already been encountered in Sect. 2.1.5 for its link with the cubic Hénon–Heiles Hamiltonian system (2.70)–(2.71). The conditions (2.99), (2.100) or (2.101) found by the Painlevé test when analyzing the cubic HH system are precisely the same values as those selected by another criterium, that the PDE (5.192) possesses non-trivial infinitesimal symmetries [160]. The corresponding PDEs, which possess N-soliton solutions, are called respectively Sawada–Kotera (SK) [387], higher-order KdV (KdV5) [274], and Kaup–Kupershmidt (KK) [246, 148] equations.

We discard here the KdV5 PDE, which is the second equation in the KdV hierarchy, because its scattering problem is identical to that of KdV, already derived by singularity analysis in Sect. 5.6.1.1.

The SK and KK equations, whose explicit writing is (5.200) and (5.202) below, are linked by a Bäcklund transformation [148] and their scattering problem has the same order three [246]. Both possess an auto-Bäcklund transformation, and the challenge is to derive, by singularity analysis only, the singular part transformation, Lax pair and auto-Bäcklund transformation for each equation, the BT between SK and KK, and the two nonlinear superposition formulae.

There is no difficulty in obtaining all these elements for SK with the procedure used in Sect. 5.6.1.2 for Boussinesq, see [316]. However, in the case of KK, a real difficulty exists, and the deep reason [319] lies in the conjunction of two features:

the scattering order is higher than two, and the link between τ and ψ is different from $\tau = \psi$.

The first feature (scattering order higher than two) implies the irrelevance of the *singular manifold equation* to obtain the Lax pair, see Sect. 5.4.1 for more details. Several authors [427, 373] have nevertheless used the SME in order to try and derive the Lax pair and the DT, in particular by uncovering a nice discrete invariance in the SME [427], but the best they could achieve was the desired result with the constraint $\lambda = 0$, thus preventing the iteration of the Darboux transformation.

The second feature (nature of the link between τ and ψ) could lead to infinitely-many possible assumptions. This is not the case, and one can prove [319] that only a finite number of such links are allowed, for a reason going back to the classification of Gambier [163]. Let us detail this quite important point [319].

5.6.3.1 Help from the Gambier Classification

Let us consider the ODE obtained by eliminating ∂_t between the two equations defining the (as yet unknown) BT. This nonlinear ODE for $Y = u - U$, with coefficients depending on U and, in the $1+1$-dimensional case, on an arbitrary constant λ, has two properties. Firstly, it is linearizable since it results from the Lax pair, a linear system, and the DT by an elimination process [60]. Secondly, it has the Painlevé property since it is linearizable. Therefore, if its order is small (at most three), it belongs to the appropriate finite list (classification) established by the Painlevé school between 1900 and 1910. This very strong property restricts the admissible choices (5.62) to a finite number of possibilities. Moreover, these very special nonlinear ODEs provide a link to both the Lax pair, *via* their linearizing transformation, and the singular part transformation, *via* an involution which leaves them invariant.

The only nonlinear ODE of first order and first degree with the PP is the Riccati equation, linearizable into a second order linear equation and defining a unique choice (5.62) for describing scattering problems which have order two.

Next, the nonlinear ODEs of order two and degree one with the Painlevé property which are also linearizable have been listed by Gambier himself, p. 21 of his thesis [163], these are the equations numbered 5, 6, 14, 24, 25, 27, in his classification of fifty equations inequivalent under the homographic group of transformations, with the respective orders for the associated linear equations, i.e. , in our context for the unknown scattering problem: 3, 2, 2, 2, 3, 2 (and 4 or 3 for the case $n = 2$ in equation 27).

Therefore, the only two generic choices for describing scattering problems of third order are the two classes of equivalence numbered 5 and 25 by Gambier. The representative equation of interest to us in each class of equivalence is what Painlevé calls the *complete equation* (in the present section, the $'$ symbol means ∂_x)

$$Y'' + 3YY' + Y^3 + rY + q = 0, \qquad \text{G5}$$
$$Y'' - 3Y'^2/(4Y) + 3YY'/2 + Y^3/4 - q'(Y' + Y^2)/(2q) - rY - q = 0, \ \text{G25}$$

in which q and r are two arbitrary functions. These two classes are equivalent under the birational group, as shown by the explicit transformation [163] between $\mathrm{G5}(y;q,r)$ and $\mathrm{G25}(Y;Q,R)$

$$Y=\frac{Q}{2z'+z^2-(Q'/Q)z-R},\ z=y+\frac{Q'}{2Q},\ 2y=\frac{Y'}{Y}+Y-\frac{Q'}{Q},\tag{5.193}$$

$$r=-R+\frac{Q''}{Q}-\frac{5Q'^2}{4Q^2},\ q=-\frac{Q}{2}-\frac{R'}{2}+\frac{Q'''}{2Q}-\frac{7Q'Q''}{4Q^2}+\frac{5Q'^3}{4Q^3}.\tag{5.194}$$

The linearizing transformations are

$$\mathrm{G5}\ \ Y=\frac{\tau'}{\tau}=\frac{\psi'}{\psi},\ \psi'''+r\psi'+q\psi=0,\tag{5.195}$$

$$\mathrm{G25}\ \begin{cases} Y=\dfrac{\tau'}{\tau}=\dfrac{q}{2z'+z^2-(q'/q)z-r},\ z=\dfrac{\psi'}{\psi}, \\ \psi'''-\dfrac{3q'}{2q}\psi''-\left(r+\dfrac{q''}{q}-\dfrac{q'^2}{q^2}\right)\psi'-\left(\dfrac{r'}{2}+\dfrac{q}{2}-\dfrac{q'r}{2q}\right)\psi=0.\end{cases}\tag{5.196}$$

The corresponding involutions

$$\mathrm{G5}\ \ (Y,q,r)\to(-Y,-q,r+6Y'),\tag{5.197}$$

$$\mathrm{G25}\ (Y,q,r)\to(-Y,-q,r-3Y'-(q'/q)Y),\tag{5.198}$$

will be seen to define the singular part transformation.

In the particular case of G25 when q is constant, the link (5.196) can be integrated explicitly

$$\tau=\psi\psi''-(1/2)(\psi')^2-(r/2)\psi^2,\ \tau'=(q/2)\psi^2.\tag{5.199}$$

To go back to the singular manifold method, in its third step, the assumption for a link (5.62) becomes, at the third scattering order, a choice between the two *and only two* above linearization formulas (5.195) and (5.196), and the associated linear third order ODEs for ψ replace the assumption (5.48). Consequently, in the determining equations of the fourth step, the unknowns (a,b) are replaced by the unknowns (q,r) of the Gambier equations G5 and G25.

These two and only two possibilities were rediscovered in 1980 in the context of scattering theory by Caudrey [56] and Kaup [246].

Let us apply this to the Sawada–Kotera and Kaup–Kupershmidt equations.

5.6.3.2 Singularity Structure of SK and KK Equations

It is convenient to handle the potential form of each equation. Denoting $u=v_x$, the conservative and potential equations are defined as

$$\mathrm{SK}(u) \equiv \beta u_t + \left(u_{xxxx} + \frac{30}{\alpha} u u_{xx} + \frac{60}{\alpha^2} u^3\right)_x = 0, \tag{5.200}$$

$$\mathrm{pSK}(v) \equiv \beta v_t + v_{xxxxx} + \frac{30}{\alpha} v_x v_{xxx} + \frac{60}{\alpha^2} v_x^3 = 0, \tag{5.201}$$

$$\mathrm{KK}(u) \equiv \beta u_t + \left(u_{xxxx} + \frac{30}{\alpha} u u_{xx} + \frac{45}{2\alpha} u_x^2 + \frac{60}{\alpha^2} u^3\right)_x = 0, \tag{5.202}$$

$$\mathrm{pKK}(v) \equiv \beta v_t + v_{xxxxx} + \frac{30}{\alpha} v_x v_{xxx} + \frac{45}{2\alpha} v_{xx}^2 + \frac{60}{\alpha^2} v_x^3 = 0. \tag{5.203}$$

Each equation has two families of movable singularities, whose leading order (in potential form) $v \sim v_0 \chi^p$ and Fuchs indices are the following [427]

$$\mathrm{pSK,F1} : p = -1,\ v_0 = \alpha,\ \text{indices } -1, 1, 2, 3, 10, \tag{5.204}$$

$$\mathrm{pSK,F2} : p = -1,\ v_0 = 2\alpha,\ \text{indices } -2, -1, 1, 5, 12, \tag{5.205}$$

$$\mathrm{pKK,F1} : p = -1,\ v_0 = \alpha/2,\ \text{indices } -1, 1, 3, 5, 7, \tag{5.206}$$

$$\mathrm{pKK,F2} : p = -1,\ v_0 = 4\alpha,\ \text{indices } -7, -1, 1, 10, 12, \tag{5.207}$$

and both the SK and KK equations pass the Painlevé test [427]. The singular part operator $\mathscr{D}$ attached to a given family is $\mathscr{D} = v_0 \partial_x$.

The existence of two families is kind of misleading, because all three integrable PDEs in the class (5.192) belong to a hierarchy, the base member of which has only one family, so one of the two families of SK and KK is irrelevant for the SMM. For KdV5, this base member is the third order KdV equation. This "minus one" member of the SK hierarchy was initially written by Hirota and Satsuma [216],

$$\mathrm{pSK}_{-1} : v_{xxt} + \frac{6}{\alpha} v_x v_t = 0, \tag{5.208}$$

and successfully processed [315, 316] by the SMM. For the KK hierarchy, the "minus one" member is defined as [436]

$$\mathrm{pKK}_{-1} : v_t v_{xxt} - \frac{3}{4} v_{xt}^2 + \frac{6}{\alpha} v_x v_t^2 = 0. \tag{5.209}$$

This equation has only one family and it can be equivalently written as [226]

$$\mathrm{pKK}_{-1} : w_{xxx} + 4 z w_x + 2 z_x w = 0,\ z_t = \frac{3}{4} \left(w^2\right)_x,\ z = v_x. \tag{5.210}$$

Therefore, in order to apply the SMM to SK and KK, the relevant family to consider is the one corresponding to the unique family of (5.208) or (5.209) under the mapping from the base member to the next member in the hierarchy. This single relevant family is F1, (5.204) and (5.206).

Practically, the SMM is applied as follows. Let us assume the one-family singular part transformation (5.33) and, successively, the second-order scalar Lax pair (5.39)–(5.40), then, for the third-order scalar one, the couple made of the t-part (5.49) and either the x-part (5.195) or the x-part (5.196).

As to the link between τ and ψ, at second order this is the identity, while at third order it can be either the linearizing transformation of the fifth Gambier equation or that of the twenty-fifth Gambier equation.

Therefore, at the fourth step of the singular manifold method, for each PDE, one has only three possibilities to examine: order two and Riccati, order three and G5, order three and G25. This is done briefly in the next two sections, the full details being available in [427, 319] and in summer school lecture notes [78, 314].

5.6.3.3 Truncation with a Second Order Lax Pair

The one-family truncation for SK and KK with the assumption

$$v = v_0 \partial_x \log \tau + V,\ \tau = \psi \tag{5.211}$$

where V is unconstrained and ψ is a solution of the second order Lax pair (5.39)–(5.40), fails to introduce an arbitrary complex constant during the resolution of the truncation equations. However, it provides in both cases the BT between the SK and KK equations. The quickest way to obtain this result is to perform the truncation of the conservative form of KK, which generates

$$\begin{cases} u = \dfrac{\alpha}{2}\partial_x^2 \log \tau + U,\ \tau = \psi, \\ E_2 \equiv U - \dfrac{\alpha S}{12} = 0, \\ E_4 \equiv S_{xx} + S^2/4 - \beta C = 0. \end{cases} \tag{5.212}$$

With the values of S and C determined by the last two equations, the cross-derivative condition (4.32) expresses that the constant level coefficient U obeys the SK equation. Next, the solutions u of KK and U of SK can be written only in terms of χ (see (4.27)),

$$U = -\alpha((\chi^{-1})_x + \chi^{-2})/6,\ u = \alpha(2(\chi^{-1})_x - \chi^{-2})/6, \tag{5.213}$$

and the elimination of χ yields the BT between SK and KK ($w = v_{pSK}$, $W = v_{pKK}$),

$$\alpha(w + W/2)_x + (w - W)^2 = 0, \tag{5.214}$$

$$\left[(w - W)(72W_x^2/\alpha^2 + 6W_{xxx}/\alpha) - 72w_x W_{xx}/\alpha - 3W_{xxxx}\right]_x + 2(w - W)_t = 0, \tag{5.215}$$

a result previously obtained by Hirota [215]. As to the inverse χ^{-1} of the expansion variable, it obeys another fifth order PDE having the Painlevé property, the Fordy–Gibbons equation [148].

Because of the failure of this truncation to introduce an arbitrary constant in the scattering problem, one has to consider a third order Lax pair.

5.6.3.4 Truncation with a Third Order Lax Pair and G5

Assuming the link $\tau = \psi$ of the G5 linearizing transformation (5.195), and the third order Lax pair defined by the x-part (5.195) and the t-part (5.49), the process is successful for SK but not for KK. It provides the scalar Lax pair of SK [427, 73]:

$$L_1 = \partial_x^3 + 6\frac{U}{\alpha}\partial_x - \lambda, \tag{5.216}$$

$$L_2 = \beta\partial_t + (18\frac{U_x}{\alpha} - 9\lambda)\partial_x^2 + (36\frac{U^2}{\alpha^2} - 6\frac{U_{xx}}{\alpha})\partial_x - 36\lambda\frac{U}{\alpha}, \tag{5.217}$$

previously obtained in the bilinear formalism by Satsuma and Kaup [386].

5.6.3.5 Truncation with a Third Order Lax Pair and G25

We finally perform the truncation of KK assuming the third order for the scattering problem, the G25 linearizing transformation (5.196) between τ and ψ, and the third order Lax pair defined by the x-part (5.196) and the t-part (5.49). The first piece of information found during the resolution [319] is

$$q_x = 0, \tag{5.218}$$

which allows one to use the integrated version (5.199) of the link between τ and ψ. The final result is the scalar Lax pair

$$L_1 = \partial_x^3 + 6\frac{U}{\alpha}\partial_x + 3\frac{U_x}{\alpha} - \lambda, \tag{5.219}$$

$$L_2 = \beta\partial_t - 9\lambda\partial_x^2 + (3\frac{U_{xx}}{\alpha} + 36\frac{U^2}{\alpha^2})\partial_x - 3\frac{U_{xxx}}{\alpha} - 72\frac{UU_x}{\alpha^2} - 36\lambda\frac{U}{\alpha}, \tag{5.220}$$

first found by Kaup [246], and the singular part transformation

$$u = (\alpha/2)\partial_x^2 \log\tau + U,\ \tau = \psi\psi_{xx} - (1/2)\psi_x^2 + 3(U/\alpha)\psi^2,\ \tau_x = \lambda\psi^2, \tag{5.221}$$

a result initially obtained in [279].

5.6.3.6 Bäcklund Transformation

The elimination leading to the BT can be reduced to a mere substitution (fifth step of the SMM), provided the third order Lax pair is represented as an equivalent non-linear first order system in two components (the so-called *pseudopotential* [424]), one of them being τ_x/τ so as to write the singular part transformation as

$$\text{SK}: Y_1 = \frac{\tau_x}{\tau} = \frac{\psi_x}{\psi} = \frac{v - V}{\alpha}, \tag{5.222}$$

$$\text{KK}: Z = \frac{\tau_x}{\tau} = 2\frac{v-V}{\alpha}. \tag{5.223}$$

The second component of this pseudopotential is chosen conveniently as

$$\text{SK}: Y_2 = \frac{\psi_{xx}}{\psi}, \tag{5.224}$$

$$\text{KK}: Y_1 = \frac{\psi_x}{\psi}. \tag{5.225}$$

The system for (Y_1, Y_2) in the SK case is the canonical projective Riccati system (5.55)–(5.59), and the system for (Y_1, Z^{-1}) is another type of linearizable Riccati system [319].

The final result for SK is the auto-BT [316]

$$(v-V)_{xx}/\alpha + 3(v-V)(v+V)_x/\alpha^2 + (v-V)^3/\alpha^3 - \lambda = 0, \tag{5.226}$$

$$\begin{aligned}
&\beta(v-V)_t/\alpha - (3/2)[(v-V)_{xxxx}/\alpha \\
&+(5(v-V)(v+V)_{xxx} + 15(v+V)_x(v-V)_{xx})/\alpha^2 \\
&+(15(v-V)^2(v-V)_{xx} + 30(v-V)(v+V)_x^2)/\alpha^3 \\
&+30(v-V)^3(v+V)_x/\alpha^4 + 6(v-V)^5/\alpha^5]_x = 0,
\end{aligned} \tag{5.227}$$

a result due to [386, 122].

For KK, the resulting auto-BT is

$$\begin{aligned}
&(v-V)_{xx}/\alpha - (3/4)(v-V)_x^2/(\alpha(v-V)) \\
&\quad + 3(v-V)(v+V)_x/\alpha^2 + (v-V)^3/\alpha^3 - \lambda = 0,
\end{aligned} \tag{5.228}$$

$$\begin{aligned}
&\beta(v-V)_t/\alpha - (3/2)[2(v-V)_{xxxx}/\alpha + 60(v-V)^3(v+V)_x/\alpha^4 \\
&\quad + 12(v-V)^5/\alpha^5 + (10(v-V)(v+V)_{xxx} + 30(v+V)_x(v-V)_{xx} \\
&\quad + 15(v-V)_x(v+V)_{xx})/\alpha^2 + (30(v-V)^2(v-V)_{xx} \\
&\quad + 60(v-V)(v+V)_x^2 + 15(v-V)(v-V)_x^2)/\alpha^3]_x = 0.
\end{aligned} \tag{5.229}$$

The x-part (5.228) was first given for $\lambda = 0$ in [373], the reason for this restriction being the (implicit) consideration of a second order scattering problem, and the result with an arbitrary λ can only be found by assuming a third order scattering problem [319].

5.6.3.7 Nonlinear Superposition Formula

For both SK and KK, the half of the BT which allows one to lower the differential order is the x-part (5.226) and (5.228). In the notation introduced in Definition 5.6 and equations like (5.106), the linear combination of the four copies of (5.226) and (5.228) which lowers the differential order is $(1, -1, 1, -1)$ for both, yielding a first order ODE for v_{n+1}.

For SK, this first order ODE has degree one, this is the Riccati equation

$$(v_{n+1}-v_{n-1})_x+\frac{1}{\alpha}(v_{n+1}^2-v_{n-1}^2)$$
$$+(v_{n+1}-v_{n-1})\left(\frac{v_{n,x}-\tilde{v}_{n,x}}{\tilde{v}_n-v_n}-\frac{1}{\alpha}(v_n+\tilde{v}_n)\right)=0. \quad (5.230)$$

For KK, this first order ODE has degree two,

$$\begin{cases} \left(v_{n+1,x}+\dfrac{2}{\alpha}v_{n+1}^2-Av_{n+1}+\alpha B\right)^2=\dfrac{\alpha^2C^2(v_n-v_{n+1})(\tilde{v}_n-v_{n+1})}{(v_n-v_{n-1})(\tilde{v}_n-v_{n-1})}, \\ A=\dfrac{2}{\alpha}(v_n+\tilde{v}_n)+\dfrac{\tilde{v}_{n,x}-v_{n,x}}{\tilde{v}_n-v_n}, \\ B=\dfrac{2}{\alpha^2}v_n\tilde{v}_n+\dfrac{\tilde{v}_{n,x}v_n-v_{n,x}\tilde{v}_n}{\alpha(\tilde{v}_n-v_n)}, \\ C=B+\dfrac{1}{\alpha}\left(v_{n-1,x}+\dfrac{2}{\alpha}v_{n-1}^2-Av_{n-1}\right). \end{cases} \quad (5.231)$$

The differential order can be lowered once more if one introduces the tau-function f,

$$\text{SK}: v=\alpha\partial_x\log f, \quad (5.232)$$
$$\text{KK}: v=\frac{\alpha}{2}\partial_x\log f. \quad (5.233)$$

Indeed, for SK, the second order linear ODE for f_{n+1} can be integrated by quadratures, leading to the NLSF

$$f_{n+1}=K_1(t)f_{n-1}\int^x\frac{D_x(f_n\cdot\tilde{f}_n)}{f_{n-1}^2}\,\mathrm{d}x, \quad (5.234)$$

in which D_x is the Hirota operator [213] and $K_1(t)$ is an irrelevant nonzero function of integration. This formula was first obtained [229] with the bilinear formalism of Hirota, shortly presented in Appendix E.

For KK, the second order ODE for f_{n+1} has degree two and it belongs to a type also integrated by quadratures by Appell [16]. Indeed, f_{n+1} obeys a third order linear ODE, and the general solution of this Appell equation is [410, 321]

$$\begin{cases} \dfrac{f_{n+1}}{f_{n-1}}=K_1^2+K_1K_2R+K_2^2\left(\dfrac{f_n\tilde{f}_n}{f_{n-1}^2}-R^2\right), \\ (\partial_xR)^2=\left(\dfrac{f_n}{f_{n-1}}\right)_{,x}\left(\dfrac{\tilde{f}_n}{f_{n-1}}\right)_{,x}, \end{cases} \quad (5.235)$$

in which $K_1(t),K_2(t)$ are two functions of integration, whose values are $K_1=0,K_2=$ arbitrary. Therefore, the NLSF is [410, 321]

$$f_{n+1}=f_{n-1}\left[\frac{f_n\tilde{f}_n}{f_{n-1}^2}-R^2\right],\ (\partial_xR)^2=\left(\frac{f_n}{f_{n-1}}\right)_{,x}\left(\frac{\tilde{f}_n}{f_{n-1}}\right)_{,x}. \quad (5.236)$$

This NLSF provides an alternative to a perturbation scheme [201] for building the N-soliton solution, whose expression for KK is [285, 314],

$$\begin{cases} u = \frac{\alpha}{2}\partial_x^2 \log \tau^{(N)},\ \tau^{(N)} = \det\left[\int^x \psi_i \psi_j \mathrm{d}x\right]_{1\le i,j\le N} \\ \forall i:\ \psi_i = A_i e^{p_i x + 9p_i^5 t} + B_i e^{r_i x + 9r_i^5 t}, \\ p_i^3 = r_i^3 = \lambda_i,\ p_i^2 + p_i r_i + r_i^2 = 0,\ p_i \ne r_i, \end{cases} \tag{5.237}$$

where λ_i, A_i, B_i are arbitrary constants. This expression for the N-soliton solution coincides with the result [285] obtained by reduction of the C-KP hierarchy.

For instance, the one-soliton solutions are, for SK,

$$u = \alpha \partial_x^2 \log\left[1 + e^{kx - k^5 t + \delta}\right], \tag{5.238}$$

for KK,

$$u = \frac{\alpha}{2}\partial_x^2 \log\left[1 + 4e^{kx - k^5 t + \delta} + e^{2(kx - k^5 t + \delta)}\right]. \tag{5.239}$$

At the two-soliton level, the tau-function $\tau^{(2)}$ is the sum of four terms for SK and nine terms for KK [201, 321].

This ends our series of examples in the integrable case. When the order of the scattering operator L_1 is higher than three, the SMM applies as well [150]. Fortunately, it may happen that the operator L_2 has an order much lower than that of L_1. Such an example is handled in [412] . If the order of L_2 is three and if L_2 is factorizable, the so-called t-part of the BT will be simple and the x-part complicated, as opposed for instance to the SK and KK cases. The reason is that $1+1$-dimensional integrable PDEs are in fact reductions [241] of $2+1$-dimensional integrable PDEs, for which none of the operators L_1 and L_2 depends on a single independent variable.

5.7 Application to Partially Integrable Equations

The methods previously introduced in this chapter still apply but, since the PDE is assumed to fail the Painlevé test, instead of yielding a Lax pair, they only provide explicit particular solutions.

5.7.1 One-Family Case: Fisher Equation

This equation [140, 255]

$$5\beta u_t - u_{xx} + \frac{6}{\delta}(u - a_1)(u - a_2) = 0,\ s_1 = a_1 + a_2,\ s_2 = a_1 a_2, \tag{5.240}$$

initially built for modelizing the evolution of mutant genes [140], also governs the neutron population in a nuclear reactor [51], or the propagation of flames, see the review [376]. The two fixed points a_1, a_2 typically represent different populations.

It admits the single family of movable singularities

$$u \sim \delta\chi^{-2}, \text{ Fuchs indices} = (-1,6),\ \mathscr{D} = \delta(\beta\partial_t - \partial_x^2), \tag{5.241}$$

leading to the Laurent series

$$\begin{aligned}\frac{u}{\delta} &= \chi^{-2} - \beta C\chi^{-1} + \beta C_x + \frac{S}{3} - \frac{\beta^2}{12}C^2 + \frac{s_1}{2\delta} \\ &+ \left(\frac{\beta^2}{12}(5C_t + 7CC_x) - \frac{\beta}{6}(CS + 3C_{xx}) - \frac{\beta^3}{12}C^3 - \frac{1}{12}S_x\right)\chi \\ &+ \mathscr{O}(\chi^2),\end{aligned} \tag{5.242}$$

and it fails the Painlevé test at index 6, i.e. the coefficient u_6 is arbitrary if and only if the following no-log condition is satisfied,

$$\begin{aligned}Q_6 \equiv &-3\beta^2\delta^{-2}(a_1 - a_2)^2C^2 + 3\beta^6C^6 - 18\beta^5C^3(C_t + CC_x) \\ &+ \frac{\beta^4}{3}(3C_t^2 + 8CC_{tt} + 14CC_tC_x + 11C^2C_x^2 + 16C^2C_{xt} + 8C^3C_{xx}) = 0.\end{aligned} \tag{5.243}$$

This expression is independent of S, and this PDE for $C(x,t)$ has not yet been integrated. In the particular case $a_1 = a_2$, its reduction $C_x = 0$ is an ODE reducible to quadratures [378]. In the even more particular case when C is a constant c, the relations $\varphi_t + C\varphi_x = 0$ and $S_t + CS_x = 0$ define the Laurent series as the traveling wave reduction $u(x,t) = U(\xi), \xi = x - ct$, of the Fisher equation, and the reduced ODE passes the Painlevé test [8] if and only if c takes one of the five values defined by $Q_6 = 0$,

$$c = 0,\ c^4 = \frac{(a_1 - a_2)^2}{\beta^4\delta^2}. \tag{5.244}$$

In such a case, the ODE for $U(\xi)$ belongs to the equivalence class of the Weierstrass equation (1.42) (equation number 3 in Gambier's list [163]), and its general solution is, for $c = 0$ the Weierstrass function itself,

$$c = 0:\ u = \frac{s_1}{2} + \delta\wp(x - x_0, g_2, g_3),\ g_2 = 3\frac{(a_1 - a_2)^2}{\delta^2}, \tag{5.245}$$

in which the two movable constants are x_0, g_3, and for $c \neq 0$ the single valued expression [8]

$$\begin{cases} c^4 = \dfrac{(a_1 - a_2)^2}{\beta^4\delta^2} :\ u = \dfrac{s_1}{2} + \delta e^{2k\xi}\wp(\dfrac{e^{k\xi} - 1}{k} - X_0, 0, g_3) - \dfrac{\delta k^2}{2}, \\ \xi = x - ct,\ k = -\beta c, \end{cases} \tag{5.246}$$

in which the two movable constants are X_0, g_3.

The singular manifold method cannot yield a Lax pair since the PDE fails the test, but it can provide particular solutions, which we now search. Assuming the one-family singular part transformation (5.86), the choice of the second-order scalar linear system (4.34)–(4.35), and the link (5.63) between τ and ψ, the truncation is defined quite similarly to that performed in Sect. 5.6.1.1,

$$\begin{aligned} u &= \delta\left(\beta\partial_t - \partial_x^2\right)\log\psi + U \\ &= \delta\chi^{-2} - \beta\delta C\chi^{-1} + \frac{\delta}{2}S + \frac{\delta\beta}{2}C_x + U,\ \chi = \frac{\psi}{\psi_x}. \end{aligned} \tag{5.247}$$

By identifying to the Laurent series (5.242), it provides the value of U,

$$U = \frac{s_1}{2} - \frac{\delta\beta^2}{12}C^2 - \frac{\delta}{6}S + \frac{\delta\beta}{2}C_x, \tag{5.248}$$

and defines the two determining equations

$$E_3 \equiv \beta^3C^3 + \beta^2(-5C_t - 7CC_x) + \beta(2SC + 6C_{xx}) + S_x = 0, \tag{5.249}$$

$$\begin{aligned} E_4 \equiv &-\frac{3(a_1-a_2)^2}{2\delta^2} + \frac{\beta^4}{24}C^4 - \frac{\beta^3}{6}(5CC_t + 6C^2C_x) \\ &+\frac{\beta^2}{6}(-17SC^2 + 37C_x^2 + 30C_{xt} - 14CC_{xx}) \\ &+\frac{\beta}{6}(16S_t + 3CS_x + 30SC_x) + \frac{S^2}{6} - \frac{1}{3}S_{xx} = 0, \end{aligned} \tag{5.250}$$

together with the cross-derivative condition (4.32). The no-log condition (5.243) is a differential consequence of these three coupled PDEs (this is easily proven by the elimination of S), but their general solution is still unknown.

The particular solutions $(S,C) =$ constant always admitted by such a system lead to a stationary wave (case $c = 0$) or a traveling wave (case $c \neq 0$) [8, (20)] which are the trigonometric degeneracies $4g_2^3 - 27g_3^2 = 0$ of the elliptic solutions (5.245) and (5.246),

$$\begin{aligned} &u = \frac{s_1}{2} + \delta\left(\beta\partial_t - \partial_x^2\right)\log\cosh\frac{k}{2}(x-ct) = \frac{s_1}{2} + \delta\left(\chi^{-2} - \beta c\chi^{-1} - \frac{\beta^2c^2}{4}\right), \\ &k = -\beta c,\ \chi^{-1} = \frac{k}{2}\tanh\frac{k}{2}(x-ct),\ c^4 = 0 \text{ or } \frac{(a_1-a_2)^2}{\beta^4\delta^2}. \end{aligned} \tag{5.251}$$

These heteroclinic solutions link the two fixed points a_1 and a_2.

In order to find the general solution of the overdetermined system (5.249), (5.250), (4.32), there exists an algorithm known as the construction of a differential Groebner basis [297, 37], a tutorial description of which can be found in [78]. One first eliminates ∂_t by solving $E_3 = 0$ for C_t and $E_4 = 0$ for S_t; next, the condition (4.32) is solved for S_{xx}, and the original system is equivalent to three new equations expressing C_t, S_t, S_{xx} as polynomials in $S_x, S, C, C_x, C_{xx}, C_{xxx}$. After elimination of S_x

and S, one obtains a sixth order ODE for C (not containing ∂_t), which we failed to integrate. This difficulty reflects the nonintegrability of the Fisher equation. There exists a particular scaling solution,

$$a_1 = a_2 : \ S = -\frac{3}{2x^2}, \ C = -\frac{3 \pm \sqrt{6}}{\beta x}, \tag{5.252}$$

but its interest is minor because the two fixed points are equal. The corresponding solution u is,

$$a_1 = a_2 : \ u = \frac{s_1}{2} + \delta \frac{2(4 \pm \sqrt{6})x^2 + 4(2 \pm \sqrt{6})\beta^{-1} t}{(x^2 + 2(3 \pm \sqrt{6})\beta^{-1} t)^2}. \tag{5.253}$$

When one assumes the one-family singular part transformation (5.86) with the link (5.63) between τ and ψ and the third order linear system (5.48)–(5.49) for ψ, one finds a unique solution,

$$a_1 = a_2 : \ U = \frac{s_1}{2}, \ \psi_{xxx} = 0, \ \psi_t = \frac{3 \pm \sqrt{6}}{\beta} \psi_{xx}, \tag{5.254}$$

which again represents the rational solution (5.253),

$$a_1 = a_2 : \ u = \frac{s_1}{2} + \delta \left(\beta \partial_t - \partial_x^2\right) \log \psi, \ \psi = x^2 + 2(3 \pm \sqrt{6})\beta^{-1} t. \tag{5.255}$$

As to sech solutions, they cannot exist because the Fisher equation has only one family.

5.7.2 Two-Family Case: KPP Equation

We restrict ourselves here to the properly defined KPP equation, i.e. without the convection term $\gamma u u_x$ introduced in [383],

$$E(u) \equiv b u_t - u_{xx} + 2d^{-2}(u - e_1)(u - e_2)(u - e_3) = 0. \tag{5.256}$$

The Painlevé test, performed in Sect. 4.4.3, fails but provides the following constructive information,

1. existence of two families with opposite singular part operators,

$$p = -1, \ u_0 = \pm d, \ \text{indices } (-1, 4), \ \mathscr{D} = \pm d \partial_x, \tag{5.257}$$

2. for each family, one necessary condition at index 4 for the absence of movable logarithms, (4.63).

Let us turn this information into exact solutions, by applying the numerous resources of the singular manifold method. The *a priori* expected solutions are

1. $u = c_1 + c_2 \tanh(k/2)(x - ct)$, with k, c, c_1, c_2 constant, because the PDE is autonomous and has the singularity degree $p = -1$,
2. $u = c_1 + c_2 \operatorname{sech} k(x - ct)$, because in addition the PDE has two families with the same $p = -1$ and opposite singular part operators,
3. $u =$ elliptic function of x, as evident if u does not depend on t.

In fact, we are going to see that the SMM can also provide other solutions, among them a degenerate two-soliton solution.

It is convenient to introduce the symmetric notation

$$\begin{cases} s_1 = e_1 + e_2 + e_3, \\ a_1 = (2e_1 - e_2 - e_3)(2e_2 - e_3 - e_1)(2e_3 - e_1 - e_1)/(27d^3), \\ a_2 = \left((e_2 - e_3)^2 + (e_3 - e_1)^2 + (e_1 - e_2)^2\right)/(18d^2). \end{cases} \tag{5.258}$$

In order to find single valued exact solutions, the various methods which can be applied are

1. enforcement of one of the two no-log conditions (4.63),
2. enforcement of the two no-log conditions,
3. one-family truncation with a second order assumption,
4. one-family truncation with a third order assumption,
5. two-family truncation with a second order assumption,
6. two-family truncation with a third order assumption.

Enforcing one or two of the no-log conditions yields intricate expressions [53, 78], unless C takes a constant value c, whose admissible values are $c = 0$ and

$$\text{one condition enforced: } c = \frac{3e_n - s_1}{bd},\ n = 1,2,3, \tag{5.259}$$

$$\text{two conditions enforced: } c = \pm\frac{3(e_2 - e_3)}{2bd},\ 2e_1 - e_2 - e_3 = 0. \tag{5.260}$$

The relation $\varphi_t + c\varphi_x = 0$ then restricts φ, the Laurent expansion for u and u itself to depend only on $x - ct$, thus defining the traveling wave reduction $u(x,t) = U(\xi), \xi = x - ct$. The reduced ODE for $U(\xi)$ then belongs to the class investigated by Painlevé and Gambier, and the conclusions are the following. For the two opposite values (5.260) of c enforcing the two no-log conditions, it has the Painlevé property and its general solution is elliptic. For the three values (5.259) of c enforcing only one no-log condition, the reduced ODE still fails the test, and all one can say at this stage is that its general solution is multivalued.

The one-family truncation (5.86) with $\tau = \psi$ yields the same result whether one chooses the second order assumption (4.34)–(4.35) [53, 72] or the third order assumption (5.48)–(5.49) [53, 71, 86]. The computation is immediate at third order or with the noninvariant WTC formalism at second order (truncation in $\varphi - \varphi_0$), while it is quite involved at second order in the invariant formalism with (S, C) as the unknowns, see [78] for full details. The common result of these truncations is a solution, displayed in Fig. 5.4, which represents the collision of two fronts (each front links two of the fixed points $u = e_j$) [247]

$$u = \frac{s_1}{3} + d\partial_x \log \Psi_3, \tag{5.261}$$

in which Ψ_3 is the entire function

$$\Psi_3 = \sum_{n=1}^{3} C_n e^{k_n(x+(3/b)k_n t)},\ k_n = \frac{3e_n - s_1}{3d},\ C_n \text{ arbitrary}, \tag{5.262}$$

i.e. the general solution of the third order linear system with constant coefficients [86]

$$\begin{cases} \psi_{xxx} - 3a_2\psi_x - a_1\psi = 0, \\ b\psi_t - 3\psi_{xx} = 0. \end{cases} \tag{5.263}$$

The degeneracy $C_1 = 0$ of this solution,

$$u = \frac{e_2+e_3}{2} + \frac{e_2-e_3}{2} \tanh\left(\frac{e_2-e_3}{2d}(x + \frac{3(e_2+e_3)}{2bd}t)\right), \tag{5.264}$$

is the single front which links the two fixed points e_2 and e_3.

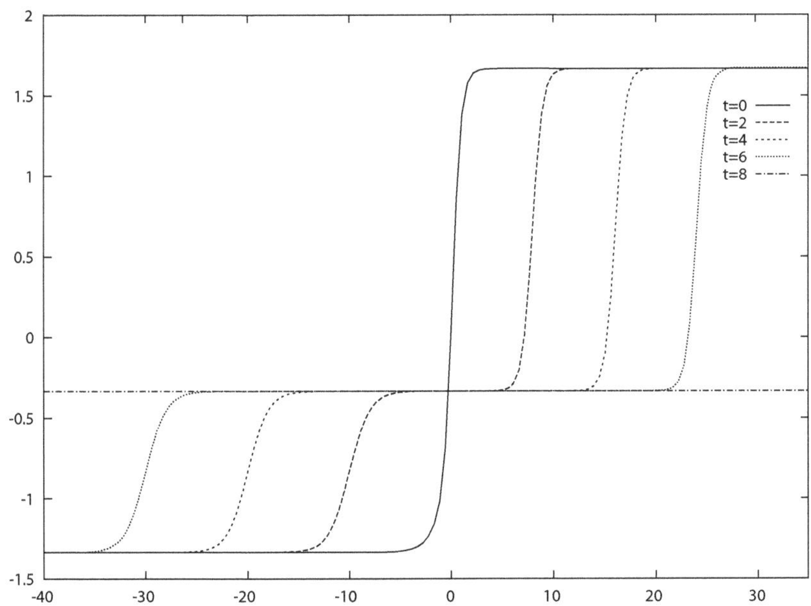

Fig. 5.4 KPP, collision of two fronts. The function $u(x,t)$ is plotted as a function of x at the successive times $t = 0,2,4,6,8$, with the parameters values $b = 1, d = 1, e_j = (0,2,-1), C_j = (1,1,1)$.

The two-family truncation with a second order assumption (5.137)–(5.140) [85, 86, 356] generates five determining equations, whose general solution is still unknown. Their particular solution for which (S,C,A,B) are constant is quite easy to find since the determining equations become algebraic, one already knows from the elementary identities (5.74) that it is polynomial in tanh and sech. The result is a stationary pulse (homoclinic orbit) which only exists for $a_1 = 0$,

$$u = e_j + \frac{e_l - e_m}{\sqrt{2}} \operatorname{sech} i \frac{e_l - e_m}{d\sqrt{2}}(x - x_0),\ a_1 = 0,\ 2e_j - e_l - e_m = 0, \tag{5.265}$$

and it depends on the arbitrary constant x_0.

The collision solution (5.261) can further be extrapolated because of a special feature of the KPP equation (shared by all PDEs having a Laurent series $u = c_0\chi^{-1} + c_1 C + c_2 + \mathcal{O}(\chi), c_j =$ constant, see (4.62)). After having defined the WTC-style truncation

$$\begin{aligned} &u = d\partial_x \log(\varphi - \varphi_0) + U,\ U = \text{constant}, \\ &E(u) = \sum_{j=0}^{3} E_j(\varphi_{xx}/\varphi_x, \varphi_t/\varphi_x) \left(\frac{\varphi - \varphi_0}{\varphi_x}\right)^{j-3},\ \forall j\ :\ E_j = 0, \end{aligned} \tag{5.266}$$

and found its solution as indicated above, $U = s_1/3, \varphi - \varphi_0 = \Psi_3$ one defines the change of function $u \mapsto f$ [54],

$$u = s_1/3 + (d\partial_x \log \Psi_3)\, f(\Psi_3), \tag{5.267}$$

which transforms (4.48) into

$$U'' - 2U^3 + 2a_1 \Psi_{3,x}^{-3} = 0,\ U(\psi) = f(\psi)/\psi. \tag{5.268}$$

This is an ODE iff $a_1 = 0$, in which case the solution is the singlevalued expression [54],

$$a_1 = 0:\ u = s_1/3 + d\psi_x\sqrt{\wp(\psi)},\ \psi = \Psi_3,\ g_3 = 0,\ g_2 \text{ arbitrary}. \tag{5.269}$$

This solution exists whenever one root e_j is at the middle of the two others, and it depends on the four arbitrary constants C_1, C_2, C_3, g_2. Its degeneracy $g_2 = 0$ (i.e. $\wp(\psi) = \psi^{-2}$) is the degeneracy $a_1 = 0$ of the collision of two fronts solution (5.261).

5.8 Reduction of the Singular Manifold Method to the ODE Case

Since the Painlevé property of a PDE is by definition preserved under a noncharacteristic reduction to an ODE, one might wonder why, nowhere in Chap. 3, have the tools of the Lax pair, singular part transformation, Bäcklund transformation and nonlinear superposition formula been used.

The reason is that they are easier to define and use at the PDE level, this is why we did not introduce them for ODEs in the first place. In fact, all these tools survive under a reduction, i.e. they can all be defined for ODEs. There is no difficulty in performing the reduction of the above mentioned tools, but there is a difficulty about the implementation of the singular manifold method to ODEs. Before examining this difficulty, let us first perform the reductions.

Consider the example of the modified Korteweg–de Vries equation (5.103) with its Lax pair in Riccati representation (5.186), singular part transformation (5.184), auto-BT (5.188)–(5.189) and NLSF (5.191). This PDE admits a reduction to the second Painlevé equation [7],

$$W = \beta + \alpha(3t/b)^{-1/3}U(\xi),\ \xi = (3t/b)^{-1/3}(x - 6\nu t/b),$$
$$(U'' - 2U^3 - \xi U)' = 0. \tag{5.270}$$

5.8.1 *From Lax Pair to Isomonodromic Deformation*

There is no difficulty in performing this reduction, and no guess work is involved. In order to reduce the Lax pair of mKdV to a Lax pair of P2 [142], it is convenient to represent the Lax pair as the one-form

$$\mathrm{d}y = y_x \mathrm{d}x + y_t \mathrm{d}t, \tag{5.271}$$

in which y is the scalar of the Riccati representation (5.186) of the Lax pair. Were the Lax pair defined in the zero-curvature representation (L, M) (5.13), the one-form would be

$$\mathrm{d}\psi = L(W,x,t,\lambda)\psi \mathrm{d}x + M(W,x,t,\lambda)\psi \mathrm{d}t, \tag{5.272}$$

in which ψ is the two-component vector defined by the correspondence $y = \psi_1/\psi_2$. The goal is to deduce a similar Lax pair for P2

$$\mathrm{d}z = (a_2 z^2 + a_1 z + a_0)\mathrm{d}\xi + (b_2 z^2 + b_1 z + b_0)\mathrm{d}\mu, \tag{5.273}$$

in which the six scalars a_j, b_j only depend on U, ξ and a scalar μ which is the ODE spectral parameter. Starting from (5.271), one first eliminates W, x, $\mathrm{d}x$ from (5.270), and U'' from $U'' = 2U^3 + \xi U + A$ to obtain

$$\mathrm{d}y = \left(-2U\mathrm{d}\xi + \left(2A + 8\lambda(\lambda - \nu\lambda^{-1})(3t/b)^{2/3}U\right)\frac{\mathrm{d}t}{3t}\right)y$$
$$-2U'(\lambda - (\lambda - \nu\lambda^{-1})y^2)\frac{\mathrm{d}t}{3t}$$

$$+(3t/b)^{1/3}\left(\mathrm{d}\xi+\left(2U^2+\xi-4\lambda(\lambda-\nu\lambda^{-1})(3t/b)^{2/3}\right)\frac{\mathrm{d}t}{3t}\right)$$
$$\times(\lambda+(\lambda-\nu\lambda^{-1})y^2). \tag{5.274}$$

This introduces μ as the additive term to either $2A$ or $2U^2+\xi$,

$$\mu=k(3t/b)^{1/3},\ k^2=\lambda\left(\lambda-\frac{\nu}{\lambda}\right),\ \frac{\mathrm{d}\mu}{\mu}=\frac{\mathrm{d}t}{3t}, \tag{5.275}$$

and allows the further elimination of t and $\mathrm{d}t$. Finally, the remaining dependence on λ/k is removed by the change

$$y=(\lambda/k)z, \tag{5.276}$$

or, in the zero-curvature representation, by

$$\psi=P\Psi,\ P=\mathrm{diag}((\lambda/k)^{1/2},(\lambda/k)^{-1/2}). \tag{5.277}$$

The final Lax pair of P2 is defined by the one-form

$$\frac{\mathrm{d}z}{z}=\left(-2U\mathrm{d}\xi+\left(\frac{2A}{\mu}+8\mu U\right)\mathrm{d}\mu\right)-2\left(z^{-1}+z\right)U'\mathrm{d}\mu$$
$$+\left(z^{-1}-z\right)\left(\mu\mathrm{d}\xi+(2U^2+\xi-4\mu^2)\mathrm{d}\mu\right), \tag{5.278}$$

i.e., in the Riccati representation

$$\frac{z_\xi}{z}=\mu\left(\frac{1}{z}-z\right)-2U, \tag{5.279}$$
$$\frac{z_\mu}{z}=\frac{2A}{\mu}+8\mu U-2\left(z^{-1}+z\right)U'+\left(z^{-1}-z\right)(2U^2+\xi-4\mu^2), \tag{5.280}$$

and in the zero-curvature traceless representation [3, 142]

$$\mathscr{L}=\begin{pmatrix}-U & \mu\\ \mu & U\end{pmatrix},$$
$$\mathscr{M}=\begin{pmatrix}A\mu^{-1}+4\mu U & 2U^2+\xi-4\mu^2-2U'\\ 2U^2+\xi-4\mu^2+2U' & -(A\mu^{-1}+4\mu U)\end{pmatrix}. \tag{5.281}$$

Another Lax pair has been established for P2 [240]

$$\mathscr{L}=\frac{1}{2}\begin{pmatrix}t+2U & 1\\ -2z & -t\end{pmatrix},\ z=U'-U^2-\frac{\xi}{2},$$
$$\mathscr{M}=\frac{1}{2}\begin{pmatrix}2t^2+2z+\xi & 2t-2U\\ -4tz-2(1-2A+2Uz) & -2t^2-2z-\xi\end{pmatrix}, \tag{5.282}$$

and the equivalence of these two Lax pairs is still an open question [338].

Conversely, given the linear differential equation

$$\frac{\mathrm{d}}{\mathrm{d}\mu}\psi=\begin{pmatrix} a_1\mu^{-1}+a_2\mu & a_3-\mu^2-a_4 \\ a_3-\mu^2+a_4 & -(a_1\mu^{-1}+a_2\mu) \end{pmatrix}\psi,\ \psi=\begin{pmatrix}\psi_1\\ \psi_2\end{pmatrix}, \tag{5.283}$$

which has the same dependence on μ as (5.281) but with unspecified coefficients a_j, it is possible to constrain these coefficients so as to generate the nonlinear P2 equation, by a method known as the *isomonodromic deformation* method, the parameters a_j being called the *deformation parameters*. For a tutorial introduction to this powerful method, we refer to summer school lecture notes [291]. This creates a very deep connection between linear ODEs and a special class of nonlinear ODEs, essentially made of those which have the Painlevé property.

The Lax pair of an ODE is quite important in itself since the Painlevé property can be proven from the Lax pair only, as opposed to the PDE case where the proof requires the knowledge of the Lax pair *and* the singular part transformation. In the case of P6 for instance, in addition to the analytic proof of Painlevé [350, ¶2], the Painlevé property has been established with the only knowledge of the Lax pair by various authors like Richard Fuchs [157], Malmquist [293, ¶30 and p. 89] and Malgrange [292].

5.8.2 From BT to Birational Transformation

In order to reduce the auto-BT (5.188)–(5.189) of mKdV, let us denote the reduced conservative fields as follows,

$$\begin{aligned} &r_x=w=\beta+\alpha(3t/b)^{-1/3}u(\xi),\ u''-2u^3-\xi u-\alpha=0,\\ &R_x=W=\beta+\alpha(3t/b)^{-1/3}U(\xi),\ U''-2U^3-\xi U-A=0. \end{aligned} \tag{5.284}$$

With the relation (5.270) between ∂_x and ∂_ξ the singular part transformation (5.184) reduces to

$$u-U=\partial_\xi \log z. \tag{5.285}$$

The two PDEs defining the auto-BT (5.188)–(5.189) reduce to a system independent of (ν,μ),

$$\begin{cases} (u'+U')^2-(u^2-U^2)^2=0,\\ u'^2-U'^2-(u-U)(\alpha+A+(u+U)(\xi+u^2+U^2))=0, \end{cases} \tag{5.286}$$

and their solution is [288]

$$u'=\varepsilon\left(u^2+\frac{\xi}{2}+\frac{\alpha+A}{2(u+U)}\right),\ U'=-\varepsilon\left(U^2+\frac{\xi}{2}+\frac{\alpha+A}{2(u+U)}\right), \tag{5.287}$$

expressions which define a birational transformation between u and U[3]. The two parameters α and A are not directly constrained by the Bäcklund transformation, and an elimination is required to find their relationship,

$$A-\alpha+\varepsilon=0,\ \varepsilon^2=1. \tag{5.288}$$

Because of the discrete invariance $(u,x,\alpha)\to(-u,x,-\alpha)$ of P2, its birational transformation depends on two signs, not one as displayed in (5.287), and the final writing is [92, (67)]

$$\begin{cases} \dfrac{N}{u-U}=U'+S_\infty D\left(U^2+\dfrac{x}{2}\right)=u'+s_\infty d\left(u^2+\dfrac{x}{2}\right), \\ u''=d^2(2u^3+xu)+\alpha,\ \alpha=-d\theta, \\ U''=D^2(2U^3+xU)+A,\ A=-D\Theta, \\ s\theta=-S\Theta+1,\ s_\infty d=S_\infty D,\ s^2=S^2=s_\infty^2=S_\infty^2=1, \\ N=(1/2)-S\Theta=s\theta-1/2=(1/2)(s\theta-S\Theta). \end{cases} \tag{5.289}$$

While the auto-Bäcklund transformation of a PDE links two solutions of the same PDE, its reduction to a birational transformation connects particular solutions of two different copies of the same ODE.

One can also reduce the BT between KdV and mKdV ((5.98)–(5.99) and (5.104)). Indeed, the KdV equation admits a reduction with the same similarity variable $xt^{-1/3}$ [143, 343] (the shift $-\xi/2$ in U is pure convenience)

$$\begin{aligned} & u=a(3t/b)^{-2/3}\left(U(\xi)-\frac{\xi}{2}\right),\ \xi=(3t/b)^{-1/3}x, \\ & U'''-6UU'+2\xi U'+U=0, \end{aligned} \tag{5.290}$$

and this third order ODE was proven by Chazy to possess the Painlevé property [59, class XIII]. The reduced BT is

$$W'+W^2-U+\frac{\xi}{2}=0, \tag{5.291}$$

$$U''-\xi W'-2U'W+(2U-\xi)W^2-2\left(U-\frac{\xi}{2}\right)^2=0, \tag{5.292}$$

$$\text{reduced mKdV}:\ W''-2W^3-\xi W-A=0,\ A=\text{arbitrary}, \tag{5.293}$$

$$\text{reduced KdV}:\ U'''-6UU'+2\xi U'+U=0, \tag{5.294}$$

and its first half (5.291) provides the general solution of (5.290) as an algebraic transform of P2. By a straightforward elimination, one generates a relation between U,U',U'' and A only which defines A as a first integral of (5.290),

[3] By a fortunate coincidence, the initials BT are common to Bäcklund transformation and birational transformation.

$$-U'' + \frac{U'^2}{2U} + 2U^2 - \xi U - \frac{\left(A+\frac{1}{2}\right)^2}{2U} = 0, \tag{5.295}$$

an ODE which bears the number 34 in Gambier's classification [163]. Finally, the elimination of W', W'' between (5.291), its derivative and (5.293) yields a polynomial relation between W, U, U', A which has first degree in W. To conclude, under the reduction $xt^{-1/3}$, the BT between KdV and mKdV reduces to a birational transformation between P2 and G34 [163, p. 31]

$$\begin{cases} U = W' + W^2 + \dfrac{\xi}{2}, \\ W = \dfrac{U' - A - \frac{1}{2}}{2U}. \end{cases} \tag{5.296}$$

As a consequence of this equivalence between P2 and G34 under the group of birational transformations, G34 could be an admissible choice to represent the equivalence class in place of P2. The reason for choosing P2 is its unique singularity $W = \infty$, while G34 has the two singularities $U = \infty, 0$[4].

An even simpler exercise, which we leave to the reader, is to consider the stationary reduction $\partial_t = 0, \xi = x$ for both KdV and mKdV, to perform the reduction of the BT between the two PDEs and to recover the well known homography (a subgroup of the group of birational transformations) between the Weierstrass equation and the other elliptic equation.

5.8.3 From NLSF to Contiguity Relation

The NLSF, which involves four solutions of mKdV, should not be reduced from the formula (5.191). Indeed, this formula results from an elimination, and the correct way to perform the reduction is to first reduce the elementary components, i.e. the BT, then to perform the elimination. The result, as we now see, is an algebraic relation between three solutions of P2, not among four solutions, and the reason is the first integral A introduced by the reduction.

Let us start from the reduction of the auto-BT, i.e. the birational transformation (5.289). The process of lowering the differential order involves the elimination of the derivatives u' and U', and this is achieved as follows [145].

1. Consider a birational transformation such as (5.289), i.e. the direct birational transformation and its inverse

$$u = f(U, U', x, \theta, \Theta), \ \theta = g(\Theta), \tag{5.297}$$
$$U = F(u, u', x, \Theta, \theta), \ \Theta = G(\theta). \tag{5.298}$$

[4] It is for exactly the same reason that, among the elliptic functions having two poles per period, the choice of Weierstrass (one function with a double pole) is much preferable to the choice of Jacobi (three functions with two simple poles), see the arguments of Halphen [193, Chap. VIII p. 253].

2. Rewrite the direct birational transformation as the forward iterate $v_{n+1} = \overline{v}$ and the inverse birational transformation as the backward iterate $v_{n-1} = \underline{v}$ of the *same* value of a discrete variable $v = u$,

$$v_{n+1} = \overline{v} = f(v, v', x, g(\theta), \theta),\ v = v_n,\ v' = v'_n, \tag{5.299}$$
$$v_{n-1} = \underline{v} = F(v, v', x, G(\theta), \theta),\ v = v_n,\ v' = v'_n. \tag{5.300}$$

3. Eliminate the single derivative v' between these two relations,

$$C(\overline{v}, \underline{v}, v, x, \theta) = 0. \tag{5.301}$$

Such a relation without any derivative is called a *contiguity relation*. As noted by Garnier [168], this is an extension to nonlinear equations of the similar notion introduced by Gauss for the (linear) hypergeometric equation.

In the case of (5.289), these two equations for $\overline{v}, \underline{v}$ are

$$\begin{cases} \dfrac{\frac{1}{2} + \theta_n}{\overline{v} - v} = v' + S_\infty D\left(v^2 + \dfrac{x}{2}\right), \\ \dfrac{\frac{1}{2} - \theta_n}{\underline{v} - v} = v' + s_\infty d\left(v^2 + \dfrac{x}{2}\right), \end{cases} \tag{5.302}$$

and their difference generates the contiguity relation

$$\frac{\frac{1}{2} + \theta_n}{\overline{v} - v} + \frac{\frac{1}{2} - \theta_n}{\underline{v} - v} + (s_\infty d - S_\infty D)(2v^2 + \xi) = 0. \tag{5.303}$$

As compared with the NLSF (5.191), this contiguity relation involves three solutions $\overline{v}, v, \underline{v}$ of three different P2 equations (against four different solutions u_{n-1}, u_n, $\tilde{u}_n$, u_{n+1} of the same PDE), the independent variable $z = nh$ is introduced by solving the recurrence relation on θ, and the variable ξ has become a mere parameter introduced by the nonautonomous reduction.

For the master equation P6, the elementary birational transformation, first found by Okamoto [341], is even simpler because it clearly displays the equivalence of the four singular points $u = \infty, 0, 1, x$ [91],

$$\frac{x(x-1)U'}{U(U-1)(U-x)} = 2\frac{s_j\theta_j - S_j\Theta_j}{u-U} - \left(\frac{S_0\Theta_0}{U} + \frac{S_1\Theta_1}{U-1} + \frac{S_x\Theta_x - 1}{U-x}\right), \tag{5.304}$$
$$\frac{x(x-1)u'}{u(u-1)(u-x)} = -2\frac{S_j\Theta_j - s_j\theta_j}{u-U} - \left(\frac{s_0\theta_0}{u} + \frac{s_1\theta_1}{u-1} + \frac{s_x\theta_x - 1}{u-x}\right). \tag{5.305}$$

In this relation, j is anyone of the four singular points $(\infty, 0, 1, x)$, the monodromy exponents θ_j, Θ_j are defined by their squares,

$$\theta_\infty^2 = 2\alpha,\ \theta_0^2 = -2\beta,\ \theta_1^2 = 2\gamma,\ \theta_x^2 = 1 - 2\delta, \tag{5.306}$$
$$\Theta_\infty^2 = 2A,\ \Theta_0^2 = -2B,\ \Theta_1^2 = 2\Gamma,\ \Theta_x^2 = 1 - 2\Delta, \tag{5.307}$$

the eight signs s_j, S_j are arbitrary and independent, and the recurrence relation among the monodromy exponents is

$$s_j\theta_j = S_j\Theta_j - \frac{1}{2}\left(\sum S_k\Theta_k\right) + \frac{1}{2}, \; S_j\Theta_j = s_j\theta_j - \frac{1}{2}\left(\sum s_k\theta_k\right) + \frac{1}{2} \cdot \tag{5.308}$$

The corresponding contiguity relation [327, (1.5)] [91] is the sum of six simple poles (including $v = \infty$)

$$\frac{\varphi(n+1/2)}{\bar{v}-v} + \frac{\varphi(n-1/2)}{\underline{v}-v} = \frac{s_0\theta_0 - S_0\Theta_0}{v} + \frac{s_1\theta_1 - S_1\Theta_1}{v-1} + \frac{s_x\theta_x - S_x\Theta_x}{v-x},$$
$$\varphi(n) = \frac{1}{2}(s_\infty\theta_\infty + s_0\theta_0 + s_1\theta_1 + s_x\theta_x - 1), \tag{5.309}$$

in which θ is taken at the center point $z = z_0 + nh$.

5.8.4 Reformulation of the Singular Manifold Method: an Additional Homography

While there is no special difficulty, as seen above, for performing the reduction on the various items of the singular manifold method, there does exist a difficulty for reducing the method itself, i.e. for generating determining equations similar to (5.66), (5.67) or (5.140).

Consider a Painlevé ODE Pn which admits a birational transformation, i.e. $n = 2,3,4,5,6$, with the goal of finding this birational transformation by a method only based on the singularities (singular manifold method). The singular part transformation assumption is [175]

$$u = u_0 Z^{-1} + U, \; u_0 \neq 0, \tag{5.310}$$
$$Z' = 1 + z_1 Z + z_2 Z^2, \; z_2 \neq 0, \tag{5.311}$$

in which u obeys a Pn equation, and let us assume for simplicity that U also obeys the same Pn equation with different parameters (as compared to the PDE case, this is an additional assumption). The unknowns (Z, z_1, z_2) are to be determined as rational functions of (x, U, U'). After this is done, the relation (5.310) represents half of the birational transformation.

This assumption (5.310)–(5.311), is indeed successful [175] when applied to Pn, $n = 2,3,4,5$ but it fails for P6 unless the following crucial point is implemented [91]. For second order first degree ODEs such as Pn, the Riccati variable Z and the derivative U' are not independent. Indeed, any N-th order, first degree ODE with the Painlevé property is necessarily [346, pp. 396–409] a Riccati equation for $U^{(N-1)}$, with coefficients depending on x and the lower derivatives of U, in our case

$$U'' = A_2(U,x)U'^2 + A_1(U,x)U' + A_0(U,x). \tag{5.312}$$

Since the group of invariance of a Riccati equation is the homographic group, the variables U' and Z are linked by a homography. Let us define it as

$$(U' + g_2)(Z^{-1} - g_1) - g_0 = 0, \ g_0 \neq 0. \tag{5.313}$$

This allows us to obtain the two coefficients z_j of the Riccati pseudopotential equation (5.311) as explicit expressions of $(g_j, \partial_U g_j, \partial_x g_j, A_2, A_1, A_0, U')$. Indeed, eliminating U' between (5.312) and (5.313) defines a first order ODE for Z, whose identification with (5.311) *modulo* (5.313) provides three relations,

$$g_0 = g_2^2 A_2 - g_2 A_1 + A_0 + \partial_x g_2 - g_2 \partial_U g_2, \tag{5.314}$$
$$z_1 = A_1 - 2g_1 + \partial_U g_2 - \partial_x \log g_0 + (2A_2 - \partial_U \log g_0) U', \tag{5.315}$$
$$z_2 = -g_1 z_1 - g_1^2 - g_0 A_2 - \partial_x g_1 - (\partial_U g_1) U'. \tag{5.316}$$

Therefore, the natural unknowns in the present problem are the two functions g_1, g_2 of the two variables (U, x), and not the two functions (z_1, z_2) of the three variables (U', U, x).

The truncation then presents no difficulty and it does provide straightforwardly [91] the birational transformation for all the Pn equations which admit one, including the one for P6 (5.305).

Table 5.1 contains a qualitative summary of the various features of a PDE which possesses the Painlevé property when the PDE is reduced to an ODE.

Table 5.1 Respective tools of integrability of a PDE having the PP and any of its noncharacteristic reductions to an ODE.

Tool	PDE	ODE
linear representation	Lax pair	isomonodromic deformation
difference of two sol.	Singular part transformation	Darboux transformation
invariance	Bäcklund transformation	birational transformation
build solutions	nonlinear superposition formula	contiguity relation
general solution	N/A	no movable critical sing.

Chapter 6
Integration of Hamiltonian Systems

Abstract In this chapter, we illustrate the various ways to "integrate" a Hamiltonian system using two examples with two degrees of freedom: the cubic (HH3) and quartic (HH4) Hénon–Heiles Hamiltonians. These various ways to integrate are

- ∘ (Liouville integrability) to find a second invariant in involution with the Hamiltonian, which is however insufficient to perform a global integration;
- ∘ (Arnol'd–Liouville integrability) to find the variables which separate the Hamilton–Jacobi equation, thus leading to a global integration;
- ∘ (Painlevé property) to find an explicit closed form single valued expression for the general solution $q_j(t), Q_j(t)$; this has been done in all seven cases (Sects. 6.2.3 and 6.3.3), *via* birational transformations to fourth order ODEs isolated and integrated by Cosgrove.

6.1 Various Integrations

When a system of nonlinear ODEs can be represented by a Hamiltonian system for $q_j(t), p_j(t)$, one can think of three types of integrability, and we will perform them all when this is possible.

The first type is the integration in the sense of Liouville, i.e. the obtention of a second constant of the motion in involution with the Hamiltonian, so as to prove the reducibility to quadratures. This is however insufficient to prove the global single valuedness, as displayed by the pendulum example in the introduction, Sect. 1.2.

The second type is the *Arnol'd–Liouville integrability*, a stronger version of the Liouville integrability introduced by Arnol'd [18, Chap. 9], which consists in finding an explicit canonical transformation converting the original Hamiltonian system to a new Hamiltonian system in which the Hamilton–Jacobi equation (see below (6.5)) is separated, i.e. has a l.h.s. equal to a sum $F(Q_1,P_1)+G(Q_2,P_2)$. These privileged new coordinates are called *Darboux coordinates* or *separating variables*.

The third type, described in Chap. 3, is to find an explicit closed form single valued expression for the general solution $q_j(t)$. For the three cases of the cubic

Hénon–Heiles system which have been selected by the Painlevé test, (2.99)–(2.101), this is done in Sect. 6.2.3.

One should keep in mind that the first and third above mentioned types are independent, see the FAQ section.

In the Arnol'd–Liouville integrability, when the second invariant has degree at most two in the momenta p_j, there exists a classical method [396] (mainly based on the decomposition of quadratic forms) to obtain explicitly the separating variables. Two of the seven cases ($\beta/\alpha = -6$ of HH3 and $A:B:C = 1:2:1$ of HH4) belong to this Stäckel class (Sects. 6.2.2.1 and 6.3.2.1). The five remaining cases (and even all the seven cases) are processed *via* a détour to soliton equations, as follows:

- ∘ each of the seven cases is a reduction of a soliton equation (SK, a higher order KdV [274] and KK for $\beta/\alpha = -1, -6, -16$, a system of two coupled NLS equations [294] for $1:2:1$, and various systems of two coupled KdV-type equations [21] for the three other HH4 cases);
- ∘ among these various soliton equations, there exist three Bäcklund transformations, a consequence of which is an equivalence (birational transformation) between the pairs $(-1,-16)$ of HH3, $(1:6:1, 1:6:8)$ and $(1:12:16, 5:9:4)$ of HH4, in which $5:9:4$ denotes a companion Hamiltonian outside the Hénon–Heiles class [21]; this restricts the remaining number of cases to three pairs;
- ∘ in each of the three pairs, one of the two elements $(-1, 1:6:1, 1:12:16)$ has a second invariant of degree two for nongeneric values of the parameters which appear in the Hamiltonian; a clever use of the birational transformation allows one to perform the separation of variables in all cases but two.

6.2 Cubic Hénon–Heiles Hamiltonians

Three cases, (2.99)–(2.101), have been selected by the Painlevé test.

6.2.1 Second Invariants

The integrability in the sense of Liouville is first established because in the three cases there exists a second constant of the motion [124, 203, 180] in involution with the Hamiltonian,

$$\text{(SK)}: K = K_0^2 + 3c_3(3p_1^2 q_2^{-2} + 4\alpha q_1 + 2\omega_2), \tag{6.1}$$

$$K_0 = 3p_1p_2 + \alpha q_2(3q_1^2 + q_2^2) + 3\omega_2 q_1 q_2,$$

$$\begin{aligned}\text{(KdV5)}: K = &\ 4\alpha p_2(q_2p_1 - q_1p_2) + (4\omega_2 - \omega_1)(p_2^2 + \omega_2 q_2^2 + c_3 q_2^{-2}) \\ &+ \alpha^2 q_2^2(4q_1^2 + q_2^2) + 4\alpha q_1(\omega_2 q_2^2 - c_3 q_2^{-2}),\end{aligned} \tag{6.2}$$

$$\begin{aligned}\text{(KK)}: K = &\ (3p_2^2 + 3\omega_2 q_2^2 + 3c_3 q_2^{-2})^2 + 12\alpha p_2 q_2^2(3q_1p_2 - q_2p_1) \\ &- 2\alpha^2 q_2^4(6q_1^2 + q_2^2) + 12\alpha q_1(-\omega_2 q_2^4 + c_3) - 12\omega_2 c_3.\end{aligned} \tag{6.3}$$

Remark. The last first integral (6.3) can be extrapolated [204] to the presence of an additional term $+c_4/(6q_2^6)$ in (2.69). This case $c_4 \neq 0$ does not have the Painlevé property, but it admits a Lax pair [225] and therefore has the weak Painlevé property. The first integral (6.2) also admits an extrapolation [225] to the presence of an additional term $+q_1/q_2^4$ in (2.69), see (6.16) below.

6.2.2 Separation of Variables

The relevant quantity for finding the separating variables is the *Hamilton–Jacobi equation* [18, Chap. 9] for the action $S(q_j,t)$,

$$\frac{\partial S}{\partial t} + H(q_1,q_2,p_1,p_2,t) = 0,\ p_1 = \frac{\partial S}{\partial q_1},\ p_2 = \frac{\partial S}{\partial q_2}. \tag{6.4}$$

In the particular case of an autonomous Hamiltonian like HH3, this nonlinear first order PDE for S reduces to

$$H(q_1,q_2,p_1,p_2) - E = 0. \tag{6.5}$$

6.2.2.1 Case $\beta/\alpha = -6$ (KdV5)

Since both invariants H,K are quadratic in the momenta p_j, the method of Stäckel [396] applies, and the separation of variables is realized by the canonical transformation to parabolic coordinates [124, 13, 433]

$$(q_1,q_2,p_1,p_2) \to (s_1,s_2,r_1,r_2), \tag{6.6}$$

$$q_1 = -(s_1+s_2+\omega_1-4\omega_2)/(4\alpha),\ q_2^2 = -s_1 s_2/(4\alpha^2), \tag{6.7}$$

$$p_1 = -4\alpha\frac{s_1 r_1 - s_2 r_2}{s_1 - s_2},\ p_2^2 = -16\alpha^2\frac{s_1 s_2 (r_1-r_2)^2}{(s_1-s_2)^2}. \tag{6.8}$$

Indeed, the two invariants H,K take the form

$$\begin{cases} H = \dfrac{f(s_1,r_1)-f(s_2,r_2)}{s_1-s_2}, \\ K = 2\dfrac{s_1 s_2}{s_1-s_2}\left(\dfrac{f(s_1,r_1)}{s_1} - \dfrac{f(s_2,r_2)}{s_2}\right), \\ f(s,r) = -\dfrac{s^2(s+\omega_1-4\omega_2)^2(s-4\omega_2)-64\alpha^4 c_3}{32\alpha^2 s} + 8\alpha^2 r^2 s, \end{cases} \tag{6.9}$$

therefore the Hamilton–Jacobi equation $H-E=0$ (6.5) allows the introduction of a separating constant,

$$f(s_1,r_1) - E s_1 = f(s_2,r_2) - E s_2 = \text{constant}, \tag{6.10}$$

the value of which is given by an elimination with the equation for K,

$$f(s_j,r_j)-Es_j+\frac{K}{2}=0,\ j=1,2. \tag{6.11}$$

The first task, to find the separating variables, is achieved.

Then, to perform the explicit integration, one builds the differential system for (s_1,s_2) by eliminating the two momenta r_1,r_2 between (6.11) and the Hamilton equations of the motion,

$$s_1'=\frac{\partial H}{\partial r_1}=16\alpha^2\frac{s_1}{s_1-s_2}r_1,\ s_2'=\frac{\partial H}{\partial r_2}=16\alpha^2\frac{s_2}{s_2-s_1}r_2. \tag{6.12}$$

The result,

$$(s_1-s_2)s_1'=\sqrt{P(s_1)},\ (s_2-s_1)s_2'=\sqrt{P(s_2)}, \tag{6.13}$$
$$P(s)=s^2(s+\omega_1-4\omega_2)^2(s-4\omega_2)+32\alpha^2Es^2-16\alpha^2Ks-64\alpha^4c_3, \tag{6.14}$$

is a classical system called *hyperelliptic system*, and the main property of interest to us is that the symmetric polynomials of s_1,s_2, i.e. the sum s_1+s_2 and the product s_1s_2, are meromorphic functions of the independent variable t. This proves the Painlevé property for this HH3 system in the variables q_1 and q_2^2. In particular, it was *a posteriori* a good idea to eliminate q_2 when looking for the dominant behaviors, since only q_2^2 is singlevalued.

The algebraic curve of the (x,y) plane defined by

$$y^2=P(x), \tag{6.15}$$

with P the above fifth degree polynomial, is called a *hyperelliptic curve*, and its *genus* g, defined by $2g+1\le \deg P\le 2g+2$, is two.

Remark. The extrapolation

$$H=\frac{1}{2}(p_1^2+p_2^2)+cq_1+\alpha q_1q_2^2+2\alpha q_1^3+\frac{c_3}{2q_2^2}+\frac{c_5q_1}{q_2^4},\ \alpha\ne 0 \tag{6.16}$$

also admits separating variables [225], however associated with a hyperelliptic curve of genus three, and the general solution presents a finite amount of movable branching (weak Painlevé property).

6.2.2.2 Cases $\beta/\alpha=-1$ (SK) and -16 (KK)

These two cases are best handled simultaneously, because there exists a canonical transformation [33, 379] between the variables of HH3.SK, which we denote here (Q_1,Q_2,P_1,P_2), and the variables of HH3.KK, denoted (q_1,q_2,p_1,p_2). With the normalization

$$H_{\mathrm{SK}} = \frac{1}{2}(P_1^2+P_2^2+\Omega_1 Q_1^2+\Omega_1 Q_2^2)+\frac{1}{2}Q_1Q_2^2+\frac{1}{6}Q_1^3-\frac{1}{8}\lambda^2 Q_2^{-2}, \quad (6.17)$$

$$H_{\mathrm{KK}} = \frac{1}{2}(p_1^2+p_2^2+16\omega_2 q_1^2+\omega_2 q_2^2)+\frac{1}{4}q_1q_2^2+\frac{4}{3}q_1^3-\frac{1}{2}\lambda^2 q_2^{-2}, \quad (6.18)$$

this canonical transformation is

$$\left\{\begin{array}{l}
Q_1 = -6\left(\frac{p_2}{q_2}+\frac{\lambda}{q_2^2}\right)^2 - q_1,\ Q_2 = \frac{\sqrt{\Omega}}{2q_2^4},\\
q_1 = -\frac{3}{2}\left(-\frac{P_2}{Q_2}+\frac{\lambda}{2Q_2^2}\right)^2 - Q_1,\ q_2 = \frac{\sqrt{\Gamma}}{Q_2},\\
P_1 = 12\frac{p_2^3}{q_2^3}+6q_1\frac{p_2}{q_2}-p_1+6\lambda\frac{q_1}{q_2^2}+36\lambda\frac{p_2^2}{q_2^4}+36\lambda^2\frac{p_2}{q_2^5}+12\frac{\lambda^3}{q_2^6},\\
p_1 = 3\frac{P_2^3}{Q_2^3}+3Q_1\frac{P_2}{Q_2}-P_1-\frac{3}{2}\lambda\frac{Q_1}{Q_2^2}-\frac{9}{2}\lambda\frac{P_2^2}{Q_2^4}+\frac{9}{4}\lambda^2\frac{P_2}{Q_2^5}-\frac{3}{8}\frac{\lambda^3}{Q_2^6},\\
P_2 = \left(-\frac{p_2}{q_2^5}+\frac{\lambda}{q_2^6}\right)\sqrt{\Omega}+\frac{\lambda q_2^4}{\sqrt{\Omega}},\\
p_2 = \left(-\frac{P_2}{2Q_2^2}+\frac{\lambda}{4Q_2^3}\right)\sqrt{\Gamma}-\frac{\lambda Q_2}{\sqrt{\Gamma}},\\
\Omega_1 = 16\omega_2,\ c = \Omega_1^2,\\
\Gamma = -12Q_2P_1P_2-6Q_1^2Q_2^2-2Q_2^4+6\Omega_1^2Q_2^2+6\lambda P_1,\\
\Omega = 48\left(3q_2^4k_{2,0}^2+6\lambda q_1q_2^5p_2+12\lambda q_2^3p_2^3-\lambda q_2^6p_1\right.\\
\qquad\left.+3\lambda^2q_1q_2^4+18\lambda^2q_2^2p_2^2+12\lambda^3q_2p_2+3\lambda^4\right),\\
k_{2,0}^2 = p_2^4-q_2^6/72-q_1^2q_2^4/12+q_1q_2^2p_2^2-q_2^3p_1p_2/3+\Omega_1^2q_2^4/12.
\end{array}\right. \quad (6.19)$$

In order to find simultaneously the separating variables of HH3.SK and HH3.KK, one proceeds as follows [372].

1. There exists a canonical transformation which trivially separates one of the two Hamilton–Jacobi equations in the particular case $\lambda = 0$, this is the rotation

$$\tilde{Q}_1 = Q_1+\Omega_2/(2\alpha)+Q_2,\ \tilde{P}_1 = (P_1+P_2)/2, \quad (6.20)$$

$$\tilde{Q}_2 = Q_1+\Omega_2/(2\alpha)-Q_2,\ \tilde{P}_2 = (P_1-P_2)/2. \quad (6.21)$$

   The separated Hamilton–Jacobi equation is that of SK,

$$\lambda = 0:\ H_{\mathrm{SK}}-E = f(\tilde{Q}_1,\tilde{P}_1)+f(\tilde{Q}_2,\tilde{P}_2)-E = 0, \quad (6.22)$$

$$f(q,p) = p^2+\frac{1}{12}q^3-4\omega_2^2 q. \quad (6.23)$$

2. One then writes the two invariants H,K of the other case (KK) in these new variables $\tilde{Q}_1,\tilde{Q}_2,\tilde{P}_1,\tilde{P}_2$. This requires applying the canonical transformation (6.19),

and the trick[1] is to take this canonical transformation for $\lambda = 0$. More precisely, the canonical transformation between q_j, p_j and $\tilde{Q}_1, \tilde{Q}_2, \tilde{P}_1, \tilde{P}_2$ is defined as [372, 413]

$$\begin{cases} q_1 = -6\left(\dfrac{\tilde{P}_1 - \tilde{P}_2}{\tilde{Q}_1 - \tilde{Q}_2}\right)^2 - \dfrac{\tilde{Q}_1 + \tilde{Q}_2}{2}, \\ q_2^2 = 24\dfrac{f(\tilde{Q}_1, \tilde{P}_1) - f(\tilde{Q}_2, \tilde{P}_2)}{\tilde{Q}_1 - \tilde{Q}_2}, \end{cases} \tag{6.24}$$

and the two invariants of the KK case take the form

$$\begin{cases} H_{\mathrm{KK}} = f(\tilde{Q}_1, \tilde{P}_1) + f(\tilde{Q}_2, \tilde{P}_2) - \dfrac{\lambda^2}{96} \dfrac{\tilde{Q}_1 - \tilde{Q}_2}{f(\tilde{Q}_1, \tilde{P}_1) - f(\tilde{Q}_2, \tilde{P}_2)}, \\ K_{\mathrm{KK}} = \dfrac{\lambda^2}{6}(\tilde{Q}_1 + \tilde{Q}_2) + 2(f(\tilde{Q}_1, \tilde{P}_1) - f(\tilde{Q}_2, \tilde{P}_2))^2 \\ \qquad + \dfrac{\lambda^4}{24^2}\left(\dfrac{\tilde{Q}_1 - \tilde{Q}_2}{f(\tilde{Q}_1, \tilde{P}_1) - f(\tilde{Q}_2, \tilde{P}_2)}\right)^2. \end{cases} \tag{6.25}$$

Therefore, the Hamilton–Jacobi equation $H_{\mathrm{KK}} - E = 0$ allows the introduction of a separating constant [413],

$$(f(\tilde{Q}_1, \tilde{P}_1) - E)^2 - \frac{\lambda^2}{4}\tilde{Q}_1 = (f(\tilde{Q}_2, \tilde{P}_2) - E)^2 - \frac{\lambda^2}{4}\tilde{Q}_2 = \text{const}, \tag{6.26}$$

the value of which is given by an elimination with the equation for K,

$$(f(\tilde{Q}_j, \tilde{P}_j) - E)^2 - \frac{\lambda^2}{4}\tilde{Q}_j + K = 0, \ j = 1, 2. \tag{6.27}$$

Since the Hamilton–Jacobi equations of the KK and SK cases are exchanged under the canonical transformation, the Hamilton–Jacobi equation of the SK case is *ipso facto* separated, by the same separating variables.

3. Next, one performs the explicit integration in the KK case. The equations of the motion for H_{KK}, (6.25), are

$$\tilde{Q}_1' = \frac{\partial H_{\mathrm{KK}}}{\partial \tilde{P}_1} = \tilde{P}_1\left(2 - \lambda^2 \frac{\tilde{Q}_1 - \tilde{Q}_2}{[f(\tilde{Q}_1, \tilde{P}_1) - f(\tilde{Q}_2, \tilde{P}_2)]^2}\right), \tag{6.28}$$

$$\tilde{Q}_2' = \frac{\partial H_{\mathrm{KK}}}{\partial \tilde{P}_2} = \tilde{P}_2\left(2 - \lambda^2 \frac{\tilde{Q}_2 - \tilde{Q}_1}{[f(\tilde{Q}_1, \tilde{P}_1) - f(\tilde{Q}_2, \tilde{P}_2)]^2}\right), \tag{6.29}$$

and the elimination of the momenta $(\tilde{P}_1, \tilde{P}_2)$ yields the system [413]

---

[1] This is not really a trick. Indeed, the present SK and KK Hamiltonian systems are the reductions of two sets of coupled KdV-like PDEs [21], and the transformation from (q_j, p_j) to $(\tilde{Q}_1, \tilde{Q}_2, \tilde{P}_1, \tilde{P}_2)$ is the reduction of some precise Bäcklund transformation.

$$(s_1 - s_2)s_1' = \sqrt{P(s_1)},\ (s_2 - s_1)s_2' = \sqrt{P(s_2)}, \tag{6.30}$$

$$P(s) = \left(s^2 - \frac{K}{\lambda^2}\right)^3 - \frac{\omega_1\omega_2}{3}\left(s^2 - \frac{K}{\lambda^2}\right) - \frac{\lambda}{9}s + \frac{2E}{9}, \tag{6.31}$$

with the notation

$$\tilde{Q}_1 = -3s_1^2 + \frac{12K}{\lambda^2},\ \tilde{Q}_2 = -3s_2^2 + \frac{12K}{\lambda^2}. \tag{6.32}$$

Therefore the general solution (q_1, q_2^2) of HH3 in the KK case is a meromorphic function of time, defined by a genus two hyperelliptic system. Its explicit expression is, in the original variables,

$$q_1 = -\frac{3}{2}\left[\left(\frac{s_1' + s_2'}{s_1 + s_2}\right)^2 - (s_1^2 + s_2^2) - \frac{2K}{\lambda^2}\right], \tag{6.33}$$

$$q_2^{-2} = \frac{s_1 + s_2}{2\lambda}. \tag{6.34}$$

4. Finally, the general solution in the SK case is obtained by carrying the general solution from the KK case with the canonical transformation (6.19). The result is

$$Q_1 = -3\left[(s_1' + s_2') + (s_1^2 + s_2^2 + s_1 s_2) - \frac{K}{\lambda^2}\right], \tag{6.35}$$

$$Q_2^2 = 18(s_1 + s_2)\left[(s_1 + s_2)\left(s_1^2 + s_2^2 - \frac{3K}{2\lambda^2}\right) - (s_1 s_1' + s_2 s_2')\right]. \tag{6.36}$$

6.2.3 Direct Integration

In order to perform the integration of the equations of Hamilton in the spirit of the theory of the integration advocated by Painlevé, it is sufficient to integrate the equivalent fourth order ODE (2.81) in the three cases isolated by the Painlevé test.

All fourth order first degree ODEs with the Painlevé property in the class

$$u^{(4)} = P(u''', u'', u', u; x), \tag{6.37}$$

in which P is polynomial in $u''', \dots, u$ with coefficients analytic in x, have been enumerated [59, 46, 104] and, at least in the autonomous case, integrated (Appendix A.3.5). In the particular case when the leading behavior of u is made of movable double poles as in (2.81), it comprises five ODEs, three nonautonomous (denoted F-I, F-V, F-VI in [104]) and two autonomous (denoted F-III, F-IV). Three out of these five equations can be readily identified with the three cases of (2.81), these are F-III, F-IV and the autonomous restriction de F-V (here denoted a-F-V). Their general solution, established in these classifications, is meromorphic and expressed with hyperelliptic functions of genus two [104] defined by the *hyperelliptic system*,

$$\begin{cases} (s_1 - s_2)s_1' = \sqrt{P(s_1)}, \ (s_2 - s_1)s_2' = \sqrt{P(s_2)}, \\ P(s) = \sum_{j=0}^{6} c_j s^j. \end{cases} \tag{6.38}$$

The canonical form of these three equations is (the fixed constants are denoted by Greek letters, the movable ones by K_1, K_2 or A, B)

$$\text{F-III} \begin{cases} u'''' = 15uu'' + \frac{45}{4}u'^2 - 15u^3 + \alpha u + \beta, \\ u = \left(\frac{s_1' + s_2'}{s_1 + s_2}\right)^2 - (s_1^2 + s_2^2) - 2A, \\ P(s) = (s^2 + A)^3 - \frac{\alpha}{3}(s^2 + A) + Bs + \frac{\beta}{3}, \end{cases} \tag{6.39}$$

$$\text{F-IV} \begin{cases} u'''' = 30uu'' - 60u^3 + \alpha u + \beta, \\ u = \frac{1}{2}\left(s_1' + s_2' + s_1^2 + s_1 s_2 + s_2^2 + A\right), \\ P(s) = \text{same as F-III}, \end{cases} \tag{6.40}$$

$$\text{a-F-V} \begin{cases} u'''' = 20uu'' + 10u'^2 - 40u^3 + \alpha u + \kappa x + \beta, \ \kappa = 0, \\ u = \frac{1}{4}(s_1 + s_2), \\ P(s) = s^5 - 2\alpha s^3 + 8\beta s^2 + 32K_1 s + 16K_2. \end{cases} \tag{6.41}$$

Therefore the correspondence is

$$\begin{aligned} &\begin{matrix}\text{KdV5}\\ \text{a-F-V}\end{matrix} \begin{cases} q_1 = -\frac{u}{\alpha} - \frac{\omega_1 + 4\omega_2}{20\alpha}, \ \kappa_{\text{V}} = 0, \ \alpha_{\text{V}} = \frac{3\omega_1^2 - 16\omega_1\omega_2 + 48\omega_2^2}{10}, \\ \beta_{\text{V}} = 2E + \frac{(\omega_1 + 4\omega_2)(\omega_1^2 - 12\omega_1\omega_2 + 16\omega_2^2)}{200\alpha^2}, \end{cases} \\ &\begin{matrix}\text{SK}\\ \text{F-IV}\end{matrix} \begin{cases} q_1 = -\frac{3u}{2\alpha} - \frac{\omega_1}{2\alpha}, \\ \alpha_{\text{IV}} = \omega_1^2, \ \beta_{\text{IV}} = \frac{E}{3} - \frac{\omega_1^3}{36\alpha^2}, \ K_{1,\text{IV}} = -\frac{c_3}{36}, \ K_{2,\text{IV}} = \frac{K}{64}, \end{cases} \\ &\begin{matrix}\text{KK}\\ \text{F-III}\end{matrix} \begin{cases} q_1 = -\frac{\alpha u}{2} - \frac{\omega_2}{2\alpha}, \ \alpha_{\text{III}} = 16\omega_2^2, \ \beta_{\text{III}} = 8E - \frac{32}{3\alpha^2}\omega_2^3. \end{cases} \end{aligned} \tag{6.42}$$

The various constants involved (c_3, E, K) are either movable or fixed depending on the considered system. In the equations of motion (2.70)–(2.71), the constant c_3 is fixed and the constants E, K are movable. In the fourth order equation (2.81), the constant E is fixed, the constants c_3, K are movable. This gives an additional insight on the HH system. Indeed, the constant c_3 is a first integral of the fourth order equation (2.81), therefore setting $c_3 = 0$ would prevent finding the general solution of (2.81), and the inverse square terms in the HH Hamiltonian (2.69) *must* be present. In fact, the particular solution for $c_3 = 0$ is much easier to obtain. For the SK case it already appears in the thesis of Chazy [59],

$$\begin{cases} -\dfrac{\alpha}{6}(Q_1 \pm Q_2) - \dfrac{\omega_2}{12} = \wp(t-t_1, g_2, g_{3,\pm}) = \wp_\pm, \\ g_2 = \dfrac{\omega_2^2}{12},\ g_{3,\pm} = -\dfrac{\alpha^2}{18}\left(E \pm \dfrac{\sqrt{K}}{3} - \dfrac{\omega_2^3}{12\alpha^2}\right), \end{cases} \tag{6.43}$$

the corresponding solution in the KK case being [372]

$$\begin{cases} \alpha q_1 + \dfrac{\omega_2}{2} = \dfrac{3}{16}\left[-4(\wp_+ + \wp_-) + (\log(\wp_+ - \wp_-))'^2\right], \\ q_2^{-2} = \alpha^2 \sqrt{K}(\wp_+ - \wp_-), \end{cases} \tag{6.44}$$

with the same values for $g_2, g_{3,\pm}$.

6.3 Quartic Hénon–Heiles Hamiltonians

We now perform the same study, however much more briefly, for the "quartic" HH Hamiltonians.

These quartic HH Hamiltonians all have the necessary form (2.78), in which the constants $A, B, C, \alpha, \beta, \gamma, \Omega_1$ and Ω_2 are further selected by the "usual" Painlevé test [369, 179], and the equations of the motion (2.79)–(2.80) pass the Painlevé test in only four cases with the result

$$\begin{cases} A:B:C = 1:2:1,\ \gamma = 0, \\ A:B:C = 1:6:1,\ \gamma = 0,\ \Omega_1 = \Omega_2, \\ A:B:C = 1:6:8,\ \alpha = 0,\ \Omega_1 = 4\Omega_2, \\ A:B:C = 1:12:16,\ \gamma = 0,\ \Omega_1 = 4\Omega_2. \end{cases} \tag{6.45}$$

6.3.1 Second Invariants

For each of these four cases there exists a second constant of motion K [204, 21, 22] in involution with the Hamiltonian,

$$\begin{aligned} 1{:}2{:}1 : K = {} & (Q_2P_1 - Q_1P_2)^2 + Q_2^2\frac{\alpha}{Q_1^2} + Q_1^2\frac{\beta}{Q_2^2} - \frac{\Omega_1 - \Omega_2}{2} \times \\ & \left(P_1^2 - P_2^2 + Q_1^4 - Q_2^4 + \Omega_1 Q_1^2 - \Omega_2 Q_2^2 + \frac{\alpha}{Q_1^2} - \frac{\beta}{Q_2^2}\right), \\ & A = \frac{1}{2}, \end{aligned} \tag{6.46}$$

$$\begin{aligned} 1{:}6{:}1 : K = {} & \left(P_1P_2 + Q_1Q_2\left(-\frac{Q_1^2 + Q_2^2}{8} + \Omega_1\right)\right)^2 \\ & -P_2^2\frac{\kappa_1^2}{Q_1^2} - P_1^2\frac{\kappa_2^2}{Q_2^2} + \frac{1}{4}\left(\kappa_1^2 Q_2^2 + \kappa_2^2 Q_1^2\right) + \frac{\kappa_1^2\kappa_2^2}{Q_1^2 Q_2^2}, \\ & \alpha = -\kappa_1^2,\ \beta = -\kappa_2^2,\ A = -\frac{1}{32}, \end{aligned} \tag{6.47}$$

$$
\begin{aligned}
1{:}6{:}8 : K = & \left(P_2^2 - \frac{Q_2^2}{16}(2Q_2^2 + 4Q_1^2 + \Omega_2) + \frac{\beta}{Q_2^2}\right)^2 \\
& -\frac{1}{4}Q_2^2(Q_2P_1 - 2Q_1P_2)^2 \\
& +\gamma\Big(-2\gamma Q_2^2 - 4Q_2P_1P_2 + \frac{1}{2}Q_1Q_2^4 + Q_1^3Q_2^2 + 4Q_1P_2^2 \\
& \qquad -4\Omega_2Q_1Q_2^2 + 4Q_1\frac{\beta}{Q_2^2}\Big),\ A = -\frac{1}{16},
\end{aligned} \tag{6.48}
$$

$$
\begin{aligned}
1{:}12{:}16 : K = & \left(8(Q_2P_1 - Q_1P_2)P_2 - Q_1Q_2^4 - 2Q_1^3Q_2^2 + 2\Omega_1Q_1Q_2^2 - 8Q_1\frac{\beta}{Q_2^2}\right)^2 \\
& +\frac{32\alpha}{5}\left(Q_2^4 + 10\frac{Q_2^2P_2^2}{Q_1^2}\right),\ A = -\frac{1}{32},
\end{aligned} \tag{6.49}
$$

therefore the Liouville integrability is established.

6.3.2 Separation of Variables

The question of finding the separating variables is much more difficult. There exist powerful methods to achieve this [393, 409], but they are too technical to be presented here, moreover their application to the full HH4 cases has, to the best of our knowledge, not yet been performed. We briefly give the results achieved to date.

6.3.2.1 Case 1:2:1 (Manakov System)

Since the two constants of the motion H and K are quadratic in the momenta, one can apply the method of Stäckel [396] to find the separating variables. The canonical transformation to elliptic coordinates [434],

$$
\begin{cases}
q_j^2 = (-1)^j \dfrac{(s_1+\omega_j)(s_2+\omega_j)}{\omega_1-\omega_2}, & j = 1,2 \\[2ex]
p_j = 2q_j \dfrac{\omega_{3-j}(r_2-r_1) - s_1r_1 + s_2r_2}{s_1-s_2}, & j = 1,2
\end{cases} \tag{6.50}
$$

transforms the invariants to

$$\begin{cases} H = -\dfrac{g(s_1,r_1) - g(s_2,r_2)}{s_1 - s_2}, \\ K = -g(s_1,r_1) - g(s_2,r_2) + (s_1 + s_2 + \omega_1 + \omega_2)\dfrac{g(s_1,r_1) - g(s_2,r_2)}{s_1 - s_2} \\ \quad -\dfrac{\alpha+\beta}{2}, \\ g(s,r) = 2(s+\omega_1)(s+\omega_2)r^2 - \dfrac{s^3}{2} - \dfrac{\omega_1+\omega_2}{2}s^2 - \dfrac{\omega_1\omega_2}{2}s \\ \quad -\dfrac{\omega_1-\omega_2}{2}\left(\dfrac{\alpha}{s+\omega_1} - \dfrac{\beta}{s+\omega_2}\right), \end{cases} \tag{6.51}$$

therefore the Hamilton–Jacobi equation $H - E = 0$ allows the introduction of a separating constant, conveniently chosen equal to the value K of the second invariant (6.46)

$$-2g(s_j,r_j) - E(2s_j + \omega_1 + \omega_2) - \frac{\alpha+\beta}{2} = K,\ j = 1,2. \tag{6.52}$$

The Hamilton equations in the new coordinates are identical to the hyperelliptic system (6.38), with

$$\begin{aligned} P(s) = {} & s(s+\omega_1)^2(s+\omega_2)^2 - \alpha(s+\omega_2)^2 - \beta(s+\omega_1)^2 \\ & -(s+\omega_1)(s+\omega_2)\left[E(2s+\omega_1+\omega_2) - K\right]. \end{aligned} \tag{6.53}$$

6.3.2.2 Cases 1:6:1 and 1:6:8

There exists a canonical transformation [21] between the two Hamiltonians 1:6:1 and 1:6:8 therefore it is sufficient to solve either case.

We need to use a different notation for the two sets of coordinates,

$$1:6:1 \begin{cases} H = \dfrac{1}{2}(P_1^2 + P_2^2) + \dfrac{\omega_1}{2}(Q_1^2 + Q_2^2) - \dfrac{1}{32}(Q_1^4 + 6Q_1^2Q_2^2 + Q_2^4) \\ \quad -\dfrac{1}{2}\left(\dfrac{\kappa_1^2}{Q_1^2} + \dfrac{\kappa_2^2}{Q_2^2}\right) = E, \\ K = \left(P_1P_2 + Q_1Q_2\left(-\dfrac{Q_1^2+Q_2^2}{8} + \omega_1\right)\right)^2 \\ \quad -P_2^2\dfrac{\kappa_1^2}{Q_1^2} - P_1^2\dfrac{\kappa_2^2}{Q_2^2} + \dfrac{1}{4}\left(\kappa_1^2Q_2^2 + \kappa_2^2Q_1^2\right) + \dfrac{\kappa_1^2\kappa_2^2}{Q_1^2Q_2^2}, \end{cases} \tag{6.54}$$

and

$$
1:6:8 \begin{cases} H = \frac{1}{2}(p_1^2+p_2^2) + \frac{\omega_2}{2}(4q_1^2+q_2^2) - \frac{1}{16}(8q_1^4+6q_1^2q_2^2+q_2^4) \\ \quad -\gamma q_1 + \frac{\beta}{2q_2^2} = E, \\ K = \left(p_2^2 - \frac{q_2^2}{16}(2q_2^2+4q_1^2+\omega_2) + \frac{\beta}{q_2^2}\right)^2 \\ \quad -\frac{1}{4}q_2^2(q_2p_1-2q_1p_2)^2 \\ \quad +\gamma\left(-4q_2p_1p_2 + \frac{1}{2}q_1q_2^4 + q_1^3q_2^2 + 4q_1p_2^2\right. \\ \qquad \left. -2\gamma q_2^2 - 4\omega_2 q_1 q_2^2 + 4q_1\frac{\beta}{q_2^2}\right). \end{cases} \tag{6.55}
$$

This transformation conserves the two invariants E and K, it reads

$$
\begin{cases} q_1 = -\frac{P_1}{Q_1} + \frac{P_2}{Q_2} + \frac{\kappa_1}{Q_1^2} + \frac{\kappa_2}{Q_2^2},\ q_2 = 2\sqrt{\Omega} \\ p_1 = \left(\frac{P_1}{Q_1}\right)^2 - \left(\frac{P_2}{Q_2}\right)^2 + \frac{Q_1^2-Q_2^2}{4} - 2\kappa_1\frac{P_1}{Q_1^3} - 2\kappa_2\frac{P_2}{Q_2^3} + \frac{\kappa_1^2}{Q_1^4} - \frac{\kappa_2^2}{Q_2^4}, \\ p_2 = \sqrt{\Omega}\left(-\frac{P_1}{Q_1} - \frac{P_2}{Q_2} + \frac{\kappa_1}{Q_1^2} - \frac{\kappa_2}{Q_2^2}\right) + \frac{\kappa_1-\kappa_2}{2\sqrt{\Omega}}, \\ \Omega = 2\frac{P_1P_2}{Q_1Q_2} - \frac{Q_1^2+Q_2^2}{4} + 2\frac{\kappa_2}{Q_2^2}\frac{P_1}{Q_1} - 2\frac{\kappa_1}{Q_1^2}\frac{P_2}{Q_2} - 2\frac{\kappa_1}{Q_1^2}\frac{\kappa_2}{Q_2^2} + 2\omega_1, \\ \omega_2 = \omega_1,\ \gamma = \frac{\kappa_1+\kappa_2}{2},\ \beta = -(\kappa_1-\kappa_2)^2, \end{cases} \tag{6.56}
$$

and its inverse is

$$
\begin{cases} Q_1 = \sqrt{\Delta_-},\ Q_2 = \sqrt{\Delta_+}, \\ P_1 = \sqrt{\Delta_-}\left(-\frac{p_2}{q_2} - \frac{q_1}{2} - \frac{\kappa_1-\kappa_2}{q_2^2}\right) + \frac{\kappa_1}{\sqrt{\Delta_-}}, \\ P_2 = \sqrt{\Delta_+}\left(-\frac{p_2}{q_2} + \frac{q_1}{2} - \frac{\kappa_1-\kappa_2}{q_2^2}\right) - \frac{\kappa_2}{\sqrt{\Delta_+}}, \\ \Delta_\pm = \mp 2p_1 - q_1^2 - \frac{q_2^2}{2} + 4\frac{p_2^2}{q_2^2} \pm q_1\frac{p_2}{q_2} \\ \qquad + 8(\kappa_1-\kappa_2)\frac{p_2}{q_2^3} \pm 4(\kappa_1-\kappa_2)\frac{q_1}{q_2^2} + 4\frac{(\kappa_1-\kappa_2)^2}{q_2^4} + 4\omega_2. \end{cases} \tag{6.57}
$$

At the present time, separating variables have only been found in the particular case $\gamma = 0$ of 1:6:8, which corresponds to $\kappa_1^2 = \kappa_2^2$ of 1:6:1. The separation of variables is achieved [411] in a manner quite similar to the two dual cases SK and KK of the cubic Hamiltonian HH3, done in Sect. 6.2.2.2.

1. The invariance of 1:6:1 under the exchange of the two particles suggests considering the canonical transformation

$$\begin{cases} \tilde{Q}_1 = \frac{1}{2}(Q_1+Q_2)^2, \ \tilde{Q}_2 = \frac{1}{2}(Q_1-Q_2)^2, \\ \tilde{P}_1 = \frac{1}{2}\frac{P_1+P_2}{Q_1+Q_2}, \ \tilde{P}_2 = \frac{1}{2}\frac{P_1-P_2}{Q_1-Q_2}, \end{cases} \tag{6.58}$$

which transforms the Hamiltonian $H_{1:6:1}$ into

$$\begin{aligned} H_{1:6:1} &= f(\tilde{Q}_1,\tilde{P}_1)+f(\tilde{Q}_2,\tilde{P}_2) \\ &\quad -\frac{\kappa_1^2}{\left(\sqrt{\tilde{Q}_1}+\sqrt{\tilde{Q}_2}\right)^2}-\frac{\kappa_2^2}{\left(\sqrt{\tilde{Q}_1}-\sqrt{\tilde{Q}_2}\right)^2}, \end{aligned} \tag{6.59}$$

$$f(q,p) = 2qp^2 - \frac{1}{16}q^2 + \frac{\omega_1}{2}q, \tag{6.60}$$

thus trivially separating the Hamilton–Jacobi equation $H_{1:6:1} - E = 0$ in the particular case $\kappa_1 = \kappa_2 = 0$,

$$\kappa_1 = \kappa_2 = 0: \ f(\tilde{Q}_j,\tilde{P}_j) - \frac{E}{2} + (-1)^j K = 0, \ j = 1,2 \tag{6.61}$$

in which the separating constant K is conveniently chosen equal to the second integral of the motion (6.54).

2. One then writes the two invariants H, K of the 1:6:8 case in these new variables $\tilde{Q}_1, \tilde{Q}_2, \tilde{P}_1, \tilde{P}_2$. This requires application of the canonical transformation (6.56), and the trick is again to take this canonical transformation for $\kappa_1 = \kappa_2 = 0$. More precisely, the canonical transformation between q_j, p_j and $\tilde{Q}_1, \tilde{Q}_2, \tilde{P}_1, \tilde{P}_2$ is defined as [414]

$$\begin{cases} \tilde{Q}_1 = 4\frac{p_2^2+z}{q_2^2} - q_1^2 - \frac{q_2^2}{2} + 4\omega_1, \\ \tilde{Q}_2 = 4\frac{p_2^2-z}{q_2^2} - q_1^2 - \frac{q_2^2}{2} + 4\omega_1, \\ \tilde{P}_1 = \frac{2q_1^3q_2^2 + q_1(q_2^4+8z) - 8\omega_1 q_1 q_2^2 - 8q_2p_1p_2 + 8q_1p_2^2}{16q_2(q_2p_1 - 2q_1p_2)}, \\ \tilde{P}_2 = \frac{2q_1^3q_2^2 + q_1(q_2^4-8z) - 8\omega_1 q_1 q_2^2 - 8q_2p_1p_2 + 8q_1p_2^2}{16q_2(q_2p_1 - 2q_1p_2)}, \\ 64z^2 = \left(2q_1^2q_2^2 + q_2^4 - 8\omega_1 q_2^2 - 8p_2^2\right)^2 - 64q_2^2(q_2p_1 - -2q_1p_2)^2. \end{cases} \tag{6.62}$$

The transformed Hamilton–Jacobi equation $H_{1:6:1} - E = 0$, which takes the form

$$\begin{aligned} & g(\tilde{Q}_1,\tilde{P}_1) - g(\tilde{Q}_2,\tilde{P}_2) \\ & -4\gamma\sqrt{\tilde{Q}_1\tilde{Q}_2}\frac{\tilde{P}_1-\tilde{P}_2}{\tilde{Q}_1-\tilde{Q}_2}\left(f(\tilde{Q}_1,\tilde{P}_1) - f(\tilde{Q}_2,\tilde{P}_2)\right) = 0, \\ & g(q,p) = (2f(q,p)-E)^2 + \frac{\delta}{8}q, \end{aligned} \tag{6.63}$$

is therefore separated in the particular case $\gamma=0$, and the value of the separating constant is given by the second integral of the motion (6.54),

$$\gamma=0,\ g(\tilde{Q}_j,\tilde{P}_j)-K=0,\ j=1,2. \tag{6.64}$$

The general solution in this particular case is

$$\begin{array}{l}1:6:8\\ \gamma=0\end{array}:\begin{cases} q_1=\dfrac{(s_2^2-C)s_1'+(s_1^2-C)s_2'}{(s_1+s_2)\sqrt{s_1^2-C}\sqrt{s_2^2-C}},\ C=\dfrac{K}{4\kappa_1^2}\\ q_2^2=\dfrac{E}{s_1+s_2},\\ P(s)=(s^2-C)\left[(s^2-C)^2+4c(s^2-C)+2(\kappa_1 s+E)\right]. \end{cases} \tag{6.65}$$

The expressions (6.65) cannot be written as rational functions of s_1,s_2,s_1',s_2' but are nevertheless meromorphic [138, 308]. The corresponding 1:6:1 solution is

$$\begin{array}{l}1:6:1\\ \kappa_1^2=\kappa_2^2\end{array}:\begin{cases} \dfrac{Q_1^2+Q_2^2}{4}=s_1^2+s_2^2+s_1s_2-(s_1'+s_2')-C,\\ \dfrac{Q_1^2-Q_2^2}{4}=\dfrac{(s_1^2-C)(s_2^2-C)-(s_2^2-C)s_1'-(s_1^2-C)s_2'}{\sqrt{s_1^2-C}\sqrt{s_2^2-C}}. \end{cases} \tag{6.66}$$

6.3.2.3 Case 1:12:16

Let us normalize the invariants as $(\alpha=-\kappa_1^2,\beta=-4\kappa_2^2)$

$$\begin{cases} H=\dfrac{1}{2}(p_1^2+p_2^2)+\dfrac{\omega}{8}(4q_1^2+q_2^2)-\dfrac{1}{32}(16q_1^4+12q_1^2q_2^2+q_2^4)\\ \qquad -\dfrac{1}{2}\left(\dfrac{\kappa_1^2}{q_1^2}+\dfrac{4\kappa_2^2}{q_2^2}\right)=E,\\ K=\left(8(q_2p_1-q_1p_2)p_2-q_1q_2^4-2q_1^3q_2^2+2\omega q_1q_2^2+32q_1\dfrac{\kappa_2^2}{q_2^2}\right)^2\\ \qquad -\dfrac{32\kappa_1^2}{5}\left(q_2^4+10\dfrac{q_2^2p_2^2}{q_1^2}\right). \end{cases} \tag{6.67}$$

The separating variables have been found only in the two particular cases $\alpha\beta=0$, these are [411]

$$\begin{cases} q_1^2 = (s_1+s_2)^2 + \alpha G_2/G_1 + 4\alpha G_2 G_3/\left(G_1^2(s_1-s_2)^2\right) \\ \quad + \sqrt{-\alpha} s_1 s_2 G_2 G_4/G_1^3 - \alpha\sqrt{-\alpha} G_2^2 G_5/\left(G_1^3(s_1^2-s_2^2)\right) \\ q_2^2 = -\dfrac{4K s_1^2 s_2^2}{G_1} \\ G_1 = -40\alpha(s_1+s_2) + K s_1 s_2 - 40\sqrt{-\alpha}(s_2^2 r_1 + s_1^2 r_2) \\ G_2 = 80 s_1 s_2 (s_1+s_2) \\ G_3 = -20\alpha(s_1^3+s_2^3) + K(s_1^2+s_2^2) + 40E s_1^2 s_2^2 (s_1+s_2) \\ \quad + 5 s_1^3 s_2^3 (s_1+s_2)(3s_1^2 - 2s_1 s_2 + 3s_2^2) \\ G_4 = K^2 s_1 s_2 (r_1 s_2 + s_1 r_2) - 80\sqrt{-\alpha} r_1 r_2 \left(K s_1 s_2 - 40(s_1 s_2)\right)(s_1+s_2) \\ G_5 = 20(s_1 r_1 - s_2 r_2)\left(K s_1 s_2 - 20\alpha(s_1+s_2)\right) \\ \quad + s_1 s_2 (r_1 s_2 - s_1 r_2)\left(K + 20(2E + s_1^2 s_2^2)(s_1+s_2)\right) \end{cases} \tag{6.68}$$

in which s_1, s_2 obey the canonical hyperelliptic system (6.38), with

$$\alpha\beta = 0: \ P(s) = s^6 - \omega s^3 + 2Es^2 + \frac{K}{20}s - \alpha - \frac{\beta}{4}. \tag{6.69}$$

The subcase $\alpha = 0$ is quite easy to integrate since both invariants H and $\sqrt{K}$ are quadratic in the momenta, a case where the method of Stäckel [396, 397] can be applied to build the separating variables. Indeed, the coordinates defined by the canonical transformation (6.68) are then parabolic,

$$\begin{cases} q_1 = s_1 + s_2, \ q_2^2 = -4 s_1 s_2, \\ p_1 = \dfrac{s_1 r_1 - s_2 r_2}{s_1 - s_2}, \ p_2 = q_2 \dfrac{r_1 - r_2}{2(s_1 - s_2)}, \end{cases} \tag{6.70}$$

the Hamiltonian becomes

$$\alpha = 0: \ H = \frac{f(s_1, r_1) - f(s_2, r_2)}{2(s_1 - s_2)}, \ f(s, r) = sr^2 - s^5 + \omega s^3 + \frac{\beta}{4s}, \tag{6.71}$$

therefore the Hamilton–Jacobi equation $H - E = 0$ allows one to introduce a separating constant

$$\alpha = 0, \ f(s_j, r_j) - 2s_j E - K = 0, \ j = 1, 2, \tag{6.72}$$

with K identical to the second constant of the motion. The Hamilton equations in the separating coordinates

$$\alpha = 0: \ s_j' = (-1)^{j+1} \frac{s_j r_j}{s_1 - s_2}, \ j = 1, 2, \tag{6.73}$$

then reduce to the canonical hyperelliptic system (6.38).

6.3.3 Painlevé Property

In the cubic case, the quickest proof of the Painlevé property is the identification of the fourth order ODE (2.81) in the three admissible cases with three classified ODEs, see Sect. 6.2.3.

The similar elimination of q_2 and $q_1'''^2$ between the three equations (2.78)–(2.80) yields a fourth order first degree ODE for q_1 [96]

$$\begin{aligned}
&-q_1'''' + 2\frac{q_1' q_1'''}{q_1} + \left(1 + 6\frac{A}{B}\right)\frac{q_1''^2}{q_1} - 2\frac{q_1'^2 q_1''}{q_1^2} + 8\left(6\frac{AC}{B} - B - C\right) q_1^2 q_1'' \\
&+ 4(B - 2C) q_1 q_1'^2 + 24C\left(4\frac{AC}{B} - B\right) q_1^5 \\
&+ \left[12\frac{A}{B}\omega_1 - 4\omega_2 + \left(1 + 12\frac{A}{B}\right)\frac{\mu}{q_1} - 4\left(1 + 3\frac{A}{B}\right)\frac{\alpha}{q_1^4}\right] q_1'' \\
&+ 6\frac{A}{B}\frac{\alpha^2}{q_1^7} + 20\frac{\alpha}{q_1^5} q_1'^2 - 12\frac{A}{B}\frac{\mu\alpha}{q_1^4} + 4\left(3\frac{A}{B}\omega_1 - \omega_2\right)\left(\mu - \frac{\alpha}{q_1^3}\right) - 2\mu\frac{q_1'^2}{q_1^2} \\
&+ 6\left(\frac{A}{B}\mu^2 + 2B\alpha - 8\frac{AC}{B}\alpha\right)\frac{1}{q_1} + \left(6\frac{A}{B}\omega_1^2 - 4\omega_1\omega_2 - 8BE\right) q_1 \\
&+ 48\frac{AC}{B}\mu q_1^2 + 4\left(12\frac{AC}{B} - B - 4C\right)\omega_1 q_1^3 = 0. \qquad (6.74)
\end{aligned}$$

Fourth order first degree ODEs with the Painlevé property have the necessary form [349]

$$u'''' = A_2(u'', u', u; x) u'''^2 + A_1(u'', u', u; x) u''' + A_0(u'', u', u; x), \qquad (6.75)$$

with A_j rational in u'', algebraic in u' and u, analytic in x, but the class of interest to us ($A_2 = 0$, which implies a polynomial dependence of A_1 and A_0 on u'') has only been studied for A_1 and A_0 polynomial in (u', u), so the results of Sect. 6.2.3 cannot be transposed to the quartic case[2].

It can nevertheless be proven [96] that all four quartic cases have the Painlevé property (general solution single valued in the complex time t). This involves building birational transformations to at least one of the classified fourth order ODEs mentioned in Sect. 6.2.3 [46, 59, 104], respectively: autonomous F-V for 1:2:1, autonomous F-VI for 1:6:1 and 1:6:8, F-III or F-IV for 1:12:16. We will not give here these developments, because they are rather lengthy and not yet optimal (some open questions are stated in [96]).

Only one case is quite easy to settle, this is the 1:2:1 case,

[2] In the 1:12:16 case with in addition $\alpha = 0$, this ODE (6.74) is the autonomous restriction of an equation considered by Kitaev [253, (5.9)] in the hierarchy of the second Painlevé equation, and reproduced by Cosgrove [106, (6.141)]. Therefore the ODE [106, (6.141)] accepts complementary terms, which represent the contribution of α in the 1:12:16 Hamiltonian.

$$H = \frac{1}{2}(p_1^2+p_2^2)+\frac{1}{2}(\omega_1 q_1^2+\omega_2 q_2^2)+\frac{1}{2}(q_1^4+2q_1^2q_2^2+q_2^4)$$
$$+\frac{1}{2}\left(\frac{\alpha}{q_1^2}+\frac{\beta}{q_2^2}\right)=E, \qquad (6.76)$$

let us process it here. The symmetries of the system suggest the change of variables

$$x=q_1^2+q_2^2,\ y=q_1^2-q_2^2. \qquad (6.77)$$

Linear combinations of the four equations (2.79), (2.80), (6.76), (6.46) generate the two ODEs,

$$\left\{\begin{array}{l}(\omega_1-\omega_2)(y''+8xy)+x'^2+2(\omega_1^2-\omega_2^2)y+4x^3\\ \quad +2(\omega_1+\omega_2)x^2+2(\omega_1-\omega_2)x-8Ex+4(\alpha+\beta+K)=0,\\ x''+2(\omega_1-\omega_2)y+6x^2+2(\omega_1+\omega_2)x-4E=0,\end{array}\right. \qquad (6.78)$$

and the elimination of y yields a fourth order ODE for the variable x in the polynomial class (6.37),

$$x''''+(20x+4\omega_1+4\omega_2)x''+10x'^2+40x^3$$
$$+8(\omega_1+\omega_2)(3x^2-E)+(16\omega_1\omega_2-E)x-8(\alpha+\beta+K)=0. \qquad (6.79)$$

This latter ODE can be identified to the autonomous F-V ODE (6.41), with

$$\left\{\begin{array}{l}u=x+\dfrac{\omega_1+\omega_2}{5},\ \kappa_{\rm V}=0,\\ \alpha_{\rm V}=16E+4(\omega_1-\omega_2)^2+\dfrac{4}{5}(\omega_1+\omega_2)^2,\\ \beta_{\rm V}=\dfrac{8}{25}(\omega_1+\omega_2)\left(2(\omega_1+\omega_2)^2-10\omega_1\omega_2-15E\right)-8(\alpha+\beta+K),\end{array}\right. \qquad (6.80)$$

therefore $q_1^2+q_2^2$ is a meromorphic function of time and, under the birational transformation between x and y,

$$\left\{\begin{array}{l}x=\left(\dfrac{1}{8}Y'^2+\dfrac{1}{16}(\omega_1-\omega_2)^2(\omega_1+\omega_2)+\dfrac{\alpha+\beta}{2}-\dfrac{K}{2}\right)\dfrac{1}{Y^2}\\ \qquad -\dfrac{Y''}{4Y}-\dfrac{\omega_1+\omega_2}{4},\\ y=Y-\dfrac{\omega_1-\omega_2}{2}=\dfrac{1}{\omega_1-\omega_2}\left(-\dfrac{1}{2}x''-3x^2-(\omega_1+\omega_2)x+2E\right),\end{array}\right. \qquad (6.81)$$

q_1^2 and q_2^2 are also meromorphic (but q_1 and q_2 are multivalued).

6.4 Final Picture for HH3 and HH4

Among the classical potentials $V(q_1, q_2)$ rational in q_1, q_2, only seven (three "cubic" ones, four "quartic") pass the Painlevé test. These Hénon–Heiles Hamiltonians have the following properties [97].

1. In the seven cases the general solution is a singlevalued function of the complex time t, whose only singularities are movable poles, explicitly expressed via a canonical hyperelliptic system with genus two.
2. Since each of the seven cases can be mapped to a fourth-order ODE which is complete in the Painlevé sense, it is impossible to add any term to the Hamiltonian without destroying the Painlevé property. The seven Hénon–Heiles Hamiltonians are *complete*.

The main open problems are to find the separating variables in three of the generic quartic cases.

Chapter 7
Discrete Nonlinear Equations

Abstract We first explain why the logistic map, which admits a continuum limit to the Riccati equation, is a bad discretization of the latter. We then present the difficulty to give an undisputed definition for the *discrete Painlevé property*, and introduce the three main methods of the discrete Painlevé test: the *singularity confinement method* [184], the *criterium of polynomial growth* [206], and the *perturbation of the continuum limit* [88]. Later, we recall the remark by Baxter and Potts that the addition formula of the Weierstrass function $\wp$ is an exact discretization of the Weierstrass equation. Finally, we introduce the two main methods able to build discrete Painlevé equations: (i) an analytic method which starts from the addition formula of the elliptic function, takes inspiration from the method of Painlevé and Gambier and produces a rather long, but incomplete, list of discrete Pn equations; (ii) a geometric method based on the theory of rational surfaces, which builds *ex abrupto* the master discrete Painlevé equation $\mathrm{e}-\mathrm{P6}$, whose coefficients have an elliptic dependence on the independent variable. The main properties of all these $\mathrm{d}-\mathrm{Pn}$ are summarized. This chapter also includes discrete Ermakov–Pinney equations and discrete nonlinear Schrödinger equations.

7.1 Generalities

When the equation for $u(x)$ is no more differential but involves a finite number of values of x, it is called a *discrete equation*. Let us start with two examples.

1. The *logistic map*

$$u_n = a u_{n-1}(1-u_{n-1}), \tag{7.1}$$

   in which u_n is short for $u(x-x_0), x=nh$.
2. the three-point mapping [184]

$$u_{n+1}+u_n+u_{n-1}-\frac{\alpha n+\beta}{u_n}=0. \tag{7.2}$$

In both examples, the points x_n are arithmetically consecutive (equispaced on a line), and one also uses the name *finite-difference equation.*

It is convenient to introduce the shorthand notation

$$u=u_n,\ \overline{u}=u_{n+1},\ \underline{u}=u_{n-1},\ \overline{\overline{u}}=u_{n+2},\ \dots \tag{7.3}$$

The logistic map (7.1) was introduced in 1845 by Verhulst to describe the evolution of a population living in a closed domain. It displays a chaotic behavior [405, 139] but, when $h\to 0$, it admits a *continuum limit* to a linearizable first order ODE. To find this limit, one must expand the u_k's as Taylor series of h up to order one only (indeed, with two discrete points, one cannot generate a second order derivative, so the continuum limit, if it exists, is necessarily a first order ODE)

$$u_n=u(x),\ u_{n-1}=u(x-h)=u(x)-hu'+\mathcal{O}(h^2), \tag{7.4}$$

leading to

$$a(u-hu')(1-u+hu')-u+\text{ h.o.t.}=0. \tag{7.5}$$

Both $u(x)$ and a must have a dependence on h for the limit to be a first order ODE. Defining the scaling transformation

$$u(x)=h^bU(x), \tag{7.6}$$

(7.5) becomes

$$U'+h^{b-1}U^2-2h^bUU'+h^{b+1}U'^2+\frac{1-a}{ah}U+\text{ h.o.t.}=0, \tag{7.7}$$

and one requires the limit to exist (this will request scaling a as well) and to be nonlinear (this determines the exponent b from the coefficient of U^2). This scaling on both the field u_n and the parameter a is called the *double scaling limit.* The final result is

$$b=1,\ a=1+hA,\ U'+U^2-AU=0. \tag{7.8}$$

Therefore the chaotic logistic map admits as its continuum limit a linearizable first order nonlinear equation, the Riccati equation, which possesses the (continuous) Painlevé property.

The second example (7.2) apparently possesses no continuum limit, but this equation is not chaotic and has some regularity properties, to be presented in Sect. 7.3.1.

If one now considers the inverse path (going from continuous to discrete), it is natural to state that the logistic map is a bad *discretization* of the Riccati equation, in the sense that the linearizability is not preserved. Therefore, two basic questions arise:

1. How to define a "good" discrete equation?
2. How to distinguish between a "good" and a "bad" discrete equation?

Remark. The word "discretization" has two meanings, which should not be confused. For numerical analysts, given an N-th order continuous differential equation, some initial data and a required precision, the question is to perform a *local* integration, and the mean is to develop a discretization scheme able to achieve the required precision. The order of the discretization scheme (e.g. fourth order Runge–Kutta) has no relation with the order N of the continuous equation. On the contrary, the question of discretization formulated above is *global* in the sense that it must preserve some global property of the equation, such as: singlevaluedness or multivaluedness of the general solution, differential order (i.e. number of arbitrary constants in the general solution), and more generally integrability (in a broad sense) or nonintegrability.

The first of the above two questions is equivalent to defining a *discrete Painlevé property*, which in the case of discrete equations admitting a continuum limit must be an extrapolation of the (continuous) Painlevé property, but which must also be defined for discrete equations without continuum limit.

The second question is equivalent to devising a *discrete Painlevé test*, in order to generate necessary conditions for a discrete equation to possess the discrete Painlevé property.

We will mainly consider three types of algebraic discrete equations, in which the discrete points (in number $N+1$, $N \geq 1$) are contiguous on a one-dimensional or a two-dimensional lattice.

1. difference equations, in which the points $x_n = x_0 + nh$ are equispaced on a line,

$$\forall x\, \forall h\ :\ E(x,h,\{u(x+kh),\ k-k_0 = 0,\ldots,N\}) = 0, \tag{7.9}$$

algebraic in the values of the field variable, with coefficients analytic in x and the stepsize h,

2. q-difference equations, in which the points $x_n = x_0 q^n$ are equispaced on a circle,

$$\forall x\, \forall q\ :\ E(x,q,\{u(xq^k),\ k-k_0 = 0,\ldots,N\}) = 0, \tag{7.10}$$

algebraic in the values of the field variable, with coefficients analytic in x and the stepsize q,

3. e-difference equations [377], in which the dependence on n of the coefficients of the equation is elliptic.

In particular, we only give one example of a discrete PDE.

The definitions of *order* and *degree* given for continuous ODEs naturally extend to discrete equations. By definition, the order is one less than the number of contiguous points in the discrete equation, and the degree is the highest of the two polynomial degrees of the l.h.s. E of the equation in the two extreme discrete points ($u(x)$

and $u(x+Nh)$ for difference equations, $u(x)$ and $u(xq^N)$ for q-difference equations), where E is assumed polynomial.

Discrete equations were in particular studied by Laguerre [266], mainly as three-term ($N=2$) recurrence relations between coefficients of orthogonal polynomials. This long remained a mathematical subject [392, 155], which then extended to topological field theory [30, 234]. Finally, the discrete equation

$$E \equiv -(\overline{u}-2u+\underline{u})/h^2+2(\overline{u}+u+\underline{u})u+x=0, \tag{7.11}$$

already considered by the authors of last five references, was again encountered by statistical physicists in two-dimensional quantum gravity [41, 123, 191] who recognized it as a discrete analogue of the first Painlevé equation P1

$$E \equiv -u''+6u^2+x=0. \tag{7.12}$$

The same happened simultaneously with a discrete analogue of the second Painlevé equation P2 [353, 326] (in the particular case $\alpha = 0$[1])

$$E \equiv -(\overline{u}-2u+\underline{u})/h^2+(\overline{u}+\underline{u})u^2+xu+\alpha=0 \tag{7.13}$$
$$E \equiv -u''+2u^3+xu+\alpha=0. \tag{7.14}$$

7.2 Discrete Painlevé Property

As opposed to the continuous case, there is not yet an undisputed definition for the *discrete Painlevé property*. Two definitions have been proposed,

1. [88] There exists a neighborhood of $h=0$ (resp. $q=1$) at every point of which the general solution $x \to u(x,h)$ (resp. $x \to u(x,q)$) has no movable critical singularities.
2. [2, p. 902] The Nevanlinna[2] order of growth of the solutions at infinity is finite, and all series representing the general solution should contain no digamma functions[3],

but none is satisfactory.

Indeed, the first one says nothing about discrete equations without continuum limit such as (7.2), and the second one is not applicable to the continuous P6 equation, whose Nevanlinna order of growth is undefined.

[1] In order to find α nonzero, physical models must get rid of the parity assumption on u, and this creates technical complications.

[2] Nevanlinna theory is presented shortly in Appendix D.

[3] The digamma functions, according to these authors, are in the discrete world the analogues of the movable logarithms in the continuous world.

Despite the lack of consensus on this definition, a *discrete Painlevé test* has been developed to generate necessary conditions for the above properties. This is the subject of the next three sections.

Of exceptional importance at this point is the *singularity confinement method* [184], which tests with great efficiency a property not yet rigorously defined, but which for sure will be an important part of the good definition of the discrete Painlevé property.

The discrete PP is invariant under the discrete analogue of (A.5), which is the group of nonlocal discrete birational transformations

$$\begin{aligned} &u = r(x, h \text{ or } q, U, \overline{U}, \underline{U}, \ldots), \\ &U = R(X, H \text{ or } Q, u, \overline{u}, \underline{u}, \ldots),\ X = \xi(x, h \text{ or } q),\ H = \eta(h),\ Q = \kappa(q), \end{aligned} \tag{7.15}$$

(r and R rational in $U, \overline{U}, \underline{U}, \ldots, u, \overline{u}, \underline{u}, \ldots$, analytic in x and the stepsize, ξ, η, κ analytic). There exist two discrete analogues of the homographic subgroup (1.43), and both may be useful to establish the discrete equivalent of the classification of Gambier. The first one is the group of transformations (7.15) which in the continuum limit reduce to the homographic transformations (1.43), where (a, b, c, d, ξ) are arbitrary analytic functions of x and of the stepsize. The second one is the group of local homographic transformations (r and R homographic in U and u, independent of $\overline{U}, \underline{U}, \ldots, \overline{u}, \underline{u}, \ldots$, analytic in x and the stepsize, ξ, η, κ analytic).

7.3 Discrete Painlevé Test

We now present the three methods which can presently yield necessary conditions for the discrete Painlevé property.

7.3.1 *Method of Singularity Confinement*

Devised by Grammaticos et al. [184], this test relies on the principle that, if a singularity is of the polar type, then far enough away from its location it is impossible to feel it. More precisely, if the field u displays a pole at a point x_0 in the complex plane x

$$u(x) \sim u_0 \chi^p,\ \chi = x - x_0 \to 0,\ u_0 \neq 0,\ -p \in \mathcal{N}, \tag{7.16}$$

it is generically regular at every point $x_0 + x_1$ in which x_1 is not infinitesimal

$$\forall x_1,\ |x_1| >> 0\ :\ u(x_0 + x_1) \neq \infty. \tag{7.17}$$

When u obeys a discrete equation of order N, the implementation of this "condition for confinement" consists in requiring the property (7.17) for N consecutive iterates, a condition which generically guarantees the same property for the next iterates; the polar behavior is then felt only during a finite number of iterations.

More specifically, one first checks (confinement in the past) the existence of an initial condition made of N regular values $u(x-Nh),\dots,u(x-h)$ ensuring a singular value for $u(x) \sim u_0/\varepsilon$, in which ε symbolizes a vanishing quantity and u_0 has a nonzero finite value in the limit $\varepsilon \to 0$. One computes the next iterates, which may be singular or not, and one requires the existence of N regular consecutive iterates, in order to also ensure confinement in the future.

Let us take as an example the second order first degree rational map [184]

$$u_{n+1}+u_n+u_{n-1}-a_n-\frac{b_n}{u_n}=0,\ b_n\neq 0, \tag{7.18}$$

which includes (7.2) as a particular case.

The only singularity to be tested for confinement is the pole $u_n=0$, therefore one assumes the initial condition

$$u_n = \text{arbitrary finite nonzero value},\ u_{n+1}=\varepsilon. \tag{7.19}$$

The successive iterates are

$$u_{n+2}=\frac{b_{n+1}}{\varepsilon}+a_{n+1}-u_n-\varepsilon, \tag{7.20}$$

$$u_{n+3}=-\frac{b_{n+1}}{\varepsilon}+a_{n+2}-a_{n+1}+u_n+\frac{b_{n+2}}{b_{n+1}}\varepsilon+\mathcal{O}(\varepsilon^2), \tag{7.21}$$

$$\begin{aligned} u_{n+4}=&\ a_{n+3}-a_{n+2}-\frac{b_{n+3}+b_{n+2}-b_{n+1}}{b_{n+1}}\varepsilon \\ &+\frac{(a_{n+1}-u_n)(b_{n+2}+b_{n+3})-a_{n+2}b_{n+3}}{b_{n+1}^2}\varepsilon^2+\mathcal{O}(\varepsilon^3), \end{aligned} \tag{7.22}$$

$$u_{n+5}=\frac{a_{n+3}-a_{n+2}+\mathcal{O}(\varepsilon)}{(a_{n+3}-a_{n+2})\varepsilon+\mathcal{O}(\varepsilon^2)}, \tag{7.23}$$

so u_{n+2} and u_{n+3} are singular, u_{n+4} is regular, and one requires u_{n+5} to be regular at $\varepsilon=0$. A first necessary condition is

$$\forall n:\ a_{n+3}-a_{n+2}=0. \tag{7.24}$$

The updated value

$$u_{n+5}=\frac{b_{n+4}-b_{n+3}-b_{n+2}+b_{n+1}+\mathcal{O}(\varepsilon)}{(-b_{n+3}-b_{n+2}+b_{n+1})\varepsilon+\mathcal{O}(\varepsilon^2)}, \tag{7.25}$$

yields the second necessary condition

$$\forall n:\ b_{n+4}-b_{n+3}-b_{n+2}+b_{n+1}=0, \tag{7.26}$$

and the new updated value

$$u_{n+5} = \frac{b_{n+4} + \mathcal{O}(\varepsilon)}{ab_{n+3} + (u_n - a)b_{n+1} + \mathcal{O}(\varepsilon)}, \tag{7.27}$$

is generically regular. Moreover, the dependence of the fifth iterate u_{n+5} on the initial condition (u_n, ε) has not been lost, so the map emerges unaltered from the singularity. This ends the method of singularity confinement.

The two necessary conditions (7.24) and (7.26) integrate as [184]

$$a_n = a,\ b_n = \alpha n + \beta + \gamma(-1)^n, \tag{7.28}$$

in which $(a, \alpha, \beta, \gamma)$ are arbitrary constants, therefore the equation selected by the test has the general form

$$u_{n+1} + u_n + u_{n-1} - a - \frac{\alpha n + \beta + \gamma(-1)^n}{u_n} = 0. \tag{7.29}$$

It will be shown in Sect. 7.5 that (7.29) has the discrete Painlevé property. Let us only investigate here its continuum limit.

In front of the $(-1)^n$ term in (7.29), a question arises: how to avoid setting $\gamma = 0$ without any good reason, i.e. how to give a meaning to the continuum limit of $(-1)^n$ (or more generally to j^n in which j is a p-th root of unity)? The solution [182, p. 461] is to split the single field u_n into two coupled fields, by denoting differently u_n according to whether n is even or odd. Therefore, when γ is nonzero, the single equation (7.29) is represented as the system

$$\begin{cases} u_{2m} = v_{m-1/4},\ u_{2m+1} = w_{m-1/4}, \\ E_1 \equiv w_m + v_m + w_{m-1} - a - \dfrac{2\alpha m - \alpha/2 + \beta + \gamma}{v_m} = 0, \\ E_2 \equiv v_{m+1} + w_m + v_m - a - \dfrac{2\alpha m + \alpha/2 + \beta - \gamma}{w_m} = 0. \end{cases} \tag{7.30}$$

It is important to remark that the order (two here) is not changed, since the map now expresses (v_{m+1}, w_{m+1}) in terms of (v_m, w_m) instead of u_{n+1} in terms of (u_n, u_{n-1}).

For $\gamma = 0, a \neq 0$, the continuum limit of (7.29) to an ODE for $U(x)$ is computed by assuming

$$nh = \frac{x - b_0}{a_0},\ u(x) = c_0(1 + c_1 U(x)),\ c_0 c_1 \neq 0, \tag{7.31}$$

in which the constants a_0, b_0, c_0 can be arbitrarily chosen. An expansion up to second order derivatives leads to

$$\begin{aligned} &\frac{x - b_0}{c_0^2 c_1} \alpha (a_0 h)^{-3} \\ &+ (1/c_1)\left(-3 + c_0^{-2}\beta + c_0^{-1} a + (a c_0^{-1} - 6) c_1 U - 3 c_1^2 U^2\right)(a_0 h)^{-2} \\ &- (1 + c_1 U) U'' + \text{h.o.t.} = 0, \end{aligned} \tag{7.32}$$

and one then requires the limit equation to have second order and to be nonlinear. The result is the P1 equation (the coefficients of U'', U^2, x, which are nonzero numerical constants, are set to their usual values $-1, 6, 1$),

$$\begin{cases} -U'' + 6U^2 + x = 0,\ c_1 = -2a_0^2 h^2,\ a = 6c_0 + Ah^3, \\ \alpha = -2a_0^5 c_0^2 h^5 + \tilde{\alpha} h^6,\ \beta = -3c_0^2 - c_0 A h^3 + \tilde{\beta} h^4, \end{cases} \tag{7.33}$$

in which (a_0, c_0) can take any numerical value and the constants $(A, \tilde{\alpha}, \tilde{\beta})$ are regular when h goes to zero. These last three constants represent the arbitrariness of the initial parameters (a, α, β), which should be preserved, so they should not be set to zero.

For $\gamma a \neq 0$, the assumption for a continuum limit is governed by the parity invariance $(v_m, w_m, h, \alpha, \gamma) \to (w_m, v_m, -h, -\alpha, -\gamma)$ of the system (7.30) [182, p. 461],

$$mh = \frac{x - b_0}{a_0},\ \frac{v_m + w_m}{2} = c_0(1 + c_2 h^2 W(x)),\ \frac{v_m - w_m}{2} = c_0 c_1 h V(x), \tag{7.34}$$

it is sufficient to expand v_{m+1} and w_{m-1} to first order in h, and the result is the first order system,

$$\begin{cases} \lim\limits_{h \to 0} \dfrac{E_1 + E_2}{h^2} = V' - \dfrac{c_1}{a_0} V^2 + 4\dfrac{c_2}{a_0 c_1} W - \dfrac{2\tilde{\alpha}}{a_0^2 c_0^2 c_1} x = 0, \\ \lim\limits_{h \to 0} \dfrac{E_1 - E_2}{h^3} = W' - \dfrac{c_1^2}{c_2} VV' - \dfrac{2c_1}{a_0} VW + \dfrac{c_1 A}{a_0 c_0 c_2} V + \dfrac{2\tilde{\gamma} - \tilde{\alpha}}{2a_0 c_0^2 c_2} = 0, \\ a = 2c_0 + Ah^2,\ \alpha = \tilde{\alpha} h^3,\ \beta = c_0^2 + (\tilde{\beta} - c_0 A) h^2, \\ \gamma = \tilde{\gamma} h^3,\ b_0 = \dfrac{a_0 \tilde{\beta}}{2\tilde{\alpha}} + Bh, \end{cases} \tag{7.35}$$

in which (a_0, c_0, c_1) are arbitrary pure numbers, and $(A, \tilde{\alpha}, \tilde{\beta}, \tilde{\gamma})$ arbitrary constants regular at $h = 0$.

This system for (V, W, x) is readily identified to the birational transformation (5.296) between a P2 equation for $V(x)$ and a G34 equation for $W + ((4c_2 + a_0 c_1)V' + (4c_2 + a_0^2)(V^2 + x/2))/(4c_2)$, with $c_1^2 = a_0^2$. Therefore the continuum limit of the single equation (7.29) when $\gamma a \neq 0$ is the second Painlevé equation, while that of the system (7.30) is richer since it is the birational transformation between G34 and P2.

Finally, for $a = 0$, no continuum limit of either (7.30) or (7.29) seems to exist [182, p. 464].

7.3.2 Method of Polynomial Growth

Let us first define the *degree* of an N-th order rational map

$$u_{n+1} = R(u_n, u_{n-1}, \ldots, u_{n-N+1}), \tag{7.36}$$

as the polynomial degree of an equivalent first order N-dimensional map when written in homogeneous coordinates.

For instance, given the second order map (7.18), one first defines two coordinates (p_n, q_n), e.g.

$$\begin{pmatrix} p_n \\ q_n \end{pmatrix} = \begin{pmatrix} u_{n-1} \\ u_n \end{pmatrix}, \tag{7.37}$$

then three homogeneous coordinates (y_n, z_n, t_n),

$$\frac{p_n}{z_n} = \frac{q_n}{y_n} = \frac{1}{t_n}, \tag{7.38}$$

so as to convert the original map to the homogeneous polynomial map

$$\begin{pmatrix} y_{n+1} \\ z_{n+1} \\ t_{n+1} \end{pmatrix} = \begin{pmatrix} -y_n^2 - y_n z_n + a_n y_n t_n + b_n t_n^2 \\ y_n^2 \\ y_n t_n \end{pmatrix}. \tag{7.39}$$

Evidently, any common factor to $(y_{n+1}, z_{n+1}, t_{n+1})$ should be divided out. The degree d_n of the map is defined as the common global polynomial degree of y_n, z_n, t_n, $n \geq 0$, in a generic initial condition (y_0, z_0, t_0) (here $d_0 = 1, d_1 = 2$).

It has been conjectured [137, 418] that any rational map which is not chaotic should have a degree d_n with a growth at most polynomial in n, therefore satisfying the necessary condition

$$\lim_{n \to +\infty} \frac{\log d_n}{n} = 0. \tag{7.40}$$

Based on this conjecture, a quite powerful indicator of integrability of rational maps has been proposed [206] for testing whether the growth is at most polynomial.

With the map (7.18), one checks that the growth is polynomial.

A much more interesting map is [206]

$$u_{n+1} + u_{n-1} - u_n - \frac{a}{u_n^2} = 0, \ a \neq 0, \tag{7.41}$$

equivalent to the homogeneous polynomial map

$$\begin{pmatrix} y_{n+1} \\ z_{n+1} \\ t_{n+1} \end{pmatrix} = \begin{pmatrix} y_n^3 - y_n^2 z_n + a t_n^3 \\ y_n^3 \\ y_n^2 t_n \end{pmatrix}. \tag{7.42}$$

Given the arbitrary initial condition

$$\begin{pmatrix} y_0 \\ z_0 \\ t_0 \end{pmatrix} \tag{7.43}$$

the degrees of the successive iterates are $1,3,9,27,\ldots$, but an accident occurs for $n = 5$. Indeed the three homogeneous polynomials y_5, z_5, t_5 have the gcd (greatest common divisor) y_0^8, so a cancellation occurs in the rational functions p_n, q_n.

In the language of the singularity confinement method, the initial singularity

$$\begin{pmatrix} y_0 \\ z_0 \\ t_0 \end{pmatrix} = \begin{pmatrix} \varepsilon \\ b \\ 1 \end{pmatrix}, \ \varepsilon \to 0, \tag{7.44}$$

is confined since the fifth iterate

$$\begin{pmatrix} y_5 \\ z_5 \\ t_5 \end{pmatrix} = \begin{pmatrix} b\varepsilon^8 + \ldots \\ -\varepsilon^9 + \ldots \\ \varepsilon^8 + \ldots \end{pmatrix}, \tag{7.45}$$

which is equivalently representable as

$$\begin{pmatrix} y_5 \\ z_5 \\ t_5 \end{pmatrix} = \begin{pmatrix} b + \ldots \\ -\varepsilon + \ldots \\ 1 + \ldots \end{pmatrix}, \tag{7.46}$$

contains the same information as the initial condition in the limit $\varepsilon \to 0$, so the map passes the test of singularity confinement.

The map is simple enough for the degrees d_n to be computed [206],

$$d_n = 1,3,9,27,73,195,513,\ldots, \ \lim_{n\to+\infty} \frac{\log d_n}{n} = \log \frac{3+\sqrt{5}}{2}. \tag{7.47}$$

The growth is exponential and indeed a numerical study displays a chaotic behavior [206, Fig. 1].

Therefore, despite its passing the singularity confinement test, the map (7.41) should be declared as not having the discrete Painlevé property, and the criterium of polynomial growth must be part of the discrete Painlevé test.

Let us remark that no continuum limit of (7.41) is known.

7.3.3 Method of Perturbation of the Continuum Limit

Consider the discrete equation (7.9) or (7.10), and assume it admits a continuum limit.

The perturbation defined by an expansion of u as a Taylor series in the lattice stepsize [88]

$$x \text{ unchanged}, \; h=\varepsilon, \; q=e^{\varepsilon}, \; u=\sum_{n=0}^{+\infty} \varepsilon^n u^{(n)}, \; a= \text{ analytic } (A,\varepsilon), \quad (7.48)$$

generates an infinite sequence of differential equations $E^{(n)}=0$

$$E=\sum_{n=0}^{+\infty} \varepsilon^n E^{(n)}(x,u^{(0)},\dots,u^{(n)}), \quad (7.49)$$

$$E^{(n)} \equiv E^{(0)}(x,u^{(0)})' u^{(n)} + R^{(n)}(x,u^{(0)},\dots,u^{(n-1)}) = 0, \; n \geq 1, \quad (7.50)$$

whose first one $n=0$ is the "continuum limit". The next ones $n \geq 1$, which are linear inhomogeneous, have the same homogeneous part $E^{(0)'}u^{(n)}=0$ independent of n, defined by the derivative of the equation of the continuum limit, while their inhomogeneous part $R^{(n)}$ ("right-hand side") comes at the same time from the nonlinearities and the discretization.

This perturbation of the continuum limit is entirely analogous to the perturbative method of the continuous case, either in its Fuchsian version (Sect. 2.2) [82] or in its nonFuchsian one (Sect. 2.3) [318], depending on the nature, Fuchsian or nonFuchsian, of the linearized equation $E^{(1)}=0$ at a singular point of $u^{(0)}$.

The simplicity of the method is best seen on the Euler scheme for the Bernoulli equation [88]

$$E \equiv \frac{u(x+h)-u(x)}{h} + u(x)^2 = 0, \quad (7.51)$$

i.e. the logistic map (7.1) of Verhulst, a paradigm of chaotic behavior which should therefore fail the test. Let us expand the terms of (7.51) according to the perturbation (7.48) up to an order in ε sufficient to build the first equation $E^{(1)}=0$ beyond the continuum limit $E^{(0)}=0$

$$u = u^{(0)} + u^{(1)}\varepsilon + \mathcal{O}(\varepsilon^2), \quad (7.52)$$

$$u^2 = u^{(0)2} + 2u^{(0)}u^{(1)}\varepsilon + \mathcal{O}(\varepsilon^2), \quad (7.53)$$

$$u(x+h) = u(x) + u'(x)h + (1/2)u''(x)h^2 + \mathcal{O}(h^3), \quad (7.54)$$

$$\frac{u(x+h)-u(x)}{h} = u^{(0)'} + (u^{(1)'} + (1/2)u^{(0)''})\varepsilon + \mathcal{O}(\varepsilon^2). \quad (7.55)$$

The equations of orders $n=0$ and $n=1$

$$E^{(0)} = u^{(0)'} + u^{(0)2} = 0, \quad (7.56)$$

$$E^{(1)} = E^{(0)'}u^{(1)} + (1/2)u^{(0)''} = 0, \; E^{(0)'} = \partial_x + 2u^{(0)}. \quad (7.57)$$

have the general solution

$$u^{(0)} = \chi^{-1}, \chi = x - x_0, \; x_0 \text{ arbitrary}, \quad (7.58)$$

$$u^{(1)} = u^{(1)}_{-1}\chi^{-2} - \chi^{-2}\log\psi, \; \psi = x - x_0, \; u^{(1)}_{-1} \text{ arbitrary}, \quad (7.59)$$

and the movable logarithm establishes the failure of the test as soon as order $n=1$, at the Fuchs index $i=-1$.

Remark. The only restriction on $u^{(0)}$ is not to be what is called a singular solution (not obtainable from the general solution by assigning values to the arbitrary data), i.e. it can be either the general solution (as above) or a particular one, it can also be either global (as above) or local (Laurent series).

As another example, one can process a straightforward discretization of P1,

$$E \equiv -(\overline{u}-2u+\underline{u})h^{-2}+3\lambda_1(\overline{u}+\underline{u})u+6\lambda_2 u^2+6\lambda_3\overline{u}\underline{u}+g=0. \qquad (7.60)$$

with $\sum\lambda_k=1$, and g an unspecified function of x. The result [88] is the selection of three values of the weights $\overrightarrow{\lambda}$, with $g=x$.

The first value $\overrightarrow{\lambda}=(2/3,1/3,0)$ corresponds to the d – P1 (7.11) found in quantum gravity and is identical, up to some translation, to the case $\gamma=0$ of (7.29); since it has a Lax pair, the condition is then sufficient. The second value $(1,0,0)$ (case $a=0$ in [184]) corresponds to a d – P1 with a second order Lax pair [88]. The third value $(1/2,1/4,1/4)$ defines an equation equivalent to that for $(1,0,0)$ under a discrete birational transformation (7.15) [370].

7.4 Discrete Riccati Equation

Consider the continuous Riccati equation

$$-\mathrm{d}u/\mathrm{d}x+\alpha u^2+2\beta u+\gamma=0, \qquad (7.61)$$

in which α,β,γ may depend on x. The property to be preserved here is the linearizability, in addition to the order one and the degree one. The discretization of $\mathrm{d}u/\mathrm{d}x$ is classical, with a second order remainder,

$$\frac{u(x+h/2)-u(x-h/2)}{h}=u'(x)+\frac{1}{24}u'''(x)h^2+O(h^4). \qquad (7.62)$$

There exists only one polynomial discretization of u^2 which preserves the order one (usage of two points, $x+h/2$ et $x-h/2$) and the degree one (in $\overline{u}=u(x+h/2)$ and $u=u(x-h/2)$), this is $\overline{u}u$, whose remainder has second order. Hence the solution [214]

$$-(\overline{u}-u)/h+\alpha\overline{u}u+(\beta_1\overline{u}+\beta_2 u)+\gamma=0, \qquad (7.63)$$

in which $(\alpha,\beta_1,\beta_2,\gamma)$ are arbitrary discretizations of (α,β,γ). It is invariant under $(h,\beta_1,\beta_2)\to(-h,\beta_2,\beta_1)$. The map $u\to\overline{u}$ is therefore homographic.

Whatever be the discretization of (α,β,γ), the equation can be linearized either into a system of two first order coupled equations,

$$u=v/w:\ (\overline{v}-v)/h=\beta_2 v+\gamma w,\ (\overline{w}-w)/h=\beta_1 w-\alpha v, \qquad (7.64)$$

or, under the transformation

$$u(x) = -\frac{(1 - h\beta_1(x))(\xi(x+h/2) - \xi(x-h/2))}{h\alpha(x+h/2)\xi(x-h/2)}, \tag{7.65}$$

into a second order equation

$$\begin{aligned}&\left[\overline{\alpha}(\xi - \underline{\xi}) - \alpha(\overline{\xi} - \xi)\right]/h^2 + \alpha(\beta_1 + \beta_2)(\overline{\xi} - \xi)/h - \overline{\alpha}\alpha\gamma\underline{\xi}\\ &-\overline{\alpha}(\beta_1 - \beta_2)(\xi - \underline{\xi})/h - \beta_2\left[\overline{\alpha}\beta_1(\xi - \underline{\xi}) + \alpha\overline{\beta_1}(\overline{\xi} - \xi)\right] = 0.\end{aligned} \tag{7.66}$$

It therefore possesses the discrete PP.

Remarks.

1. The Euler scheme
$$-(\overline{u} - u)/h + \alpha u^2 + 2\beta u + \gamma = 0 \tag{7.67}$$
is identical, at least for constant values of α, β, γ, to the map of Verhulst (logistic map) whose behavior is generically chaotic. It possesses no symmetry, as opposed to the well discretized equation (7.63), invariant under $(h, \beta_1, \beta_2) \to (-h, \beta_2, \beta_1)$.
2. From the singularities point of view, the homographic correspondence between $\overline{u}$ and u in (7.63) remedies a key omission of (7.67), which is the absence of a pole depending on the initial condition $u(0)$ (movable pole).

7.5 Discrete Lax Pairs

After the discrete Painlevé test has generated some necessary conditions for the discrete equation under study to possess the discrete PP, one must prove or disprove the Painlevé property.

In the continuous ODE case described in Sect. 5.8, it has been recalled that, from the (continuous) Lax pair, one can prove the (continuous) Painlevé property. Following most researchers, we will admit that, in the discrete case, the existence of a discrete Lax pair is a constructive proof of the discrete Painlevé property.

Let us first define a discrete Lax pair. Let us consider the Lax pair of a (continuous) ODE $E(x,u) = 0$, (5.13)

$$\begin{cases}\partial_x \psi = L\psi,\ \partial_\mu \psi = M\psi,\\ (C \equiv \partial_\mu L - \partial_x M + LM - ML = 0) \Leftrightarrow (E = 0),\end{cases} \tag{7.68}$$

in which μ is the spectral parameter, and let us restrict here to the difference type equations (7.9).

One requires the differential order in the column vector ψ to be conserved, therefore ψ must be discretized with two points, which we denote $\overline{\psi} = \psi(x+h/2)$ and $\underline{\psi} = \psi(x-h/2)$. As to the operator L, it must be discretized with as many points

u as required by the differential order of the equation under consideration, points which we denote $\overline{u} = u(x+h), u = u(x), \underline{u} = u(x-h)$ for a second order equation. In order to keep a linear correspondence between the continuous operators (L, M) and their discrete counterparts, it is then convenient to discretize $\partial_x \psi = L\psi$ in a dissymmetric-looking way and to introduce [32] the linear operator A linking ψ to $\overline{\psi}$, thus defining the *discrete Lax pair* (A, B, z, ψ, h) as [233]

$$\begin{cases} \overline{\psi} = A\psi, \ \partial_z \psi = B\psi, \\ (K \equiv \partial_z A + AB - \overline{B}A = 0) \Leftrightarrow (E = 0) \end{cases} \tag{7.69}$$

(it is an easy exercise to establish the expression for K).

The continuum limit is then

$$\begin{cases} \dfrac{A-1}{h} \to L, \ (\mathrm{d}z/\mathrm{d}\mu)B \to M, \\ (\mathrm{d}z/\mathrm{d}\mu)(\partial_z A + AB - \overline{B}A)/h \to \partial_\mu L - \partial_x M + LM - ML, \end{cases} \tag{7.70}$$

with some link $F(\mu, z, h) = 0$ between the spectral parameters μ and z.

For a second order equation $E(\overline{u}, u, \underline{u}, x, h) = 0$, the operators A and B must have the u–dependences $A(\overline{u}, u, \underline{u}), B(u, \underline{u})$.

Remark. The definition is invariant under the stepsize reversal

$$(E, A, B, x, h, \overline{u}, u, \underline{u}) \to (E, A^{-1}, B, x, -h, \underline{u}, u, \overline{u}). \tag{7.71}$$

No discrete equivalent is known to the singular manifold method, i.e. one does not know how to take advantage of the singularity structure of a discrete equation in order to build a discrete Lax pair and the discrete analogues of the tools presented in Sect. 5.2. Therefore, given a discrete equation which passes the discrete test, its Lax pair is often guessed from the knowledge of the continuous Lax pair.

For the equation (7.29) previously isolated by the test of singularity confinement, such a discrete Lax pair is [182, p. 494]

$$A = \begin{pmatrix} (\beta - \alpha n/2 - \gamma(-1)^n)/u & 1 & 0 \\ 0 & 0 & 1 \\ z & 0 & 0 \end{pmatrix}, \tag{7.72}$$

$$B = \begin{pmatrix} \lambda_1 & u & 1 \\ z & \alpha n/2 + \gamma(-1)^n & a - u - \underline{u} \\ z\underline{u} & z & 1 + \alpha(n-1)/2 - \gamma(-1)^n \end{pmatrix}, \tag{7.73}$$

in which λ_1 is constant. This Lax pair is valid even in the case $a = 0$ where no continuum limit is known.

7.6 Exact Discretizations

We now give two examples of discrete equations in which the discrete field also obeys the (continuous) ODE which is the continuum limit of the considered discrete equation.

7.6.1 Ermakov–Pinney Equation

If one assumes that the field $u(x)$ which satisfies the linear Schrödinger (Sturm–Liouville) equation

$$\psi'' + f(x)\psi = 0, \tag{7.74}$$

is a complex field $\psi = ve^{iS}$ and the potential f remains real, like in quantum mechanics, the modulus v obeys a second order nonlinear ODE,

$$v'' + fv + c^2 v^{-3} = 0,\ c = \text{ constant},$$

first introduced by Ermakov [133] and later considered by Pinney [358], therefore often called *Ermakov–Pinney equation*. Its movable branching $v \sim v_0(x - x_0)^{1/2}, v_0^2 = 2c$, can be removed, thanks to the parity, by considering the ODE for $w = v^2$,

$$ww'' - \frac{1}{2}w'^2 + 2f(x)w^2 + 2c^2 = 0. \tag{7.75}$$

In the continuous case, the Sturm–Liouville equation enjoys many beautiful properties, among them its equivalence [346, p. 230] (see details in [75]) with the Riccati equation and the Schwarz equation. It is therefore natural to investigate whether this is still the case after discretization, and this section summarizes the main results achieved so far.

The properties of the (continuous) ODE (7.75) are

1. its general solution is a nonlinear superposition of two solutions of a linear equation; indeed, given any two linearly independent solutions ψ_+, ψ_- of the Schrödinger equation

$$\psi'' + f\psi = 0, \tag{7.76}$$

with a Wronskian equal to $2c$, the general solution of (7.75) is

$$w = \psi_+\psi_- = \frac{2c}{(\log(\psi_+/\psi_-))'},\ \psi_+'\psi_- - \psi_-'\psi_+ = 2c; \tag{7.77}$$

This proves the Painlevé property for (7.75).

2. as a member of the class number 22 of Gambier [162], it is linearizable by elimination of c into the third order equation

$$w''' + 4fw' + 2f'w = 0; \tag{7.78}$$

3. the equation for $u = w^{-1}$ is transformed into a Schwarzian equation under the singular part transformation [74]

$$u = w^{-1} = \frac{1}{2c}(\log\varphi)', \tag{7.79}$$
$$\{\varphi;x\} - 2f = 0. \tag{7.80}$$

The two formulae (7.77) and (7.79) are identical under the correspondence

$$\varphi = \frac{\psi_+}{\psi_-}. \tag{7.81}$$

Consider now the problem of discretizing (7.75) while conserving as many of the above properties as possible.

From the Darboux transformation (5.22) rewritten in matrix form,

$$\begin{pmatrix}\Psi\\ \Psi'\end{pmatrix} = \kappa^{-1}\begin{pmatrix}-y & 1\\ y^2-\kappa^2 & -y\end{pmatrix}\begin{pmatrix}\psi\\ \psi'\end{pmatrix},\ y = \frac{\theta'}{\theta}, \tag{7.82}$$

in which κ is an arbitrary nonzero constant, Hone [223] found a birational transformation which maps (7.75) to an equation of the same form

$$WW'' - \frac{1}{2}W'^2 + 2FW^2 + 2C^2 = 0. \tag{7.83}$$

This is

$$w = \frac{(yW-(1/2)W')^2 - C^2}{\kappa^2 W},\ W = \frac{(yw+(1/2)w')^2 - c^2}{\kappa^2 w}, \tag{7.84}$$
$$y^2 = \kappa^2 - \frac{f+F}{2}, \tag{7.85}$$
$$C^2 - c^2 = 0,\ (f'+F')^2 + 2(f-F)^2(f+F-2\kappa^2) = 0. \tag{7.86}$$

Let us now define discrete variables by

$$\Psi = \overline{\psi},\ W = \overline{w},\ F = \overline{f}. \tag{7.87}$$

As indicated in Sect. 5.8, one then builds [223] the contiguity relation

$$\sqrt{c^2+\kappa^2 w\overline{w}} + \sqrt{c^2+\underline{\kappa}^2 w\underline{w}} + w\Omega = 0, \tag{7.88}$$
$$\Omega = -\sqrt{\kappa^2-(f+\overline{f})/2} - \sqrt{\underline{\kappa}^2-(f+\underline{f})/2},$$

a second order discrete equation for w, which has second degree after removal of the square roots involving w.

In the continuous variable

$$Z = \frac{w'}{2} - yw, \tag{7.89}$$

the degree of the contiguity relation becomes unity [223],

$$\kappa^2(\overline{Z}+Z)(\underline{Z}+Z) - \Omega\overline{\Omega}(Z^2-c^2) = 0. \tag{7.90}$$

The two discrete equations (7.88) and (7.90) both admit a continuum limit $h \to 0$ to the Ermakov–Pinney equation, an exercise which is left to the reader. Since the discrete field w in (7.88) also obeys the continuous Ermakov–Pinney equation (7.75), one says that the discretization (7.88) is *exact*.

Remark. The two discrete equations (7.88) and (7.90) were first given in [68] in the particular case

$$\kappa = \frac{1}{h},\ \Omega = -\frac{2}{h} + hf. \tag{7.91}$$

Equations (7.88) and (7.90) have the discrete Painlevé property. Indeed, the two functions ψ_+, ψ_- of (7.77) which satisfy the continuous Schrödinger equation (7.76) also satisfy, as seen from the Darboux transformation and its inverse, the same second order discrete linear equation

$$\underline{\kappa\psi} + \Omega\psi + \kappa\overline{\psi} = 0, \tag{7.92}$$

with the discrete Wronskian

$$\psi_-\overline{\psi_+} - \psi_+\overline{\psi_-} = \frac{2c}{\kappa}. \tag{7.93}$$

From Definition (7.89) of Z and the Darboux transformation (7.82), the elimination of derivatives ψ'_+, ψ'_- leads to a formula expressing Z in terms of two solutions of the same linear equation,

$$Z = \frac{\kappa}{2}\left(\overline{\psi_+}\psi_- + \overline{\psi_-}\psi_+\right), \tag{7.94}$$

identical to that in [68].

The elimination of $\varphi = \psi_+/\psi_-$ between (7.90), (7.94), and (7.93) provides the discrete analogue of (7.80)

$$\frac{(\overline{\varphi}-\underline{\varphi})(\overline{\overline{\varphi}}-\varphi)}{(\varphi-\underline{\varphi})(\overline{\overline{\varphi}}-\overline{\varphi})} - \frac{\Omega\overline{\Omega}}{\kappa^2} = 0. \tag{7.95}$$

This discrete Schwarzian was first given by Faddeev and Takhtajan [136].

Finally, the elimination of c by discrete differentiation linearizes both (7.88) and (7.90) into two different third order discrete linear equations admitting (7.78) as a continuum limit.

7.6.2 Elliptic Equation

The (continuous) elliptic equation has two usual kinds of normalized forms (Appendix C), the one of Weierstrass and the twelve ones of Jacobi (to which we will prefer the three ones $\mathrm{h}_1, \mathrm{h}_2, \mathrm{h}_3$ of Halphen, Appendix Sect. C.2), defined by the first and second order equations

$$\wp'^2 = 4\wp^3 - g_2\wp - g_3, \tag{7.96}$$

$$\wp'' = 6\wp^2 - g_2/2, \tag{7.97}$$

$$\mathrm{h}'_\alpha = -\mathrm{h}_\beta\,\mathrm{h}_\gamma, \tag{7.98}$$

$$\mathrm{h}''_\alpha = \mathrm{h}_\alpha(2\mathrm{h}^2_\alpha - 3e_\alpha), \tag{7.99}$$

in which g_2, g_3, e_α are constants and (α, β, γ) is an arbitrary permutation of (1,2,3).

As noticed in the context of discretization by Baxter [24] and Potts [364], the addition formulae (C.7) and (C.19) *ipso facto* define an exact discretization scheme for the first order equations (7.96) and (7.98). This is obvious by renaming (x_1, x_2) as $(x - h/2, h)$. A scheme for the second order equations (7.97) and (7.99) then results from the difference of the discrete first order equations taken for $(x_1, x_2) = (x, h)$ and $(x - h, h)$. The results are [364, 365, 366],

$$\begin{aligned}(\overline{u} - u)^2\wp(h) &= 2\overline{u}u(\overline{u} + u) - (g_2/2)(\overline{u} + u) - g_3\\ &\quad -[(\overline{u}u + g_2/4)^2 + g_3(\overline{u} + u)]\wp^{-1}(h),\end{aligned} \tag{7.100}$$

$$\begin{aligned}(\overline{u} - 2u + \underline{u})\wp(h) &= 2u(\overline{u} + u + \underline{u}) - g_2/2\\ &\quad -[u^2(\overline{u} + \underline{u}) + (g_2/2)u + g_3]\wp^{-1}(h),\end{aligned} \tag{7.101}$$

$$(\overline{u} - u)^2\mathrm{h}^2_\alpha(h) = (\overline{u}u)^2 - 2(\mathrm{h}'_\alpha(h) + \mathrm{h}^2_\alpha(h))\overline{u}u + 9e_\beta e_\gamma, \tag{7.102}$$

$$(\overline{u} - 2u + \underline{u})\,\mathrm{h}^2_\alpha(h) = u^2(\overline{u} + \underline{u}) - 2(\mathrm{h}'_\alpha(h) + \mathrm{h}^2_\alpha(h))u. \tag{7.103}$$

Remarks.

1. The general solution of (7.100) and (7.102) is by construction $\wp(x - x_0, g_2, g_3)$ and $\mathrm{h}_\alpha(x - x_0, g_2, g_3)$, where the step h is arbitrary, i.e. *not necessarily small.* These equations therefore possess the discrete PP. Equations (7.101) and (7.103) also possess the discrete PP since they admit a discrete Lax pair [89].
2. Order and degree are conserved by the four discretizations.
3. From these schemes of infinite order, the Laurent expansion of $\wp(h)$ around its double pole $h = 0$, or of $\mathrm{h}_\alpha(h)$ around its simple pole, defines the second order schemes

$$(\overline{u}-2u+\underline{u})h^{-2} = 2u(\overline{u}+u+\underline{u})-g_2/2, \tag{7.104}$$
$$(\overline{u}-2u+\underline{u})h^{-2} = u^2(\overline{u}+\underline{u})+3e_\alpha u, \tag{7.105}$$

which possess a discrete Lax pair [89] and therefore have the discrete PP.

4. If one replaces in (7.102) (or (7.100)) the three fixed coefficients by arbitrary constants, then applies to the resulting equation

$$a(\overline{u}-u)^2 = (\overline{u}u)^2+b\overline{u}u+c, \tag{7.106}$$

an arbitrary homography, one generates a first order second degree discrete equation

$$P(\overline{u},u,c_j)=0, \tag{7.107}$$

depending on six arbitrary constants c_j and having second degree separately in $\overline{u}$ and u[4]. The latter equation was considered and integrated (with elliptic functions of course!) in 1973 by Baxter [24] in the eight-vertex model, as a commutation condition of the Yang–Baxter, or star–triangle relations. These Yang–Baxter relations [237], which are second order discrete tensorial equations, play in the discrete domain a role as central as the one played by the Yang–Mills equations in the continuous domain.

The exceptional importance of these exact discretizations, and in particular of the six-parameter equation (7.107), will be emphasized in Sect. 7.8.

7.7 Discrete Versions of NLS

The discrete version of the nonlinear Schrödinger equation which naturally occurs in most physical problems is *not* derived as a discretization of the NLS, but as the output of some assumption not involving a stepsize, and the resulting equation is generically (the denominator h^2 is pure convenience)

$$iu_{n,t}+p\frac{u_{n+1}+u_{n-1}-2u_n}{h^2}+q|u_n|^2u_n=0,\ pq\neq 0, \tag{7.108}$$

with p and q real. This is the case in Bose–Einstein condensation [359] and in ideal optical fibers [10]. Its only analytic solution known so far is the plane wave

$$u_n = ae^{i(knh-\omega t)},\ \omega-4p\sin^2\frac{kh}{2}+q|a|^2=0. \tag{7.109}$$

[4] This freedom can be increased [230, 251]. For instance [251], if the starting equation $u=\mathrm{h}_\alpha(x)$ or $u=\wp(x)$ is replaced by $u=\mathrm{h}_\alpha(x+(-1)^n\gamma)$ or $u=\wp(x+(-1)^n\gamma)$ and the system is split according to the parity of n as done in (7.30), the degrees in $\overline{u}$ and u become four.

One very interesting and open question is to find, if it exists, a closed form expression for the bright and/or dark solitary wave of (7.108).

From the point of view of the Painlevé property, (7.108) fails the three methods of the Painlevé test (Sects. 7.3.1, 7.3.2 and 7.3.3), therefore, in the language of Sect. 7.1, it is a bad discretization of the NLS equation.

An inverse method to find a good discretization of the NLS is to discretize the Lax pair of the (continuous) NLS equation, which leads to the Ablowitz–Ladik [4] discrete NLS

$$iu_{n,t} + p\frac{u_{n+1}+u_{n-1}-2u_n}{h^2} + qu_n\overline{u}_n\frac{u_{n+1}+u_{n-1}}{2} = 0. \tag{7.110}$$

This equation possesses by construction the discrete Painlevé property. A discretization of the NLS equation itself in the polynomial class could not isolate [265] another integrable discrete NLS (in contrast for instance with the discretization (7.60) which admits three solutions).

An instructive approach [141] is to consider the NLS-type equation

$$iu_{n,t} + p\frac{u_{n+1}+u_{n-1}-2u_n}{h^2} + q\left(\frac{u_{n+1}+u_{n-1}}{2}F(|u_n|) + u_nG(|u_n|)\right) = 0, \tag{7.111}$$

and to determine the couples of functions (F,G) which allow the solutions sech and tanh of the continuous NLS. One such physically relevant equation is the *saturable discrete nonlinear Schrödinger equation*

$$iu_t + p\frac{u(x+h,t)+u(x-h,t)-2u(x)}{h^2} + q\frac{|u|^2u}{1+\nu(qh^2/p)|u|^2} = 0,\ pq\nu \neq 0, \tag{7.112}$$

which admits various elliptic solitary waves [250, 62]; in the continuum limit, these solutions require that $\nu \to 1/2$, making the limit of (7.112) equal to the NLS equation.

7.8 A Sketch of Discrete Painlevé Equations

A major problem is to find new discrete transcendental functions, i.e. to extend to the discrete world the discovery of the six transcendents of Painlevé. This amounts to establishing a discrete analog of the classification of all second order first degree nonlinear ODEs which possess the PP, see Appendix Sect. A.3.2. For each discrete type (difference equation, q-difference, e-difference), one has

1. to build a discrete analog of the α-method so as to forget no equation;
2. to define groups of invariance for the discrete PP so as to arrange the equations in distinct equivalence classes. The discrete birational group has been defined in

(7.15), but proving the equivalence of two given discrete equations under this group is quite difficult;
3. to prove the PP for an equation which passes the discrete Painlevé test. As said in Sect. 7.5, we will admit that it is sufficient to exhibit a discrete Lax pair;
4. to define a discrete analog of the notion of irreducibility (Appendix Sect. A.2). This has not been done yet.

This huge task has been tackled from two quite different approaches, an analytic one and a geometric one, which we now briefly present.

7.8.1 Analytic Approach

This relies on the following remark. In the list of 50 equations of Gambier [163] (Appendix Sect. A.3.2), all autonomous equations either are linearizable or admit a first integral defining a first order second degree elliptic ODE, while all nonautonomous equations admit a coalescence to an autonomous equation of the list. Therefore, starting *only* from the most general first order second degree elliptic ODE, one should be able, in principle, to generate the full classification. For instance, the elimination of the constant K between the equation

$$u'^2 = 2\left[\alpha u - \frac{\beta}{u} - \frac{\gamma}{u-1} - \frac{\delta}{u-a} + K\right] u(u-1)(u-a) \tag{7.113}$$

and its derivative yields an equation (numbered 48 in Gambier's list) which is the autonomous limit of the sixth Painlevé equation P6. To proceed from the 48-th Gambier equation to P6, one merely replaces the constant coefficients by arbitrary functions of x, which are then selected by requiring this nonautonomous equation to pass the Painlevé test.

This idea has been extended to the discrete equations [371] with great success. The starting point is the six-parameter first order second degree discrete autonomous equation (7.107), obtained by eliminating $\mathrm{h}_\alpha(x)$ between $u = \mathrm{h}_\alpha(x)$ and $\overline{u} = \mathrm{h}_\alpha(x+h)$ then by performing a homography. By eliminating one of the six parameters c_j between $P(\overline{u}, u, c_j) = 0$ and $P(u, \underline{u}, c_j) = 0$, one generates a family of five-parameter second order first degree discrete autonomous equations

$$Q(\overline{u}, u, \underline{u}, c_j) = 0, \tag{7.114}$$

(the degree of the polynomial Q in its first three arguments is $1, 4, 1$). This family of equations, first introduced in [302, 367, 368] and sometimes called the QRT mapping, is a direct consequence of the addition formula of elliptic functions. After changing the constants in (7.114) to functions of x and requiring the nonautonomous equation to pass the discrete Painlevé test, the selected equations are likely to possess the discrete Painlevé property, and a search for their Lax pair may settle this final point. Many $\mathrm{d}-\mathrm{Pn}$ and $\mathrm{q}-\mathrm{Pn}$ equations have been isolated by this procedure (detailed in lecture notes [182]), in particular a q − P6 equation with a Lax pair (discovered in [181] and identified as such in [242])

$$\begin{cases} y_n y_{n-1} = \dfrac{st(x_n-a)(x_n-b)}{(x_n-c)(x_n-d)}, \\ x_n x_{n+1} = \dfrac{cd(y_n-p)(y_n-r)}{(y_n-s)(y_n-t)}, \quad \dfrac{prcd}{abst} = q, \end{cases} \qquad (7.115)$$

in which c,d,s,t are constants and a,b,p,r are proportional to q^n. Both x_n and y_n admit a continuum limit to P6 and the relation between $\lim x_n$ and $\lim y_n$ is the elementary birational transformation (5.305) of P6 [182, p. 463]. However, the above procedure cannot guarantee that all the discrete Painlevé equations have been found since it is only sufficient.

7.8.2 Geometric Approach

The Painlevé transcendents were isolated from two different points of view: the construction of new functions defined by nonlinear ODEs (Painlevé and Gambier), and the monodromy which preserves the deformation of linear differential equations [156, 240]. By exploring a third, geometric approach, Sakai [377] uncovered in two features of the continuous Pn the key to the construction of *all* the discrete Pn (additive d − Pn, multiplicative q − Pn and elliptic e − Pn).

These two features are

1. the group of symmetries which leave invariant each Pn, detailed in Sect. B.4;
2. another group arising from the space of initial conditions as constructed in [339] by the blowing-up procedure (éclatement in French).

Both groups can be represented by diagrams of the Dynkin type (Table 7.1), and their comparison is quite instructive. In terms of complexity of the groups ($E_8^{(1)}$ is the richest, $A_1^{(1)}$ the poorest), one list is increasing, the other is decreasing, and P6 is the only common point. Most of all, this provides the information that the richest group $E_8^{(1)}$ should play a quite important role.

Table 7.1 Comparison of two sets of Dynkin diagrams. The first line lists the diagrams of the affine Weyl groups representing the elementary birational transformations. The second line shows the diagrams describing the configurations of irreducible components of the "vertical leaves".

Feature	P1	P2	P3	P4	P5	P6
Birational transfo	None	$A_1^{(1)}$	$(2A_1)^{(1)}$	$A_2^{(1)}$	$A_3^{(1)}$	$D_4^{(1)}$
Vertical leaves	$E_8^{(1)}$	$E_7^{(1)}$	$D_6^{(1)}$	$E_6^{(1)}$	$D_5^{(1)}$	$D_4^{(1)}$

This role was uncovered by a very beautiful construction starting from the theory of rational surfaces and ending in the master discrete Painlevé equation [377]. Pedagogical details can be found in [331] and [245, §6 and 7]. In one of its equivalent geometric constructions, one starts from the most general cubic curve in the plane. Such a curve is characterized by nine parameters (like the group $E_8^{(1)}$) and has genus one, its natural parametric representation in homogeneous coordinates (x_1, x_2, x_3) is

$$\begin{cases} x_1 = -\sigma(u)\sigma(u+c_2)\sigma(u-c_2), \\ x_2 = -\sigma(u)\sigma(u+c_1)\sigma(u-c_1), \\ x_3 = -\sigma(u-c_0)\sigma(u-c_1)\sigma(u-c_2), \\ c_1 = \dfrac{\omega_1}{2},\ c_2 = \dfrac{\omega_2}{2},\ c_0 = -c_1 - c_2, \end{cases} \tag{7.116}$$

in which σ is the entire function of Weierstrass and ω_1, ω_2 the associated half-periods. Then the master equation $\mathrm{e}-\mathrm{P6}$ can be represented as a rational relation between ten points: eight belonging to the same cubic curve (the so-called "parameters" of $\mathrm{e}-\mathrm{P6}$), and two others (the so-called "independent" and "dependent" variables). Its affine Weyl group is $E_8^{(1)}$ by construction, but its Lax pair is still unknown. As compared to P6, it depends on four more parameters.

The explicit expression for $\mathrm{e}-\mathrm{P6}$ is far from short, but some codimension one $\mathrm{e}-\mathrm{P6}$ (i.e. depending on one less arbitrary parameter) can be written in a few lines, such as [336, (12)]

$$\begin{cases} \dfrac{x_{n-1} - \wp(\lambda n + a_n)}{x_{n-1} - \wp(\lambda n + b_n)} \times \dfrac{x_n - \wp(-2\lambda n + c_n)}{x_n - \wp(-2\lambda n + d_n)} \\ \qquad \times \dfrac{x_{n+1} - \wp(\lambda n + e_n)}{x_{n+1} - \wp(\lambda n + f_n)} = F(n, (-1)^n). \end{cases} \tag{7.117}$$

Here, $a_n, \dots, f_n$ are linear combinations of a constant, a period-3 function $\varphi(n)$, a period-4 function $\omega(n)$, and finally a parity-dependent term $(-1)^n\gamma$ with γ constant, and the r.h.s. F is independent of x_n and involves similar values of the elliptic function $\wp$.

7.8.3 Summary of Discrete Painlevé Equations

It is outside the scope of the present volume to give a complete description of all the discrete analogues of the Painlevé equations. The interested reader can refer to the thesis of Sakai [377] and to subsequent reviews [245, 183].

Some of their important properties are

1. The coefficients of their equation have a dependence on n which is elliptic (e-discrete) or degenerate of elliptic (trigonometric $e^{n \log q}$, rational), in full analogy to the first order autonomous ODEs which have the Painlevé property (Appendix Sect. A.3.1).

2. The master elliptic-difference equation can generate by confluence all the discrete Painlevé equations. Naturally, the e-type degenerates to the q-type then to the difference type. This can be described by a coalescence tree with e – P6 at the top and the continuous Pn at the bottom, see diagram in [377].
3. e – P6 and all the q – P6 and d – P6 admit P6 as their continuum limit.
4. e – P6 admits classical solutions (as defined in Sect. A.2). The analogue of the hypergeometric function for P6 is the elliptic hypergeometric function ${}_{10}E_9$ [244] introduced in [154] and studied in [394]. By coalescence, this generates various generalizations of the hypergeometric function, such as the Askey–Wilson polynomials. See [332] for a review.
5. The best description of all these discrete equations is made by their associated affine Weyl groups.

Chapter 8
FAQ (Frequently Asked Questions)

Abstract This chapter contains frequently asked questions or frequently encountered incorrect statements.

Q. *Why is the solution of $u' + u^2 + S/2 = 0$, with S constant, always presented as a* tanh, *whatever the sign of S?*

A. This is a matter of taste. Indeed, in the complex domain, there is no need to introduce four functions (tanh, cotanh, tan, cotan) to represent the solution of this equation, one is enough. We prefer tanh for it is the only one to be bounded on the real axis, a feature appreciated for the solitary waves of physics.

Q. *Linear ODEs have no dominant balance, therefore they fail the P test.*

A. Wrong. All linear ODEs have by definition the PP, therefore the test must be formulated so as to let them pass.

Q. *What is the general solution of the* P2 *equation? It is written nowhere.*

A. Each Pn equation has for general solution the Pn function, by construction! This tautology only reflects the original problem, to define functions from nonlinear ODEs. There exist nongeneric values of $\alpha, \beta, \gamma, \delta$ for which the Pn equations possess solutions (called classical, Appendix Sect. A.2) expressible with classical functions only (elliptic function and solutions of linear equations). In the P2 case, these are $\alpha - 1/2$ equal to an integer (there then exists a one-parameter solution which is an algebraic transform of the Airy function), and α equal to an integer (for which there exists a zero-parameter solution rational in x).

Q. *When an ODE is reducible to quadratures, is it integrable in the Painlevé sense?*

A. The two properties are independent, here are examples of the four situations.

1. reducible to quadratures, integrable in the Painlevé sense: elliptic equation,
2. reducible to quadratures, not integrable in the Painlevé sense: $u' = u^3$,
3. not reducible to quadratures, integrable in the Painlevé sense: P1 equation,

4. not reducible to quadratures, not integrable in the Painlevé sense: KS ODE $u''' + (1/2)u^2 + K = 0$.

Integrating by quadratures only was the old way, and Jacobi proved its insufficiency on the pendulum example.

Q. *Is the Riccati equation integrable?*
A. First define "integrable". If "integrable" means "reducible to a quadrature", the answer is no, unless a particular solution u_0 is known. If "integrable" means "possessing the Painlevé property", the answer is yes, since the equation can be mapped by a closed form transformation to a linear equation.
Note that, if three particular solutions are known, the general solution is given by the crossratio formula (5.36).

Q. *The PP of an ODE is defined as all solutions having no other movable singularities than poles.*
A. Two words are wrong, "all" and "poles". "All" because, out of the three kinds of solutions admitted by an ODE (general, particular, singular, see details in [76, p. 100]), only the general solution is involved in the definition of the PP. For instance, the third order second degree ODE [59, p. 360]

$$(u''' - 2u'u'')^2 + 4u''^2(u'' - u'^2 - 1) = 0, \tag{8.1}$$

admits the single valued general solution,

$$u = \frac{e^{c_1 x + c_2}}{c_1} + \frac{c_1^2 - 4}{4c_1} x + c_3, \tag{8.2}$$

and also a singular solution (envelope solution) with a movable critical singularity, defined by canceling the discriminant of (8.1)

$$u'' - u'^2 - 1 = 0, \; u = C_2 - \log\cos(x - C_1). \tag{8.3}$$

For first and second order ODEs having the PP, it is true that the movable singularities are only poles. But, from third order on, the ODEs having the PP may present various types of noncritical movable singularities, and "une discussion qui écarterait d'avance certaines singularités comme invraisemblables serait inexistante."[1] [349, p. 6]. The best known example is the celebrated Chazy's equation of class III

$$u''' - 2uu'' + 3u'^2 = 0, \tag{8.4}$$

whose general solution is only defined inside or outside a circle characterized by the three initial conditions (two for the center, one for the radius); this solution is holomorphic in its domain of definition and cannot be analytically continued beyond it. This equation therefore has the PP, and the only singularity is a movable

[1] A discussion which would from the beginning discard some singularities as being unlikely would be nonexistent.

analytic essential singular line which is a natural boundary. For more details, see the arguments of Painlevé [76, §2.6] and Chazy [76, §5.1].

Q. *This equation passes the Painlevé test, therefore it has the Painlevé property.*
A. The word "therefore" is wrong since passing the test is only necessary, not sufficient, for the PP. As an algorithm, the test must terminate after a finite number of steps, and Picard [355] built the beautiful example of an ODE with the general solution $u = \wp(\lambda \log(x - c_1) + c_2, g_2, g_3)$, namely

$$u'' - \frac{u'^2}{4u^3 - g_2 u - g_3}\left(6u^2 - \frac{g_2}{2}\right) - \frac{u'^2}{\lambda\sqrt{4u^3 - g_2 u - g_3}} = 0, \tag{8.5}$$

which has the PP iff $2\pi i\lambda$ is a period of the elliptic function $\wp$. This is a transcendental condition on (λ, g_2, g_3) impossible to obtain in a finite number of algebraic steps such as the Painlevé test.

Q. *For the PP, movable essential singularities are forbidden.*
A. Wrong. Essential singularities may be critical or noncritical, and only the movable essential critical singularities are forbidden. The ODE with the general solution $c_1 e^{1/(x-c_2)}$ has the Painlevé property, but the ODE with the general solution [349, p. 5]

$$u = \tan\log(c_1 x + c_2), \tag{8.6}$$

which is

$$u'' = u'^2 \frac{2u - 1}{u^2 + 1} \tag{8.7}$$

has not the PP since the movable point $x = -c_2/c_1$ is essential and critical. The difficulty with essential singularities is that they cannot be uncovered by algebraic procedures such as the test of Kowalevski and Gambier. Details on this difficulty are well documented in Painlevé [349, §4 and 5].

Q. *Resonance -1 is always compatible.*
A. (In our vocabulary, *The no-log condition at Fuchs index -1 is always satisfied.*) This is wrong. A counterexample is given in Sect. 3.2.7. Other examples can be found in [82].

Q. *A negative Fuchs index indicates a movable essential singularity.*
A. This is wrong. The meromorphic expression

$$u = \frac{1}{x-a} + \frac{1}{x-b}, \ (a, b) \text{ arbitrary}, \tag{8.8}$$

is the general solution of

$$u'' + 3uu' + u^3 = 0, \tag{8.9}$$

an ODE in the class of Painlevé and Gambier which has the two families,

(F1) $p=-1, u_0=1$, indices $(-1,1)$,
(F2) $p=-1, u_0=2$, indices $(-2,-1)$.

See [82] for the associated Laurent series.

Q. *In the WTC truncation (Sect. 5.5.2), since the "constant level coefficient"* u_{-p} *is another solution of the PDE, one has obtained a Bäcklund transformation.*

A. Wrong, for two reasons. Firstly, this feature of u_{-p} is true for any PDE, integrable or not. Secondly, for PDEs which do admit a BT, after completion of the singular manifold method the coefficient u_{-p} is *never* the second solution involved in the BT. More in [84, 77].

Appendix A
The Classical Results of Painlevé and Followers

In the process of searching for new functions defined by a differential equation (cf. text on p. 8), one must input a class of ODEs, e.g. the most general n-th order m-th degree ODE,

$$P(u^{(n)}, u^{(n-1)}, \ldots, u', u; x) = 0, \tag{A.1}$$

in which P is a polynomial of all its arguments but the last one, of degree m in the highest derivative, with an analytic dependence on x. Since this is a formidable task, these classes are often restricted to simpler classes such as (6.37) ($n = 4, m = 1$ and coefficient unity for the term $u^{(4)}$), but there is no good reason for that. Naturally (see the FAQ section), linear ODEs are excluded since they already have the Painlevé property.

The output of such a search is twofold (this is the "double interest" of differential equations):

1. some *new functions* (defined from the general solution of an ODE which is irreducible to a lower differential order or to a linear equation),
2. an exhaustive list (called *classification*) of ODEs which includes the ones defining new functions, and which are explicitly integrated with functions found at a lower order.

The importance of these classifications is worth being emphasized. If one has an ODE in such an already studied class (e.g. second order second degree binomial-type ODEs [107] $u''^2 = F(u', u, x)$ with F rational in u' and u, analytic in x), and which is suspected to have the PP (for instance because one has been unable to detect any movable critical singularity, see Chap. 2), then two cases are possible: either there exists a transformation (1.43) mapping it to a listed equation, in which case the ODE has the PP and is explicitly integrated, or such a transformation does not exist, and the ODE does not have the PP.

For instance, the class $n = 1, m = 1$ of first order first degree equations

$$u' = F(u, x), \tag{A.2}$$

with F rational in u and analytic in x, only yields the Riccati equation

$$u' = a_2(x)u^2 + a_1(x)u + a_0(x), \tag{A.3}$$

and defines no new function since the Riccati equation is linearizable,

$$u = -\frac{\varphi'}{a_2\varphi},\ \varphi'' - \left(\frac{a_2'}{a_2} + a_1\right)\varphi' + a_0a_2\varphi = 0. \tag{A.4}$$

A.1 Groups of Invariance of the Painlevé Property

In these classifications, it is sufficient to take one representative equation by class of equivalence of the PP. There exist two such relations of equivalence.

The first group of invariance is the *homographic group* defined in the text, (1.43). This is precisely the group of invariance of the Riccati equation. Since it conserves the crossratio of four points, its main practical use is, given an ODE with three polar singularities $U = p_1(x), p_2(x), p_3(x)$, to change it to an equivalent ODE whose poles u are set at predefined locations, e.g. $(\infty, 0, 1)$, *via* the homography $(U, p_1(x), p_2(x), p_3(x)) = (u, \infty, 0, 1)$. This is what has been done for choosing a representative of P6 in its equivalence class, see p. 8.

The second group [346, 163] is the group of birational transformations, in short the *birational group* $(u,x) \leftrightarrow (U,X)$

$$\begin{cases} u = r(x, U, \mathrm{d}U/\mathrm{d}X, \dots, \mathrm{d}^{n-1}U/\mathrm{d}X^{n-1}) = 0,\ x = \Xi(X), \\ U = R(X, u, \mathrm{d}u/\mathrm{d}x, \dots, \mathrm{d}^{n-1}u/\mathrm{d}x^{n-1}) = 0,\ X = \xi(x), \end{cases} \tag{A.5}$$

(n order of the equation, r and R rational in U, u and their derivatives, analytic in x, X). This group admits as a subgroup the group of homographic transformations.

Example. Given the ODE $u'' - 2u^3 = 0$ and the new dependent variable $U = u' + u^2$, the algebraic elimination of (u', u'') among these two equations and the derivative of the second one yields the inverse transformation $u = U'/(2U)$, which, once inserted in the direct transformation, yields the transformed equation $UU'' - U'^2/2 - 2U^3 = 0$.

There exist other groups of transformations of nonlinear ODEs which do not conserve the PP. The most important one is the group of *point transformations* which considers the variables u and x as equivalent geometric coordinates,

$$u = f(X, U),\ x = g(X, U),\ U = F(x, u),\ X = G(x, u). \tag{A.6}$$

The class of second order first degree ODEs invariant under (A.6) is

$$u'' + \sum_{j=0}^{3} A_j(u, x)u'^j = 0 \tag{A.7}$$

(as compared to the class (3.16), it contains an additional coefficient A_3), and it has been extensively studied by Roger Liouville [282], Tresse [404] and Cartan [55]. The subgroup of *"fiber-preserving" transformations*

$$u = f(X,U),\ x = g(X),\ U = F(x,u),\ X = G(x), \tag{A.8}$$

also violates the PP but it conserves the integrability in the sense of Poincaré as defined on p. 6, and its equivalence classes are called *Cartan equivalence classes* [55].

These last two groups provide an additional insight to that of Painlevé, which forbids the exchange of the dependent and independent variables as done for instance in Sect. 3.2.6.

A.2 Irreducibility. Classical Solutions

In order to be declared new, a function (general solution of some nonlinear ODE with the PP) must be shown to obey neither a nonlinear ODE of lower differential order nor a linear ODE of any order. To understand the difficulty, it is sufficient to consider the fourth order ODE for $u(x)$ defined by

$$u = u_1 + u_2,\ u_1'' = 6u_1^2 + x,\ u_2'' = 6u_2^2 + x. \tag{A.9}$$

This ODE (easy to write by the elimination of u_1, u_2) has a general solution which depends transcendentally on the four constants of integration, and it is reducible to P1, a second order equation.

The initial definition of irreducibility as given by the "groupe de rationalité" of Jules Drach [351, vol. III p. 14], [348, p. 246], [363] relied on the infinite dimensional differential Galois theory, which Picard and Vessiot continued to develop, and is so difficult that it is not yet finished. Painlevé believed that the theory could be completed soon, and so used unproven results of Drach in his argumentation. This was pointed out by Roger (not Joseph) Liouville in a passionate discussion with Painlevé in the *Comptes rendus* [351, vol. III, pp. 81–114]. A precise, purely algebraic, definition of irreducibility has been given by Umemura [407]. This is the following.

Given a class of differential equations, e.g. the linear ODEs and the first order ODEs, a function is called *classical* with respect to that class if it can be built from solutions of ODEs in that class by the following operations,

addition,
multiplication, division,
composition of functions,
derivation,
integration (once, i.e. taking the primitive),

algebraic function (i.e. root of algebraic equations whose coefficients are functions in the considered class).

The two notions "classical solution" and "irreducibility" are of course equivalent. The above definition shares many features with the algorithm of Risch and Norman in computer algebra (which decides whether the primitive of a class of expressions, e.g., rational functions, is inside or outside the class).

Examples.

1. With respect to the linear equations, the solution of the Weierstrass equation is nonclassical.
2. With respect to the set made of linear equations plus the Weierstrass equation, the solution of Picard (B.3) for P6 is nonclassical, because g_2 and g_3 are not constant.
3. With respect to the set made of the linear equations plus the first order equations having the Painlevé property, the general solution of any of the six Painlevé equations is nonclassical (except for nongeneric values of the parameters $\alpha,\beta,\gamma,\delta$ listed in Sect. B.8).

Even if an ODE is irreducible, for nongeneric values of its fixed parameters it may admit particular solutions which are also solutions of either a lower order nonlinear ODE or a linear ODE. For instance, P6 (defined on p. 8) admits for $2\alpha=-2\beta=\lambda^2, 2\gamma=1-2\delta=\mu^2$ the zero-parameter solution $u=\sqrt{x}$.

A.3 Classifications

Each equation is characterized by one representative in its equivalence class. For instance, under the homographic group, the ODE (3.20) for $x(t)$ in the case $b=0, \sigma=1/3$ of the Lorenz model is not distinct from the P3 equation with $\alpha=\beta=0,\ \gamma=\delta=1$. A recent list of achieved classifications can be found in Cosgrove [104, 106].

A.3.1 First Order Higher Degree ODEs

First order algebraic ODEs

$$F(u',u,x)=0, \tag{A.10}$$

with F polynomial in u,u', analytic in x, define only one function, the *Weierstrass elliptic function* $\wp$, new in the sense that its ODE

$$u'^2-4u^3+g_2u+g_3=0,\ (g_2,g_3) \text{ arbitrary complex constants,} \tag{A.11}$$

is not reducible to a linear ODE. Its singularities are movable double poles.

For the classifications, one must distinguish [361, Vol. III p. 26] the autonomous case (F independent of x) of the nonautonomous case. Let g be the genus of the curve $(u,u') \to F(u,u',x) = 0$ for a generic value of x.

In the autonomous case, the genus is necessarily 1 or 0. When $g = 1$, there is only one class of equivalence of the birational group, and a suitable representative is the Weierstrass equation (A.11). When $g = 0$, the general solution is rational in either $(e^{a(x-x_0)} - 1)/a$ or in its limit $x - x_0$, therefore the ODE is linearizable into $u' - au - 1 = 0$, and neither group is relevant.

In the nonautonomous case, the genus can be arbitrary, see details in Poincaré [361, vol. III p. 26], Valiron [408, vol. II, §141 p. 286] and [309].

A.3.2 Second Order First Degree ODEs

The class

$$u'' = F(u',u,x), \tag{A.12}$$

with F rational in u', algebraic in u, analytic in x, has been fully processed by Painlevé [348] and Gambier [163].

The classification contains 53 equivalence classes under the homographic group (50 with a rational dependence of F in u and 3 with an algebraic dependence), and 24 equivalence classes under the birational group. Their list can be found in various locations [163, 232, 116]. Among these 24, six classes are irreducible [to a first order or linear equation] and define the six Painlevé transcendents listed on p. 8. Given an ODE in the class (A.12) and having the PP, the method to identify it to one of the 53 classes is well explained in [116] and can be found in Sect. 3.1.1.2.

The reduction of the 53 classes has also been done under the fiber preserving point group (A.8) [228] and under the point group (A.6) [19]. A beautiful result is that all six Pn equations are equivalent to [19]

$$\frac{\mathrm{d}^2U}{\mathrm{d}X^2} + f(U,X) = 0, \tag{A.13}$$

i.e. without contribution of the first derivative, e.g. for P6

$$\frac{\mathrm{d}^2U}{\mathrm{d}X^2} + \sum_{j=\infty,0,1,x} \theta_j^2 \frac{\mathrm{d}}{\mathrm{d}U} \wp(U + c_j\omega_1 + d_j\omega_2) = 0, \tag{A.14}$$

in which $2\omega_1, 2\omega_2$ are the periods of the Weierstrass elliptic function $\wp$ and $(c_j,d_j) = (0,0),(1,0),(0,1),(1,1)$. This compact writing was initially found in 1906 by Painlevé [350] by extending the representation of Picard [354] to the generic case $(\theta_\infty,\theta_0,\theta_1,\theta_x) \neq (0,0,0,0)$, then later retrieved by geometrical considerations [295, 19].

It should also be mentioned that Garnier [169] classified the set of two-dimensional systems

$$u' = F(u,v), v' = G(u,v), \tag{A.15}$$

with F and G homogeneous polynomials.

A.3.3 Second Order Higher Degree ODEs

General necessary conditions have been established by Painlevé [348, p. 252] [349, p. 67] for an arbitrary order and a degree one.

Only partial classifications exist at second order and a degree higher than one. For second degree, preliminary investigations have been carried out by Bureau [47]. For an arbitrary degree the subset of binomial equations

$$u''^n = F(x,u,u'), n \geq 2, \tag{A.16}$$

with F rational in u' and u, has been classified for $n = 2$ [107, 103] and for $n \geq 2$ [102].

A.3.4 Third Order First Degree ODEs

The classification of third order first degree ODEs

$$u''' = \sum_{j=0}^{2} A_j(u',u,x)u''^j, \tag{A.17}$$

has been nearly finished by Chazy [59]. The subset $A_2 = 0$ comprises 13 classes under the homographic group, its classification has been continued by Bureau [46], Exton [135] and completed by Cosgrove [105].

A.3.5 Fourth Order First Degree ODEs

Only the subclass

$$u'''' = F(u''',u'',u',u;x), \tag{A.18}$$

with F polynomial in u''', u'', u', rational in u, analytic in x, has been investigated. Several examples were encountered in the text, Sect. 6.2.3. After some preliminary results obtained by Chazy [59], Bureau [46], Exton [135] and Martynov [299], with the further restriction F polynomial in u[1], Cosgrove finished the classification, finding 12 classes inequivalent under the homographic group when u has simple poles [106], and 5 classes when it has double poles [104].

Four of the 17 isolated ODEs could not be integrated (with linear equations, elliptic functions or Painlevé transcendents). They have a transcendental dependence on the four constants of integration [262], but their irreducibility is not proven, therefore one cannot yet claim that they define new functions. These are (we have added the Fif-IV fifth order equation [104] which shares the same properties)

$$\text{F-V} \quad u'''' = 20uu'' + 10u'^2 - 40u^3 + \alpha u + \kappa x + \beta, \tag{A.19}$$

$$\text{F-VI} \quad u'''' = 18uu'' + 9u'^2 - 24u^3 + \alpha u^2 + \frac{\alpha^2}{9}u + \kappa x + \beta, \tag{A.20}$$

$$\begin{aligned}\text{F-XVII} \quad u'''' &= 10u^2u'' + 10uu'^2 - 6u^5 \\ &\quad - 10\delta(u'' - 2u^3) + (\lambda x + \mu)u + \gamma,\end{aligned} \tag{A.21}$$

$$\text{F-XVIII} \quad u'''' = -5u'u'' + 5u^2u'' + 5uu'^2 - u^5 + (\lambda x + \alpha)u + \gamma, \tag{A.22}$$

$$\begin{aligned}\text{Fif-IV} \quad u''''' &= 18uu''' + 36u'u'' - 72u^2u' + 3\lambda u'' \\ &\quad + \frac{1}{x}\left\{u'''' - 18uu'' - 9u'^2 + 24u^3 - 3\lambda u' + \kappa\right\} \\ &\quad + \frac{\lambda x}{2}\left(5u''' - 36uu'\right) - \frac{\lambda^2 x}{2}\left(2xu' + u\right).\end{aligned} \tag{A.23}$$

Remark. The same feature (transcendental dependence but unproven irreducibility) is also observed in the Garnier system [167], which admits P6 as a degeneracy.

These five equations are reductions of various soliton equations, F-V of KdV5 [7, 261], F-VI of a coupled KdV system [126], F-XVII of modified KdV5 [142] F-XVIII of a fifth-order PDE [148].

At least three soliton systems of two equations, which are all reductions of the KP hierarchy, admit a reduction to F-VI. One system has order 6+6 [176], another the order 5+3 [121], and the shortest system the order 3+3,

$$\begin{cases} u_t - \dfrac{1}{2}\left(u_{xx} + au^2 + 6v\right)_x = 0, \\ v_t + v_{xxx} + auv_x = 0. \end{cases} \tag{A.24}$$

The latter system, isolated in [126], admits the reduction

$$\begin{cases} au = -6U + \dfrac{\alpha}{12},\ av = V - \kappa t,\ \xi = x + \dfrac{\alpha}{4}t, \\ \left(U'' - 6U^2 - \dfrac{\alpha}{3}U - V\right)' = 0,\ V''' - 6UV' + \dfrac{\alpha}{3}V' - \kappa = 0, \\ \left(-U'''' + 18UU'' + 9U'^2 - 24U^3 + \alpha U^2 + \dfrac{\alpha^2}{9}U + \kappa\xi\right)' = 0. \end{cases} \tag{A.25}$$

[1] This excludes (6.74).

A.3.6 Higher Order First Degree ODEs

Some results exist for fifth [135, 104, 106] (see e.g. (A.23) in the previous section) and sixth [135] order equations in the class

$$u^{(n)} = P(u^{(n-1)},...,u;x), \tag{A.26}$$

with P analytic in x and polynomial in its other arguments.

A.3.7 Second Order First Degree PDEs

After preliminary indications by Hlavatý [221], second order semilinear PDEs have been classified, both in the hyperbolic/elliptic case [100] (no such distinction is made by the Painlevé property) and in the parabolic case [101]. No new PDE has been isolated, and all the previously known PDEs were found (Burgers, Liouville, sine-Gordon, Tzitzéica, etc).

Appendix B
More on the Painlevé Transcendents

The classical results are well documented in [235]. For an up to date summary of the (mainly geometric) properties of the Painlevé transcendents, we refer the reader to the nice review [231].

The master equation P6 was first written by Picard [354] in 1889 in a particular case. Let φ be the elliptic function defined by

$$\varphi\ :\ y \mapsto \varphi(y,x),\ y = \int_\infty^\varphi \frac{\mathrm{d}z}{\sqrt{z(z-1)(z-x)}}, \tag{B.1}$$

and let $\omega_1(x), \omega_2(x)$ be its two half-periods. Then the function

$$u\ :\ x \mapsto u(x) = \varphi(c_1\omega_1(x) + c_2\omega_2(x), x), \tag{B.2}$$

with (c_1, c_2) arbitrary constants, has no movable critical singularities, and it obeys a second order ODE which is P6 in the particular case $\alpha = \beta = \gamma = \delta - 1/2 = 0$. Its explicit expression is [346, pp. 506–508, 512–513]

$$\begin{cases} u = \dfrac{x+1}{3} + 4\wp(\omega, g_2, g_3), \\ x(x-1)\dfrac{\mathrm{d}^2\omega}{\mathrm{d}x^2} + (2x-1)\dfrac{\mathrm{d}\omega}{\mathrm{d}x} + \dfrac{\omega}{4} = 0, \\ g_2 = \dfrac{x^2 - x + 1}{12},\ g_3 = \dfrac{2x^3 - 3x^2 - 3x + 2}{432}, \end{cases} \tag{B.3}$$

therefore ω is any linear combination of the two linearly independent half-periods $\omega_1(x), \omega_2(x)$.

Despite its integration with elliptic functions, this "Picard case" is not a classical solution in the sense defined in Sect. A.2. As noticed by Painlevé [349, §19 p. 26], it is part of the differential equations describing the dependence of the elliptic functions on their invariants, established by Halphen [193, vol. I, pp. 252–253 and 291–331].

The complete P6 is absent in [349] because of an omission in the tables of Painlevé. It was first established by R. Fuchs [156], who considered a second order linear ODE for $\psi(t)$ with four Fuchsian singularities of crossratio x (located for instance at $t=\infty,0,1,x$), with in addition, as prescribed by Poincaré [360, pp. 217–220] for the isomonodromy problem, one apparent singularity $t=u$,

$$
\begin{aligned}
-\frac{2}{\psi}\frac{\mathrm{d}^2\psi}{\mathrm{d}t^2} &= \frac{A}{t^2}+\frac{B}{(t-1)^2}+\frac{C}{(t-x)^2}+\frac{D}{t(t-1)}+\frac{3}{4(t-u)^2} \\
&\quad +\frac{a}{t(t-1)(t-x)}+\frac{b}{t(t-1)(t-u)},
\end{aligned}
\tag{B.4}
$$

(A,B,C,D) being constants and (a,b) parameters. The requirement that the monodromy matrix (which transforms two independent solutions ψ_1,ψ_2 when t goes around a singularity) be independent of the nonapparent singularity x (isomonodromy condition) results in the constraint that u, as a function of x, obeys P6.

B.1 Coalescence Cascade

Most results on the Pn equations can be deduced from the sole consideration of P6 by a confluence process [350]. This process is an extension of the confluence (or coalescence, or degeneracy, these are all synonyms) among second order linear ODEs which, starting from the hypergeometric equation of Gauss, yields successively the equations of Whittaker, Weber–Hermite, Bessel, Airy.

The P6 equation defined on p. 8 depends on four parameters $\alpha,\beta,\gamma,\delta$ and the coefficient of u'^2 possesses four poles $(\infty,0,1,x)$ having the same residue $1/2$. The coalescence scheme of these four poles

$$
(1/2,1/2,1/2,1/2) \;\to\; (1/2,1,1/2)
\begin{array}{c}
\nearrow\ (1/2,3/2)\ \searrow \\
\\
\searrow\ (1,1)\ \nearrow
\end{array}
(2)
\tag{B.5}
$$

defines from P6$(u,x,\alpha,\beta,\gamma,\delta)$ four equations with four parameters, chosen as follows by Garnier [167][1]

$$
\begin{aligned}
\text{P6}\ :\ u'' &= \frac{1}{2}\left[\frac{1}{u}+\frac{1}{u-1}+\frac{1}{u-x}\right]u'^2-\left[\frac{1}{x}+\frac{1}{x-1}+\frac{1}{u-x}\right]u' \\
&\quad +\frac{u(u-1)(u-x)}{x^2(x-1)^2}\left[\alpha+\beta\frac{x}{u^2}+\gamma\frac{x-1}{(u-1)^2}+\delta\frac{x(x-1)}{(u-x)^2}\right],
\end{aligned}
$$

[1] This choice is not unique, and another P4′ has been proposed [338] which enjoys the same properties.

$$
\begin{aligned}
\text{P5} &: u'' = \left[\frac{1}{2u} + \frac{1}{u-1}\right] u'^2 - \frac{u'}{x} + \frac{(u-1)^2}{x^2}\left[\alpha u + \frac{\beta}{u}\right] + \gamma\frac{u}{x} + \delta\frac{u(u+1)}{u-1},\\
\text{P4}' &: u'' = \frac{u'^2}{2u} + \gamma\left(\frac{3}{2}u^3 + 4xu^2 + 2x^2u\right) + 4\delta(u^2 + xu) - 2\alpha u + \frac{\beta}{u},\\
\text{P3} &: u'' = \frac{u'^2}{u} - \frac{u'}{x} + \frac{\alpha u^2 + \gamma u^3}{4x^2} + \frac{\beta}{4x} + \frac{\delta}{4u},\\
\text{P2}' &: u'' = \delta(2u^3 + xu) + \gamma(6u^2 + x) + \beta u + \alpha,\\
\text{J} &: u'' = 2\delta u^3 + 6\gamma u^2 + \beta u + \alpha.
\end{aligned}
$$

The added equation J (like “Jacobi”) is the autonomous limit of P2′. As compared to the usual choice listed on p. 8, the additional symbols γ in P4 and δ in P2 have been added to represent the signs of the two opposite residues $\pm u_0$ of each family of movable simple poles, and the coefficient δ in P4 can be set to zero when $\gamma \neq 0$.

Table B.1 collects various basic data on the above Pn equations, such as leading behavior, Fuchs indices and monodromy exponents.

Table B.1 [Reproduced from [92]] Useful data for the Pn equations. We follow the notation of Okamoto [340], rather than the one of Jimbo and Miwa [240], and each monodromy exponent θ_j, including θ_∞, has a square rational in $\alpha, \beta, \gamma, \delta$. The successive lines are : the singularity order q of the Pn equation and the positive Fuchs index i, the value of the first coefficient u_0 of the Laurent series for u, the notation for the square root of u_0, the definition of the monodromy exponents θ, the components of the column vector θ.

	P2′	P3	P4′	P5	P6
q,i	$-3,4$	$-4,2$	$-4,3$	$-5,1$	$-6,1$
u_0	d^{-1}	$2c^{-1}x$	c^{-1}	$\theta_\infty^{-1}x$	$\theta_\infty^{-1}x(1-x)$
$\sqrt{}$	$d^2=\delta$	$c^2=\gamma$	$c^2=\gamma$	$\theta_\infty^2=2\alpha$	$\theta_\infty^2=2\alpha$
α	$-d\theta_\infty$	$2c\theta_\infty$	$2c\theta_\infty$	$\theta_\infty^2/2$	$\theta_\infty^2/2$
β		$-2d\theta_0$	$-8\theta_0^2$	$-\theta_0^2/2$	$-\theta_0^2/2$
γ		c^2	c^2	$-d\theta_1$	$\theta_1^2/2$
δ	d^2	$-d^2$		$-d^2/2$	$(1-\theta_x^2)/2$
θ	(θ_∞)	$\begin{pmatrix}\theta_\infty\\ \theta_0\end{pmatrix}$	$\begin{pmatrix}\theta_\infty\\ \theta_0\end{pmatrix}$	$\begin{pmatrix}\theta_\infty\\ \theta_0\\ \theta_1\end{pmatrix}$	$\begin{pmatrix}\theta_\infty\\ \theta_0\\ \theta_1\\ \theta_x\end{pmatrix}$

When studying Lax pairs, birational transformations, etc, the above sequence of equations is more advantageous to consider than the usual sequence on p. 8. Indeed, the latter emphasizes the irreducibility at the expense of the conservation of the number of parameters, while the former allows one to make full use of the four parameters $\alpha, \beta, \gamma, \delta$. The successive coalescences

$$\begin{array}{ccccccccc} & & & & \mathrm{P4}' & & & & \\ & & & \nearrow & & \searrow & & & \\ \mathrm{P6} & \rightarrow & \mathrm{P5} & & & & \mathrm{P2}' & \rightarrow & \mathrm{J} \\ & & & \searrow & & \nearrow & & & \\ & & & & \mathrm{P3} & & & & \end{array} \tag{B.6}$$

of an equation $E(x,u,\alpha,\beta,\gamma,\delta)=0$ to another equation $E(X,U,A,B,C,D)=0$ are described by affine transformations $(x,u,\alpha,\beta,\gamma,\delta)\rightarrow(X,U,A,B,C,D,\varepsilon)$,

$$\begin{aligned}
6\rightarrow 5: \ (x,u,\alpha,\beta,\gamma,\delta) &= (1+\varepsilon X, U, A, B, \varepsilon^{-1}C-\varepsilon^{-2}D, \varepsilon^{-2}D),\\
5\rightarrow 4': \ (x,u,\alpha,\beta,\gamma,\delta) &= (1+\varepsilon X, \varepsilon U/2, 2C\varepsilon^{-4}+28D\varepsilon^{-3}, B/4,\\
&\quad -4C\varepsilon^{-4}-60D\varepsilon^{-3},\\
&\quad 2A\varepsilon^{-2}-2C\varepsilon^{-4}-32D\varepsilon^{-3}),\\
5\rightarrow 3: \ (x,u,\alpha,\beta,\gamma,\delta) &= (X, 1+\varepsilon U, \varepsilon^{-1}A/4+\varepsilon^{-2}C/8,\\
&\quad -\varepsilon^{-2}C/8, \varepsilon B/4, \varepsilon^2 D/8),\\
4'\rightarrow 2': \ (x,u,\alpha,\beta,\gamma,\delta) &= (\varepsilon^2 X/4, 1+\varepsilon U, -4B\varepsilon^{-4}+96C\varepsilon^{-5}-24D\varepsilon^{-6},\\
&\quad 16A\varepsilon^{-3}-8B\varepsilon^{-4}+48C\varepsilon^{-5}-8D\varepsilon^{-6},\\
&\quad -32C\varepsilon^{-5}+16D\varepsilon^{-6}, 48C\varepsilon^{-5}-16D\varepsilon^{-6}),
\end{aligned}$$

$$
\begin{aligned}
3 \to 2' : \ (x,u,\alpha,\beta,\gamma,\delta) = (&1+\varepsilon^2 X/2, 1+\varepsilon U,\\
&16B\varepsilon^{-4} - 64C\varepsilon^{-5} - 32D\varepsilon^{-6},\\
&32D\varepsilon^{-6}, -8B\varepsilon^{-4} + 48C\varepsilon^{-5} + 16D\varepsilon^{-6},\\
&16A\varepsilon^{-3} - 8B\varepsilon^{-4} + 16C\varepsilon^{-5} - 16D\varepsilon^{-6}),\\
2' \to \mathrm{J} : \ (x,u,\alpha,\beta,\gamma,\delta) = (&\varepsilon X, \varepsilon^{-1}U, \varepsilon^{-3}A, \varepsilon^{-2}B, \varepsilon^{-1}C, D)
\end{aligned}
$$

in which ε goes to zero. The classical coalescence among the Pn listed on p. 8 can be found in [350, 163]. The confluence for the monodromy exponents θ, established in [340], is recalled in Table B.2. All these transformations are affine, and this is certainly the reason why in 1906 Painlevé changed his original choice of P3[2].

Table B.2 [Reproduced from [92]] Confluence of the monodromy exponents. The parameters c, d (which essentially represent signs) also participate to the confluence. The choice of square roots is such that there are only $+$ signs in the successive values $\theta_\infty + \theta_0 + \theta_1 + \theta_x$, $\theta_\infty + \theta_0 + \theta_1$, $2\theta_\infty + 2\theta_0$, $\theta_\infty + \theta_0$, $2\theta_\infty$.

	x	u	θ_∞	θ_0	θ_1	θ_x	c	d
$6 \to 5$	$1+\varepsilon X$	U	θ_∞	θ_0	$\theta_1 - \varepsilon^{-1}d$	$\varepsilon^{-1}d + \varepsilon$		
$5 \to 4$	$1+\varepsilon X$	$\varepsilon U/2$	$-2c\varepsilon^{-2}$	$2\theta_0$	$2\theta_\infty + 2c\varepsilon^{-2}$			$2c\varepsilon^{-2} - 2\theta_\infty$
$5 \to 3$	X	$1+\varepsilon U$	$\theta_\infty + \varepsilon^{-1}c/2$	$-\varepsilon^{-1}c/2$	θ_0			$\varepsilon d/2$
$4 \to 2$	$\varepsilon X/4 - \varepsilon^{-1}$	$\varepsilon^{-1} + U$	$-\varepsilon^{-3}d$	$\theta_\infty + \varepsilon^{-3}d$			$4\varepsilon^{-1}d$	
$3 \to 2$	$1+\varepsilon^2 X/2$	$1+\varepsilon U$	$4d\varepsilon^{-3}$	$2\theta_\infty - 4d\varepsilon^{-3}$			$-4d\varepsilon^{-3}$	$2\theta_\infty + 4d\varepsilon^{-3}$

B.2 Invariance Under Homographies

The permutations of the four singular points $\infty, 0, 1, x$ of P6 which leave this equation invariant act as homographies on x and u. These 24 permutations are generated by the three elements $(\theta, x, u) \to (\Theta, X, U)$,

$$\mathrm{H_{bacd}} : \Theta = (\theta_0, \theta_\infty, \theta_1, \theta_x), \ xX = 1, \ uU = 1, \tag{B.7}$$

$$\mathrm{H_{bcad}} : \Theta = (\theta_0, \theta_1, \theta_\infty, \theta_x), \ x(1-X) = 1, \ (u-1)U = 1, \tag{B.8}$$

$$\mathrm{H_{badc}} : \Theta = (\theta_0, \theta_\infty, \theta_x, \theta_1), \ x = X, \ uU = x. \tag{B.9}$$

Four of these twenty-four also conserve x, the identity and

$$\mathrm{H_{dcba}} : \Theta = (\theta_x, \theta_1, \theta_0, \theta_\infty), \ (u-x)(U-x) = x(x-1), \tag{B.10}$$

$$\mathrm{H_{badc}} : \Theta = (\theta_0, \theta_\infty, \theta_x, \theta_1), \ uU = x, \tag{B.11}$$

[2] The old choice of P3 is probably responsible for the absence of the Hamiltonian of P3 in Malmquist [293].

$$\text{H}_{\text{cdab}} : \Theta = (\theta_1, \theta_x, \theta_\infty, \theta_0), \ (u-1)(U-1) = 1 - x. \tag{B.12}$$

For generic values of $\alpha, \beta, \gamma, \delta$, the confluence defines the following homographies leaving invariant the $\text{Pn}(\alpha, \beta, \gamma, \delta)$,

$$\begin{aligned}
\text{P6} \ & (u,x,\alpha,\beta,\gamma,\delta) \to (1/u, 1/x, -\beta, -\alpha, \gamma, \delta), \\
& \qquad \to (1-1/u, 1-1/x, -\beta, -\gamma, \alpha, \delta), \\
& \qquad \to (x/u, x, -\beta, -\alpha, -\delta+1/2, -\gamma+1/2), \\
\text{P5} \ & (u,x,\alpha,\beta,\gamma,\delta) \to (u, -x, \alpha, \beta, -\gamma, \delta), \\
& \qquad \to (1/u, x, -\beta, -\alpha, -\gamma, \delta), \\
\text{P4}' \ & (u,x,\alpha,\beta,\gamma,\delta) \to (ku, kx, k^2\alpha, \beta, k^4\gamma, k^3\delta), \\
\text{P4} \ & (u,x,\alpha,\beta) \to (ku, kx, k^2\alpha, \beta), \ k^4 = 1, \\
\text{P3} \ & (u,x,\alpha,\beta,\gamma,\delta) \to (\lambda u, \mu x, \lambda^{-1}\mu^{-1}\alpha, \lambda\mu^{-1}\beta, \lambda^{-2}\mu^{-2}\gamma, \lambda^2\mu^{-2}\delta), \\
& \qquad \to (u^{-1}, x, -\beta, -\alpha, -\delta, -\gamma), \qquad \text{(B.13)} \\
\text{P2}' \ & (u,x,\alpha,\beta,\gamma,\delta) \to (ku, k^2x, k^{-3}\alpha, k^{-4}\beta, k^{-5}\gamma, k^{-6}\delta), \\
\text{P2} \ & (u,x,\alpha) \to (ku, k^2x, k^{-3}\alpha), \ k^6 = 1, \\
\text{P1} \ & (u,x) \to (ku, k^2x), \ k^5 = 1,
\end{aligned}$$

(B.14)

in which k, λ, μ are arbitrary nonzero constants.

B.3 Invariance Under Birational Transformations

There exists another group of invariance, which maps a Pn equation to itself while changing its parameters (this therefore does not apply to P1). Already encountered in the text for P2 (5.289) and P6 (5.305), this group acts on the equation for u as a birational transformation and on the monodromy exponents as an *affine transformation* (relation (5.308) in the case of P6).

Before applying the confluence to the elementary birational transformation T_6 of P6, one should be aware of the noncommutativity of the permutation of the four singularities of P6 on one hand, the convention adopted for defining the lower Pn equations on the other hand (e.g., at the P5 level, two equivalent singularities plus another one, chosen once and for all as $(\infty, 0)$ and 1). This results in two distinct sequences [92].

B.3.1 Normal Sequence

The choice of $(1,x)$ as the coalescing pair (adopted in Table B.2) leads to

$$\text{P6}: \begin{pmatrix} s_\infty\theta_\infty \\ s_0\theta_0 \\ s_1\theta_1 \\ s_x\theta_x \end{pmatrix} = \frac{1}{2}\begin{pmatrix} 1 & -1 & -1 & -1 \\ -1 & 1 & -1 & -1 \\ -1 & -1 & 1 & -1 \\ -1 & -1 & -1 & 1 \end{pmatrix}\begin{pmatrix} S_\infty\Theta_\infty \\ S_0\Theta_0 \\ S_1\Theta_1 \\ S_x\Theta_x \end{pmatrix} + \frac{1}{2}\begin{pmatrix} 1 \\ 1 \\ 1 \\ 1 \end{pmatrix}, \tag{B.15}$$

$$\text{P5}: \begin{pmatrix} s_\infty\theta_\infty \\ s_0\theta_0 \\ s_1\theta_1 \end{pmatrix} = \frac{1}{2}\begin{pmatrix} 1 & -1 & -1 \\ -1 & 1 & -1 \\ -2 & -2 & 0 \end{pmatrix}\begin{pmatrix} S_\infty\Theta_\infty \\ S_0\Theta_0 \\ S_1\Theta_1 \end{pmatrix} + \frac{1}{2}\begin{pmatrix} 1 \\ 1 \\ 2 \end{pmatrix}, s_\infty d = S_\infty D, \tag{B.16}$$

$$\text{P4}: \begin{pmatrix} s_\infty\theta_\infty \\ s_0\theta_0 \end{pmatrix} = \frac{1}{2}\begin{pmatrix} -1 & -3 \\ -1 & 1 \end{pmatrix}\begin{pmatrix} S_\infty\Theta_\infty \\ S_0\Theta_0 \end{pmatrix} + \frac{1}{4}\begin{pmatrix} 3 \\ 1 \end{pmatrix}, s_\infty c = S_\infty C, \tag{B.17}$$

$$\text{P3}: \begin{pmatrix} s_\infty\theta_\infty \\ s_0\theta_0 \end{pmatrix} = \begin{pmatrix} 0 & -1 \\ -1 & 0 \end{pmatrix}\begin{pmatrix} S_\infty\Theta_\infty \\ S_0\Theta_0 \end{pmatrix} + \begin{pmatrix} 1 \\ 1 \end{pmatrix}, s_\infty c = S_\infty C,\ s_0 d = S_0 D, \tag{B.18}$$

$$\text{P2}: (s\theta_\infty) = -(S\Theta_\infty) + (1), s_\infty d = S_\infty D, \tag{B.19}$$

and (we omit the signs, they can easily be restored)[3]

$$\text{P6}: \frac{N}{u-U} = \frac{x(x-1)U'}{U(U-1)(U-x)} + \frac{\Theta_0}{U} + \frac{\Theta_1}{U-1} + \frac{\Theta_x - 1}{U-x}, \tag{B.20}$$

$$\text{P5}: \frac{N}{u-U} = \frac{xU'}{U(U-1)^2} + \frac{\Theta_0}{U} + \frac{\Theta_1 - 1}{U-1} + \frac{Dx}{(U-1)^2}, \tag{B.21}$$

$$\text{P4}: \frac{N}{u-U} = \frac{U'}{U} + \frac{4\Theta_0}{U} + CU + 2Cx, \tag{B.22}$$

$$\text{P3}: \frac{N}{u-U} = \frac{xU'}{U^2} + \frac{\Theta_0 - 1}{U} + \frac{Dx}{2U^2} - \frac{C}{2}, \tag{B.23}$$

$$\text{P2}: \frac{N}{u-U} = U' + DU^2 + D\frac{x}{2}, \tag{B.24}$$

with

$$\text{P6}: N = 1 - \Theta_\infty - \Theta_0 - \Theta_1 - \Theta_x = (1/2)\sum(\theta_j - \Theta_j), \tag{B.25}$$

$$\text{P5}: N = 1 - \Theta_\infty - \Theta_0 - \Theta_1 = (1/2)\sum(\theta_j - \Theta_j), \tag{B.26}$$

$$\text{P4}: N = -2(1 - 2\Theta_\infty - 2\Theta_0) = 2\sum(\theta_j - \Theta_j), \tag{B.27}$$

$$\text{P3}: N = 1 - \Theta_\infty - \Theta_0 = (1/2)\sum(\theta_j - \Theta_j), \tag{B.28}$$

$$\text{P2}: N = \frac{1}{2} - \Theta_\infty = (1/2)(\theta_\infty - \Theta_\infty). \tag{B.29}$$

For the signs $s_j = 1$, all the shifts are positive, and, for the signs $s_j = S_j$, the linear part has determinant -1. The sum of the shifts remains equal to two (except for P4 and P2 because of a global rescaling, see Table B.1).

[3] The second half of each birational transformation, not written, is deduced from the first half by exchanging the lowercase and uppercase notation, as done in the text (5.289). With such a convention, each transformation is an involution.

Since the affine representations (B.15)–(B.19) are involutions when the signs satisfy $s_j = S_j$, only half of the birational representation needs to be written, the second half resulting from the permutation of (u,θ,c,d) and (U,Θ,C,D).

These transformations were first found respectively, for P5 by Okamoto [342], for P4 by Murata [310], for P3 by Gromak [185, Eqs. (14)–(15)], for P2 by Lukashevich [288].

B.3.2 Biased Sequence

After performing on T_6 the permutation $\mathrm{H}_{\mathrm{dcba}}$ (B.10) and suitable sign reversals so as to conserve the involution property ($\mathrm{S_a}, \mathrm{S_b}, \mathrm{S_c}, \mathrm{S_d}$ denote the operators which change the sign of, respectively, θ_∞, θ_0, θ_1, θ_x),

$$\mathrm{T}_{6,\mathrm{b}} = \mathrm{S_a S_d T_6 H_{dcba} S_d S_a}, \tag{B.30}$$

the application of the confluence to the involution $\mathrm{T}_{6,\mathrm{b}}$ yields

$$\mathrm{P6}: \begin{pmatrix} s_\infty\theta_\infty \\ s_0\theta_0 \\ s_1\theta_1 \\ s_x\theta_x \end{pmatrix} = -\frac{1}{2}\begin{pmatrix} 1 & -1 & -1 & -1 \\ -1 & 1 & -1 & -1 \\ -1 & -1 & 1 & -1 \\ -1 & -1 & -1 & 1 \end{pmatrix}\begin{pmatrix} S_\infty\Theta_\infty \\ S_0\Theta_0 \\ S_1\Theta_1 \\ S_x\Theta_x \end{pmatrix} + \frac{1}{2}\begin{pmatrix} -1 \\ 1 \\ 1 \\ -1 \end{pmatrix}, \tag{B.31}$$

$$\mathrm{P5}: \begin{pmatrix} s_\infty\theta_\infty \\ s_0\theta_0 \\ s_1\theta_1 \end{pmatrix} = -\frac{1}{2}\begin{pmatrix} 1 & -1 & -1 \\ -1 & 1 & -1 \\ -2 & -2 & 0 \end{pmatrix}\begin{pmatrix} S_\infty\Theta_\infty \\ S_0\Theta_0 \\ S_1\Theta_1 \end{pmatrix} + \frac{1}{2}\begin{pmatrix} -1 \\ 1 \\ 0 \end{pmatrix},$$
$$s_\infty d = -S_\infty D, \tag{B.32}$$

$$\mathrm{P4}: \begin{pmatrix} s_\infty\theta_\infty \\ s_0\theta_0 \end{pmatrix} = -\frac{1}{2}\begin{pmatrix} -1 & -3 \\ -1 & 1 \end{pmatrix}\begin{pmatrix} S_\infty\Theta_\infty \\ S_0\Theta_0 \end{pmatrix} + \frac{1}{4}\begin{pmatrix} -1 \\ 1 \end{pmatrix}, \; s_\infty c = -S_\infty C, \tag{B.33}$$

$$\mathrm{P3}: \begin{pmatrix} s_\infty\theta_\infty \\ s_0\theta_0 \end{pmatrix} = -\begin{pmatrix} 0 & -1 \\ -1 & 0 \end{pmatrix}\begin{pmatrix} S_\infty\Theta_\infty \\ S_0\Theta_0 \end{pmatrix}, \; s_\infty c = -S_\infty C, \; s_0 d = -S_0 D, \tag{B.34}$$

$$\mathrm{P2}: (s\theta_\infty) = (S\Theta_\infty), \; s_\infty d = -S_\infty D, \tag{B.35}$$

and

$$\mathrm{P6}: \frac{-Nx(x-1)}{(u-x)(U-x)-x(x-1)} = (U-x) \times\left(\frac{x(x-1)U'}{U(U-1)(U-x)} + \frac{\Theta_0}{U} + \frac{\Theta_1}{U-1} + \frac{\Theta_\infty - \Theta_0 - \Theta_1}{U-x}\right), \tag{B.36}$$

$$\mathrm{P5}: \frac{-2Dx}{(u-1)(U-1)} = (U-1)\left(\frac{xU'}{U(U-1)^2} + \frac{\Theta_0}{U} + \frac{\Theta_\infty - \Theta_0}{U-1} + \frac{Dx}{(U-1)^2}\right), \tag{B.37}$$

$$\mathrm{P4}: 2C(u+U) = \frac{U'}{U} + \frac{4\Theta_0}{U} + CU - 2Cx, \tag{B.38}$$

$$\text{P3}: \frac{Dx}{uU} = -C, \tag{B.39}$$

$$\text{P2}: u + U = 0. \tag{B.40}$$

In (B.36), the constant N is any expression among

$$N = \sum(\theta_k^2 - \Theta_k^2) \tag{B.41}$$

$$= 1 + S_\infty\Theta_\infty - S_0\Theta_0 - S_1\Theta_1 + S_x\Theta_x \tag{B.42}$$

$$= -1 - s_\infty\theta_\infty + s_0\theta_0 + s_1\theta_1 - s_x\theta_x \tag{B.43}$$

$$= -2(s_\infty\theta_\infty - S_x\Theta_x) = 2(s_0\theta_0 - S_1\Theta_1)$$

$$= 2(s_1\theta_1 - S_0\Theta_0) = -2(s_x\theta_x - S_\infty\Theta_\infty). \tag{B.44}$$

At the P3 level, the transformation reduces to the permutation of the two singular points $(\infty, 0)$, a homography on u which leaves P3 invariant. Therefore, at the P2 level this is just the parity.

The biased affine representations (B.31)–(B.35) and the normal ones (B.15)–(B.19) have opposite linear parts (this results from our involution convention), but the sum of the biased shifts is zero.

The transformation for P5 has first been obtained by Gromak [186, (10)–(11)], and the one for P4 by Lukashevich [287].

If one denotes $\mathrm{T}_{6,\mathrm{b}}$, $\mathrm{T}_{5,\mathrm{b}}$ and $\mathrm{T}_{4,\mathrm{b}}$ the biased transformations, $\mathrm{T}_{6,\mathrm{u}}$, $\mathrm{T}_{5,\mathrm{u}}$ and $\mathrm{T}_{4,\mathrm{u}}$ the normal ones, and H the unique homography of P5 which conserves x,

$$\text{P5}: \mathrm{H}(x, u, \theta_\infty, \theta_0, \theta_1) = (x, u^{-1}, \theta_0, \theta_\infty, \theta_1), \tag{B.45}$$

the relations

$$\mathrm{T}_{6,\mathrm{u}} = \mathrm{S_a S_d T_{6,b} H_{dcba} S_d S_a}, \tag{B.46}$$

$$\mathrm{T}_{5,\mathrm{u}} = \mathrm{S_a T_{5,b} S_a S_c T_{5,b} S_a H}, \tag{B.47}$$

$$\mathrm{T}_{4,\mathrm{u}} = \mathrm{S_a T_{4,b} S_b T_{4,b} S_a}, \tag{B.48}$$

(the relation (B.48) is due to [23]) show that, at the P5 and P4 levels, the biased birational transformations are more elementary than the unbiased ones.

B.4 Invariance Under Affine Weyl Groups

Under the birational group, the Pn equations define six classes of equivalence, leading to the historical labelling as P1, ..., P6. This labelling can be refined with the classes of equivalence of another group, the group of *affine transformations* which acts on the monodromy exponents in the birational transformations, see Sect. B.3, (B.31)–(B.35) and (B.15)–(B.19). Indeed, these affine transformations generate groups known as *affine Weyl groups*, whose generators are reflections and translations, well introduced for instance in [435].

Let us present the affine Weyl group of P6 [339]. As a consequence of the affine transformations (B.15), there exists a group of reflections (a synonym is Coxeter group) leaving P6 invariant. Indeed, if one denotes

$$\kappa_0 = \frac{1-\theta_\infty-\theta_0-\theta_1-\theta_x}{2},\ (\kappa_1,\kappa_2,\kappa_3,\kappa_4) = (\theta_\infty,\theta_0,\theta_1,\theta_x), \quad \text{(B.49)}$$

this group is generated by the five reflections σ_i through the planes $\kappa_i = 0$,

$$\sigma_i(\kappa_j) = \kappa_j - \kappa_i c_{ij},\ C = (c_{ij}) = \begin{pmatrix} 2 & -1 & -1 & -1 & -1 \\ -1 & 2 & 0 & 0 & 0 \\ -1 & 0 & 2 & 0 & 0 \\ -1 & 0 & 0 & 2 & 0 \\ -1 & 0 & 0 & 0 & 2 \end{pmatrix}. \quad \text{(B.50)}$$

This group is of a particular type called *affine Weyl group* of type $D_4^{(1)}$, denoted $W(D_4^{(1)})$ or simply $D_4^{(1)}$, and it is characterized by its *Cartan matrix* $C = (c_{ij})$.

All the affine transformations (B.31)–(B.35) and (B.15)–(B.19) can similarly be represented by some affine Weyl group and, since the definition of the monodromy exponents Table B.1 assumes some constants to be nonzero, one must split the P5 and P3 equations into cases represented by different affine Weyl groups.

In order to classify the Pn equations under both criteria (irreducibility, affine Weyl groups), one must discard (i) reducible equations (the only one is P3 for $\alpha = \beta = \gamma = \delta = 0$, which integrates as $u = c_1 x^{c_2}$) and (ii) any equation birationally equivalent to an equation already in the list (the only one is P5 for $\delta = 0$, birationally equivalent to P3-$D_6^{(1)}$ [185]). The result is a list of eight (instead of six) equations [377] enumerated in Table B.3.

Table B.3 List of the eight Painlevé equations which are irreducible, inequivalent and with distinct affine Weyl groups. The last column indicates the number of essential parameters in the equation.

Dynkin diagram	Affine Weyl group	Painlevé equation	Param
$D_4^{(1)}$	$D_4^{(1)}$	P6	4
$D_5^{(1)}$	$A_3^{(1)}$	P5, $\delta \neq 0$	3
$E_6^{(1)}$	$A_2^{(1)}$	P4	2
$D_6^{(1)}$	$(2A_1)^{(1)}$	P3, $\gamma\delta \neq 0$	2
$D_7^{(1)}$	$A_1^{(1)}$	P3, $(\gamma \neq 0, \delta = 0)$ or $(\gamma = 0, \delta \neq 0)$	1
$D_8^{(1)}$	Σ_2	P3, $\gamma = \delta = 0, (\alpha,\beta) \neq (0,0)$	0
$E_7^{(1)}$	$A_1^{(1)}$	P2	1
$E_8^{(1)}$	None	P1	0

B.5 Invariance Under Nonbirational Transformations

For nongeneric values of the parameters $\alpha, \beta, \gamma, \delta$, the eight Painlevé equations (except P1 and P3 with $D_8^{(1)}$ Dynkin diagram, see Table B.3) are invariant under an algebraic, nonbirational, noncanonical[4] transformations [406]. These transformations are extensions of the Goursat transformation of the hypergeometric equation,

$$F\left(a,b,\frac{a+b+1}{2};x\right) = F\left(\frac{a}{2},\frac{b}{2},\frac{a+b+1}{2};4x(1-x)\right). \tag{B.51}$$

For instance, when all monodromy exponents θ_j of P6(u,x,θ_j) are ± 1, this equation is invariant under $(u,x,\theta_j) \to (U,X,\Theta_j)$, with

$$\begin{aligned} &x = \frac{1}{2} + \frac{X^{1/2}+X^{-1/2}}{4},\ u = \frac{1}{2} + \frac{X^{1/2}U^{-1}+X^{-1/2}U}{4}, \\ &\theta_j = (\varepsilon_1,\varepsilon_1,\varepsilon_2,\varepsilon_2),\ \Theta_j = (2\varepsilon_1,0,0,2\varepsilon_2),\ \varepsilon_1^2 = \varepsilon_2^2 = 1. \end{aligned} \tag{B.52}$$

The exhaustive list of such transformations can be found in [406].

B.6 Hamiltonian Structure

Each Pn possesses a Hamiltonian structure, with $H(q,p,x)$ polynomial in the canonical variables q and p,

$$\frac{\mathrm{d}q}{\mathrm{d}x} = -\frac{\partial H}{\partial p},\ \frac{\mathrm{d}p}{\mathrm{d}x} = \frac{\partial H}{\partial q},\ u = q. \tag{B.53}$$

These Hamiltonians obey the coalescence and their expression is [293],

---

[4] Under this transformation, the symplectic form $\mathrm{d}p \wedge \mathrm{d}q - \mathrm{d}H \wedge \mathrm{d}x$ is multiplied by a small integer, 2 for (B.52).

$$
\begin{cases}
\text{P6}: \; H = \dfrac{q(q-1)(q-x)}{x(x-1)} \left[p^2 - \left(\dfrac{\theta_0}{q} + \dfrac{\theta_1}{q-1} + \dfrac{\theta_x - 1}{q-x} \right) p \right. \\
\qquad\qquad \left. - \dfrac{\theta_\infty^2 - (\theta_0 + \theta_1 + \theta_x - 1)^2}{4q(q-1)(q-x)} \right], \\
\text{P5}, \; \delta \neq 0: \; H = \dfrac{q(q-1)^2}{x} \left[p^2 + \left(\dfrac{\theta_0}{q} + \dfrac{\theta_1 + 1}{q-1} + \dfrac{dx}{(q-1)^2} \right) p \right. \\
\qquad\qquad \left. - \dfrac{\theta_\infty^2 - (\theta_0 + \theta_1 + 1)^2}{4q(q-1)} \right], \\
\text{P4}': \; H = qp^2 + \left[cq^2 + 2 \left(cx + \dfrac{\delta}{c} \right) q + 4\theta_0 \right] p \\
\qquad\qquad + \left[c(2\theta_\infty + 2\theta_0 + 1) + \dfrac{\delta^2}{c^2} \right] q, \\
\text{P3}: \; H = \dfrac{q^2}{x} \left[p^2 + \left(\dfrac{a_0}{q} + \dfrac{dx}{2q^2} + \dfrac{a_1}{2} \right) p + \dfrac{a_2}{q^2} \right. \\
\qquad\qquad \left. + \dfrac{2a_0 a_1 - \alpha}{8q} + \dfrac{a_1^2 - \gamma}{16} + \dfrac{(\beta - 2d + 2a_0 d)x}{q^3} \right], \; \delta = -d^2, \\
\text{P2}': \; H = \dfrac{p^2}{2} + \dfrac{d}{2}(2q^3 + xq + q) - \alpha q - \dfrac{\beta}{2} q^2 - \gamma(2q^3 + xq),
\end{cases}
\tag{B.54}
$$

in which the a_j in P3 are arbitrary constants which can be used to match the three varieties of P3, see Table B.3.

B.7 Lax Pairs

All Pn equations admit second order Lax pairs.

In scalar form, this second order Lax pair of P6 [156] easily defines Lax pairs for the lower Pn equations under action of the coalescence [167]. These scalar Lax pairs are linear in $\alpha, \beta, \gamma, \delta$ and characterized by the two homographic invariants (S, C), with the cross-derivative condition $Z = 0$

$$\partial_t^2 \psi + (S/2)\psi = 0, \tag{B.55}$$

$$\partial_x \psi + C\partial_t \psi - (1/2)C_t \psi = 0, \tag{B.56}$$

$$Z \equiv S_x + C_{ttt} + CS_t + 2C_t S = 0. \tag{B.57}$$

The value of S is given by

$$-\frac{S}{2} = \frac{3/4}{(t-u)^2} + \frac{\beta_1 u' + \beta_0}{(t-u)e_1} + \frac{[(\beta_1 u')^2 - \beta_0^2]e_0 + f_{\mathrm{G}}(u)}{e_2} + f_{\mathrm{G}}(t), \tag{B.58}$$

in which the function f_{G} and the various scalars are defined in Table B.4. However, these scalar Lax pairs do not unveil the invariance under Weyl groups (Appendix B.4).

Table B.4 Coalescence of the scalar Lax pairs of the Painlevé equations

	P6	P5	P4′	P3	P2′
β_1	$-\dfrac{x(x-1)}{2(u-x)}$	$-\dfrac{x}{2(u-1)}$	$1/4$	$-\dfrac{x}{2u}$	-1
β_0	$-u+1/2$	$-u+1/2$	$1/2$	$-1/2$	0
$f_G(z)$	$\dfrac{a}{z^2}+\dfrac{b}{(z-1)^2}+\dfrac{c}{(z-x)^2}+\dfrac{d}{z(z-1)}$	$\dfrac{a}{z^2}-\dfrac{bx}{(z-1)^3}+\dfrac{cx^2}{(z-1)^4}+\dfrac{d}{(z-1)^2}$	$-\dfrac{a}{z^2}-\dfrac{\delta(z+2x)}{4}-b-\dfrac{\gamma(z+2x)^2}{16}$	$a-\dfrac{bx}{z^3}+\dfrac{cx^2}{z^4}+\dfrac{d}{z}$	$2\alpha z+\beta z^2+2\gamma(2z^3+xz)+\delta(z^4+xz^2)$
α	$2(a+b+c+d+1)$	$2(a+d+1)$	$-4b$	$8d$	
β	$-2(a+1/4)$	$-2(a+1/4)$	$-8a-2$	$8b$	
γ	$2(b+1/4)$	$2b$		$16a$	
δ	$-2c$	$-2c$		$-16c$	
e_0	$\dfrac{u-x}{u(u-1)}$	$1/u$	$1/u$	1	1
e_1	$t(t-1)$	$t(t-1)$	$-t$	t	1
e_2	$t(t-1)(t-x)$	$t(t-1)^2$	t	t^2	1
$-C$	$\dfrac{t(t-1)(u-x)}{(t-u)x(x-1)}$	$\dfrac{t(t-1)(u-1)}{(t-u)x}$	$2\dfrac{t}{t-u}$	$\dfrac{tu}{(t-u)x}$	$\dfrac{1/2}{t-u}$

As to the matrix form of the Lax pair of P6 [240], it presents two main advantages over the scalar form: (i) to remove the apparent singularity $t=u$ unavoidable in the scalar form; (ii) to display the invariance under affine Weyl groups. However it also presents various minor drawbacks [281], including: (i) meromorphic dependence on one of the four monodromy exponents θ_j, (ii) uneasy confluence, and they seem difficult to remove [80]. The original Lax pair [240] can be made traceless and with elements L_{ij}, M_{ij} rational in u', u, x. Introducing the Pauli matrices σ_k

$$\sigma_1=\begin{pmatrix}0&1\\1&0\end{pmatrix},\ \sigma_2=\begin{pmatrix}0&-i\\i&0\end{pmatrix},\ \sigma_3=\begin{pmatrix}1&0\\0&-1\end{pmatrix}, \qquad \text{(B.59)}$$

$$\sigma^+=\begin{pmatrix}0&1\\0&0\end{pmatrix},\ \sigma^-=\begin{pmatrix}0&0\\1&0\end{pmatrix}, \sigma_j\sigma_k=\delta_{jk}+i\varepsilon_{jkl}\sigma_l,$$

this Lax pair is [291, 281],

$$\partial_x\psi=L\psi,\quad \partial_t\psi=M\psi, \qquad \text{(B.60)}$$

with

$$L=-\frac{M_x}{t-x}+L_\infty,\ M=\frac{M_0}{t}+\frac{M_1}{t-1}+\frac{M_x}{t-x}, \qquad \text{(B.61)}$$

$$L_\infty=-\frac{(\Theta_\infty-1)(u-x)}{2x(x-1)}\sigma_3, \qquad \text{(B.62)}$$

$$2M_\infty=\Theta_\infty\sigma_3, \qquad \text{(B.63)}$$

$$2M_0 = z_0\sigma_3 - \frac{u}{x}\sigma^+ + (z_0^2 - \theta_0^2)\frac{x}{u}\sigma^-, \tag{B.64}$$

$$2M_1 = z_1\sigma_3 + \frac{u-1}{x-1}\sigma^+ - (z_1^2 - \theta_1^2)\frac{x-1}{u-1}\sigma^-, \tag{B.65}$$

$$2M_x = \left((\theta_0^2 - z_0^2)\frac{x}{u} - (\theta_1^2 - z_1^2)\frac{x-1}{u-1}\right)\sigma^- - \frac{u-x}{x(x-1)}\sigma^+$$
$$-(\Theta_\infty + z_0 + z_1)\sigma_3, \tag{B.66}$$

$$z_0 = \frac{1}{2\Theta_\infty x(u-1)(u-x)}\Big[\left(x(x-1)u' - (u-1)(u - \Theta_\infty(u-x))\right)^2$$
$$-(\Theta_\infty^2 + \theta_0^2)x(u-1)(u-x) + \theta_1^2(x-1)u(u-x) - \theta_x^2 x(x-1)u(u-1)\Big],$$

$$z_1 = \frac{-1}{2\Theta_\infty(x-1)u(u-x)}\Big[\left(x(x-1)u' - u(u-1-\Theta_\infty(u-x))\right)^2$$
$$+(\Theta_\infty^2 + \theta_1^2)(x-1)u(u-x) - \theta_0^2 x(u-1)(u-x) - \theta_x^2 x(x-1)u(u-1)\Big],$$

$$(2\alpha, -2\beta, 2\gamma, 1-2\delta) = ((\Theta_\infty - 1)^2, \theta_0^2, \theta_1^2, \theta_x^2),\ M_\infty = -M_0 - M_1 - M_x.$$

The meromorphic dependence (denominator Θ_∞) comes from the assumption $M_\infty = (\Theta_\infty/2)\sigma_3$.

B.8 Classical Solutions

The classical solutions (defined in Appendix A.2) admitted by the eight irreducible *Pn* (Table B.3) have all been found, except for P6 where the list is not yet exhaustive. P1 admits no such solution. Classical solutions have a dependence on the two constants of integration which is less than twice transcendental.

A first set depends rationally on one arbitrary constant, its seed is the Riccati equation defined by equating to zero the l.h.s. N and the r.h.s. of the birational transformations (B.20)–(B.24). All the elements in the set are generated from the seed by repeated applications, in any order, of (i) the birational transformation, (ii) the homographies leaving invariant the Pn equation, see details in [190]. For the respective Pn equations $n = 6,5,4,3,2$, they depend on 3,2,1,1,0 arbitrary complex constants θ_j, and 4,3,2,2,1 arbitrary integer constants. Naturally, this sequence can be generated from the Riccati subequation of P6 [157] by the coalescence. All these solutions obey a first order, any degree, genus zero ODE which is, by construction, linearizable into, respectively, the Gauss hypergeometric, Whittaker, Weber–Hermite, Bessel and Airy equations.

The second set depends on no constant at all. If one excludes particular solutions of the above Riccati equations, these solutions are rational functions for P2–P5 (more precisely rational in x for P2 [437, 419], P4 [287], P5 [286], rational in $x^{1/2}$ for the $D_6^{(1)}$ – P3 [311], rational in $x^{1/3}$ for the $D_7^{(1)}$ – P3 [187]) and algebraic functions for P6. Their list cannot be generated by the confluence. The rational ones only exist for the following constraints among the monodromy exponents

$$\begin{cases} \text{P2}: \ \theta_\infty = n, \\ \text{P3}\ D_6^{(1)} \gamma\delta \neq 0: \ \dfrac{\theta_\infty - \theta_0}{2} = n, \\ \text{P3}\ D_7^{(1)} \gamma = 0,\ \alpha\delta \neq 0: \ \dfrac{\beta}{2\sqrt{-\delta}} = n, \\ \text{P4}: \ 2\theta_\infty = m,\ 2\theta_0 = 2n - m + 1, \\ \text{P4}: \ 2\theta_\infty = m,\ 2\theta_0 = 2n - m + 1/3, \\ \text{P5},\ \delta \neq 0: \ (\theta_\infty = n,\ \theta_0 + \theta_1 = m) \text{ or } (\theta_0 = n,\ \theta_\infty + \theta_1 = m),\ m+n \text{ odd}, \end{cases} \tag{B.67}$$

in which m and n denote arbitrary integers, and all of them are generated by the birational and homographic transformations from the following seeds,

$$\begin{cases} \text{P2}: \ u = 0,\ n = 0, \\ \text{P3}\ D_6^{(1)} \gamma\delta \neq 0: \ u = z,\ z^2 = \dfrac{d}{c}x,\ n = 0, \\ \text{P3}\ D_7^{(1)} \gamma = 0,\ \alpha\delta \neq 0: \ u = z^2,\ z^3 = \alpha^{-1/2} dx,\ n = 0, \\ \text{P4}: \ u = -2x,\ \alpha = 0,\ \beta = -2,\ m = 0,\ n = 0, \\ \text{P4}: \ u = -\dfrac{2}{3}x,\ \alpha = 0,\ \beta = -\dfrac{2}{9},\ m = 0,\ n = 0, \\ \text{P5}, \delta \neq 0: \ u = -1,\ \alpha + \beta = 0,\ \gamma = 0. \end{cases} \tag{B.68}$$

The classical solutions of P6 disctinct from the Riccati ones are necessarily algebraic. The simplest one,

$$u = \sqrt{x},\ \theta_j^2 = (\lambda^2, \lambda^2, \mu^2, \mu^2), \tag{B.69}$$

with λ, μ arbitrary, is equivalently described by an algebraic curve $P(u,x) = 0$ of degree two and genus zero. Under birational transformations, the genus is conserved but the degree changes, therefore, in order to represent the algebraic solutions, it is important to minimize the degree with the help of birational transformations. Since these solutions have not yet all been found, we will not list them all, the interested reader can refer to [130, 220, 389, 131, 254, 34]. Some solutions [130, 131] are associated with the five regular three-dimensional polyhedra of Plato (cube, tetrahedron, octahaedron, dodecahedron, icosahedron), such as for the cube [130, 254, 188] (genus zero, degree three)

$$x - 6xu + 3u^2 + 3xu^2 + x^2 - 2u^3 = 0,\ \theta = (2a, 1/3, a, a),\ a \text{ arbitrary}, \tag{B.70}$$

for the tetrahedron [130, 254] (genus zero, degree six)

$$\begin{aligned} &2xu^3 - u^4 - 6x^2u^2 + 2x^2u + 2x^2u^3 + xu^4 - x^2u^4 - x^3 + 2x^3u = 0, \\ &\theta = (1/2, a, a, a),\ a \text{ arbitrary}. \end{aligned} \tag{B.71}$$

As another example, the solution [34]

$$x = \frac{(5s^2-8s+5)(7s^2-7s+4)}{s(s-2)(s+1)(2s-1)(4s^2-7s+7)}, \quad u = \frac{(7s^2-7s+4)^2}{s^3(4s^2-7s+7)^2},$$
$$\theta = \frac{(2,2,2,4)}{7}, \tag{B.72}$$

with genus zero and degree ten, can be linked to a group of Klein.

Appendix C
Brief Presentation of the Elliptic Functions

The trigonometric functions $u = \sin, \cos, \tan, \mathrm{cotan}, \sinh, \cosh, \tanh, \coth$ all obey a nonlinear first order ODE of the type

$$u'^n + a + bu^2 = 0,\ n = 1 \text{ or } 2, \tag{C.1}$$

and Abel and Jacobi have proven that the first order second degree ODE

$$u'^2 + A + Bu^2 + Cu^4 = 0, \tag{C.2}$$

which includes as particular cases all the equations (C.1), has a general solution which is single valued and therefore defines a function. This function, known as the "cnoidal wave" by physicists and as the "Jacobi elliptic function" by mathematicians, realizes an extrapolation of the above trigonometric functions to one more parameter.

There exist several equivalent definitions of the *elliptic functions* $u(x)$, for instance

1. doubly periodic functions having two incommensurate periods T_1, T_2,

$$\exists T_1, T_2,\ T_1/T_2 \notin \mathcal{Q},\ \forall n_1, n_2 \in \mathcal{Z} :\ u(x + n_1 T_1 + n_2 T_2) = u(x), \tag{C.3}$$

2. general solution of the first order m-th degree ODE with constant coefficients,

$$F(u, u') \equiv \sum_{k=0}^{m} \sum_{j=0}^{2m-2k} a_{j,k} u^j u'^k = 0,\ a_{0,m} \neq 0, \tag{C.4}$$

when it has no movable critical singularities and the algebraic curve $F(X, Y) = 0$ has genus one.

C.1 The Notation of Jacobi and Weierstrass

Jacobi has defined twelve "Jacobi functions" $\mathrm{pq}(z,k)$ as the general solution of equations (C.2) in which A,B,C only depend on one adimensional complex constant called the *modulus* k, and p,q are four letters among s,c,d,n. By moving to ∞ one of the four zeros u_j of u' in (C.2) with a homography $u(z) \to u_j + b/(\wp(x) - a)$, with x/z constant, Weierstrass has defined a very convenient normalized form, called the *Weierstrass elliptic equation*

$$\left(\frac{\mathrm{d}\wp}{\mathrm{d}x}\right)^2 = 4\wp^3 - g_2\wp - g_3 = 4(\wp - e_1)(\wp - e_2)(\wp - e_3), \tag{C.5}$$

a differential consequence of which is

$$\frac{\mathrm{d}^2\wp}{\mathrm{d}x^2} = 6\wp^2 - \frac{g_2}{2}. \tag{C.6}$$

Elliptic functions possess an addition formula, i.e. an algebraic relation between the values of the function at the three points $(x_1, x_2, x_1 + x_2)$, with (x_1, x_2) arbitrary.

$$\wp(x_1 + x_2) + \wp(x_1) + \wp(x_2) = \frac{1}{4}\left(\frac{\wp'(x_1) - \wp'(x_2)}{\wp(x_1) - \wp(x_2)}\right)^2, \tag{C.7}$$

The link between the Jacobi modulus k and the constants of Weierstrass

$$k^2 = m = \frac{e_2 - e_3}{e_1 - e_3}, \tag{C.8}$$

unfortunately breaks the invariance under permutation of the roots e_j, this follows from the requirement of Jacobi that two of his twelve functions $(\mathrm{cn}, \mathrm{sn})$ should extrapolate the cos and sin functions. As a consequence, computations involving the Jacobi functions are not very practical because they need to use a large number of formulae [9, Chap. 16] [289]. For reference only, we give the basic ones involving the *copolar trio* $(\mathrm{sn}, \mathrm{cn}, \mathrm{dn})$ (preferred by physicists, for they are bounded on the real axis, see Fig. C.1) and the *copolar trio* $(\mathrm{cs}, \mathrm{ds}, \mathrm{ns})$ (preferred by mathematicians, for they share the same simple pole at the origin)

$$\mathrm{sn}'(z) = \mathrm{cn}(z)\,\mathrm{dn}(z),\ \mathrm{cn}'(z) = -\mathrm{dn}(z)\,\mathrm{sn}(z),\ \mathrm{dn}'(z) = -m\,\mathrm{sn}(z)\,\mathrm{cn}(z), \tag{C.9}$$
$$-\mathrm{dn}^2 + 1 - m = -m\,\mathrm{cn}^2 = m\,\mathrm{sn}^2 - m, \tag{C.10}$$
$$\mathrm{cs}'(z) = -\mathrm{ds}(z)\,\mathrm{ns}(z),\ \mathrm{ds}'(z) = -\mathrm{ns}(z)\,\mathrm{cs}(z),\ \mathrm{ns}'(z) = -\mathrm{cs}(z)\,\mathrm{ds}(z), \tag{C.11}$$
$$\mathrm{cs}^2 + 1 - m = \mathrm{ds}^2 = \mathrm{ns}^2 - m. \tag{C.12}$$

Exercise. Prove the equivalence of the twelve Jacobi equations under the class of transformations

$$u(z,k) \to U(Z,K),\ u = \lambda U,\ Z = \mu z,\ (\lambda, \mu) = \text{constant}. \tag{C.13}$$

Hint: since (C.2) admits AC/B^2 as an invariant under this transformation, it is sufficient to check that the induced relation between the two moduli k and K indeed has a solution (complex of course). For instance, with $u = \mathrm{sn}, U = \mathrm{cn}$, there are two possibilities,

$$u = \mathrm{cn},\ U = \mathrm{sn},\ \begin{cases} \mu^2 = -\dfrac{1}{m},\ \lambda^2 = \dfrac{m}{m-1},\ m(1-M) = 1, \\ \mu^2 = \dfrac{1}{1-m},\ \lambda^2 = 1,\ (1-m)(1-M) = 1. \end{cases} \tag{C.14}$$

An important practical consequence is the possibility to represent the solution $\mathrm{sn}(z,k)$ (for instance) of some nonlinear ODE as twelve equivalent complex expressions.

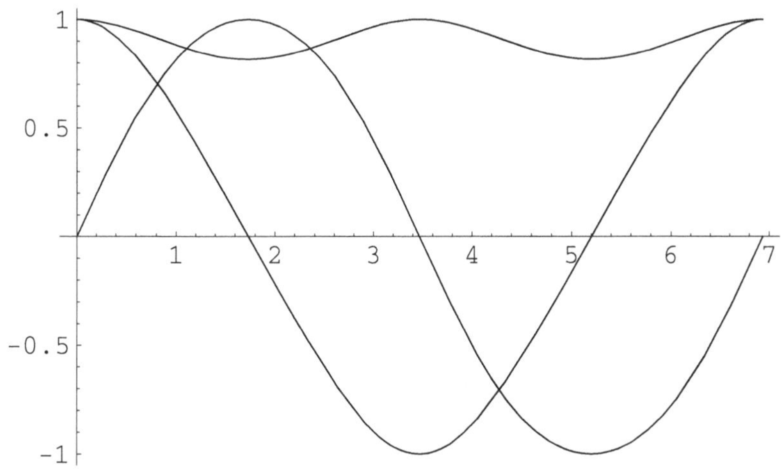

Fig. C.1 The copolar trio $(\mathrm{sn}, \mathrm{cn}, \mathrm{dn})$ of Jacobi, over one real period, for the value $k = 1/3$. The curves of sn, cn are close to those of sin, cos, and dn never vanishes.

C.2 The Symmetric Notation of Halphen

Since $\wp$ is single valued, the three functions

$$\mathrm{h}_\alpha(x) = \sqrt{\wp(x) - e_\alpha},\ \alpha = 1,2,3, \tag{C.15}$$

are single valued, and Halphen [193, Chap. VIII p. 253] has emphasized their numerous advantages over the notation pq of Jacobi. The square roots in (C.15) are chosen so that each $\mathrm{h}_\alpha(x)$ has a residue $+1$ at its simple pole $x = 0$. This copolar trio is closely related to the copolar trio (cs, ds, ns) of Jacobi [193, Chap II, (16) p. 46],

$$\frac{\mathrm{cs}(z)}{\mathrm{h}_1(x)} = \frac{\mathrm{ds}(z)}{\mathrm{h}_2(x)} = \frac{\mathrm{ns}(z)}{\mathrm{h}_3(x)} = \frac{x}{z} = \frac{1}{\sqrt{e_1 - e_3}},\ k^2 = m = \frac{e_2 - e_3}{e_1 - e_3}, \tag{C.16}$$

and its advantage is the invariance under any permutation of $(1,2,3)$ of all the formulae involving h_α, including the solutions of physical nonlinear ODEs [63].

When dealing with an ODE or a discrete equation, the only formulae required are
the derivation formula

$$\mathrm{h}'_\alpha(x) = -\mathrm{h}_\beta(x)\,\mathrm{h}_\gamma(x), \tag{C.17}$$

the algebraic dependence relations

$$\mathrm{h}^2_\beta(x) + e_\beta = \mathrm{h}^2_\gamma(x) + e_\gamma, \tag{C.18}$$

the addition formula

$$\mathrm{h}_\alpha(x+y) = \frac{\mathrm{h}^2_\alpha(x)\,\mathrm{h}^2_\alpha(y) - (e_\alpha - e_\beta)(e_\alpha - e_\gamma)}{\mathrm{h}_\alpha(x)\,\mathrm{h}_\beta(y)\,\mathrm{h}_\gamma(y) + \mathrm{h}_\alpha(y)\,\mathrm{h}_\beta(x)\,\mathrm{h}_\gamma(x)}, \tag{C.19}$$

the Laurent expansion at the origin [193, Chap VII p. 237] [193, Chap IX p. 304]

$$\mathrm{h}_\alpha(x) = \frac{1}{x} - e_\alpha\frac{x}{2} + (g_2 - 5e^2_\alpha)\frac{x^3}{40} + (5g_3 - 7e_\alpha g_2)\frac{x^5}{2^6.5.7} + O(x^7), \tag{C.20}$$

and the degeneracy to trigonometric functions [193, Chap VIII (75) p. 289]

$$\begin{cases} g_2 = \dfrac{4}{3}\left(\dfrac{\pi}{2\omega}\right)^4,\ g_3 = \dfrac{8}{27}\left(\dfrac{\pi}{2\omega}\right)^6,\ -\dfrac{e_\alpha}{2} = e_\beta = e_\gamma = -\dfrac{3g_3}{2g_2}, \\ \sigma_\beta(x) = \sigma_\gamma(x) = \exp\left(\dfrac{1}{6}\left(\dfrac{\pi x}{2\omega}\right)^2\right),\ \sigma_\alpha(x) = \sigma_\beta(x)\cos\dfrac{\pi x}{2\omega}, \\ \sigma(x) = \dfrac{2\omega}{\pi}\sigma_\beta(x)\sin\dfrac{\pi x}{2\omega},\ \mathrm{h}_\alpha(x) = \dfrac{\pi}{2\omega}\operatorname{cotg}\dfrac{\pi x}{2\omega}. \end{cases} \tag{C.21}$$

From (C.17) and (C.18), one deduces the differential equation

$$\left(\mathrm{h}'_\alpha(x)\right)^2 = (\mathrm{h}^2_\alpha(x) + e_\beta - e_\alpha)(\mathrm{h}^2_\alpha(x) + e_\gamma - e_\alpha), \tag{C.22}$$

and from (C.19) the expression of the three elliptic functions h_α as rational functions of $\wp$ and $\wp'$ (a characteristic property of the elliptic functions),

$$\mathrm{h}_\alpha(2u) = -\frac{(\wp(u,g_2,g_3) - e_\alpha)^2 - (e_\alpha - e_\beta)(e_\alpha - e_\gamma)}{\wp'(u,g_2,g_3)}. \tag{C.23}$$

Appendix D
Basic Introduction to the Nevanlinna Theory

There exist two theories, respectively due to the Nevanlinna brothers [323, 324] for the first one, and to Wiman and Valiron [408, §207] for the second one, which may give an additional insight to the solutions of nonlinear differential equations.

However, while the scope of the theory of Painlevé is the *singlevaluedness* (at the movable singularities) of the *general* solution of nonlinear differential equations, the theory of Nevanlinna deals with *meromorphic* functions of one complex variable and the theory of Wiman and Valiron with *entire* functions of one complex variable (for extension to meromorphic functions with a finite number of poles, see e.g.[29]), i.e. two subclasses of the class of singlevalued functions. Moreover, when these functions obey a nonlinear differential equation, they are usually only *particular* solutions. It is quite important to keep these main differences in mind. Let us now briefly introduce the Nevanlinna theory. Complete expositions can be found in [432, 207, 267], and additional results on the Painlevé transcendents in [189, 195].

Let us first precisely define the word *meromorphic*, because it is sometimes used with two meanings.

A function $f(x)$ is said to be *locally meromorphic* near $x = a$ iff its Laurent series near $x = a$ contains a finite number of negative powers.

A function $f(x)$ is said to be *globally meromorphic* iff its only singularities at a finite distance are poles.

Examples.

1. The function $x \to 1/x + \log(x-1)$ is locally meromorphic near $x = 0$ but not globally meromorphic.
2. The functions $x \to e^{1/x}, x \to \wp(e^x, g_2, g_3)$ are singlevalued but not (globally) meromorphic.

In the remaining part of this section, f denotes a globally meromorphic (and we will simply say meromorphic) function whose value at $x = 0$ is neither 0 nor ∞. Rational functions are particular meromorphic functions whose growth at infinity is measured by their *degree* (defined as the maximum value of the degrees of the numerator and denominator). Let us first define a similar notion for meromorphic functions.

1. The function (sometimes called unintegrated counting function) $n(r,f)$ is defined as the number of poles of f (counting multiplicity) in the domain $|z|<r$.
2. The *counting function* (also called integrated counting function) $N(r,f)$

$$N(r,f) = \int_0^r \frac{n(t,f)-n(0,f)}{t} \mathrm{d}t + n(0,f)\log r. \tag{D.1}$$

3. The *proximity function* $m(r,f)$ measures how close f is to ∞. It is defined as

$$m(r,f) = \frac{1}{2\pi}\int_{\theta=0}^{2\pi} \log^+ |f(re^{i\theta})| \mathrm{d}\theta, \tag{D.2}$$

where, for any positive number α, one denotes $\log^+ \alpha = \max\{0, \log \alpha\}$.
4. The main quantity is the *Nevanlinna characteristic*. It is defined as the sum

$$T(r,f) = m(r,f) + N(r,f). \tag{D.3}$$

This characteristic function $T(r,f)$ generalizes to meromorphic functions the notion of degree of a rational function. Indeed, it enjoys the properties, for any meromorphic functions f,g and any positive integer n,

$$T(r,fg) \le T(r,f) + T(r,g), \tag{D.4}$$
$$T(r,f^n) = nT(r,f), \tag{D.5}$$
$$T(r,f+g) \le T(r,f) + T(r,g) + O(1), \tag{D.6}$$
$$T(r,1/f) = T(r,f) + O(1). \tag{D.7}$$

These properties are consequences of similar properties of $N(r,f)$ and $m(r,f)$, for instance

$$m(r,fg) \le m(r,f) + m(r,g). \tag{D.8}$$

The *order* of growth of a meromorphic function is defined as

$$\sigma(f) = \limsup_{r\to+\infty} \frac{\log T(r,f)}{\log r}. \tag{D.9}$$

When f is entire, this coincides with the usual definition

$$\sigma(f) = \limsup_{r\to+\infty} \frac{\log\log M(r,f)}{\log r}, \ M(r,f) = \max_{|z|=r} |f(z)|. \tag{D.10}$$

For instance, the entire functions $z^n, e^z, \sin \pi z, \exp(z^n), \exp(\exp z)$ have the respective orders $0, 1, 1, n, \infty$. The rational functions are characterized by the property $T(r,f) = O(\log r)$, therefore their order is zero. The meromorphic functions P1, P2, P4 have the orders $5/2, 3, 4$. The P6 function being nonmeromorphic, its order is not defined and the Nevanlinna theory is not applicable to it.

This ends the main definitions.

For applications to differential equations, the most important property is the following lemma.

Lemma of the logarithmic derivative (Nevanlinna).

If f is meromorphic, then

$$m(r,f'/f) = O(\log T(r,f) + \log r), \quad r \to \infty, \, r \notin E, \tag{D.11}$$

where E is some exceptional set of finite length.

If f has no essential singularity at infinity, the term $\log r$ can be omitted. The exceptional set E may indeed occur but it makes no harm in most applications.

If f has a pole of order k, hence f' a pole of order $k+1$, the upper bound 2 for $|(k+1)/k|$ yields the inequality

$$N(r,f') \le 2N(r,f). \tag{D.12}$$

Combining the above lemma (D.11) and the property (D.8), one obtains

$$\begin{aligned} T(r,f') &= N(r,f') + m(r,f') \le 2N(r,f) + m(r,ff'/f) \\ &\le 2N(r,f) + m(r,f) + m(r,f'/f) \le (2+o(1))T(r,f). \end{aligned}$$

Thus

$$\forall n \in \mathcal{N} : \, T(r,f^{(n)}) = O(T(r,f)). \tag{D.13}$$

If f has no poles, we obtain the upper bound

$$\begin{aligned} T(r,f') = m(r,f') &= m(r,ff'/f) \le m(r,f) + m(r,f'/f) \\ &\le (1+o(1))T(r,f), \end{aligned} \tag{D.14}$$

and, by recurrence,

$$\forall n \in \mathcal{N} : \, T(r,f^{(n)}) \le (1+o(1))T(r,f). \tag{D.15}$$

Similarly, if $L(f)$ is a linear differential polynomial of f with rational coefficients, and if f has only a finite number of poles, one has the upper bound

$$T(r,L(f)) \le (1+o(1))T(r,f). \tag{D.16}$$

The inequalities (D.15) and (D.16) are the key ingredients to the proof that, for some particular nonlinear ODEs, any meromorphic solution if it exists can only be elliptic. An elegant application to the KS ODE (2.27) has been made by Eremenko [132]; its results are presented in Sect. 3.2.8.

Another application of the Nevanlinna theory in the context of the present book is its probable relation with a good definition of the discrete Painlevé property [2], see Sect. 7.2. Indeed, the upper bound (D.14) also exists in the discrete case [2]

$$r\varepsilon \geq 1: \; T(r, f(z \pm 1)) \leq (1+\varepsilon)T(r+1, f(z)) + \kappa, \tag{D.17}$$

with κ equal to some constant.

Appendix E
The Bilinear Formalism

For all soliton equations, most expressions (the equation itself, its Bäcklund transformation, etc) can be considerably shortened if one uses a compact notation introduced by Hirota [208][1]. The *bilinear operator* D is a first order differential operator acting on a couple of functions f and g, defined by

$$D_{x_j} f.g = \left(\partial_{x_j} - \partial_{x'_j}\right) f(x_1, \cdots, x_l) f(x'_1, \cdots, x'_l)|_{x'_j = x_j}. \tag{E.1}$$

For instance,

$$\begin{cases} D_x f.g = f_x g - f g_x, \\ D_x^2 f.g = f_{xx} g - 2 f_x g_x + f g_{xx}, \\ D_x^3 f.g = f_{xxx} g - 3 f_{xx} g_x + 3 f_x g_{xx} - f g_{xxx}, \\ D_x^n f.g = D_x^{n-1} \left(f_x g - f g_x\right), \\ D_x D_t f.g = f_{xt} g - f_x g_t - f_t g_x + f g_{xt}. \end{cases} \tag{E.2}$$

Any sum of such monomials whose coefficients are independent of the usual derivatives ∂_{x_j} is called by definition a (Hirota) *bilinear form* (even if it is a quadratic form in case $g = f$). This formalism presents a strong analogy [219] with a variety of determinant known as *Pfaffian*.

E.1 Bilinear Representation of a PDE

Given a 1+1-dimensional soliton PDE defined by the differential polynomial equation

$$P(\partial_x, \partial_t, u) = 0, \tag{E.3}$$

[1] This notation can be traced back to Clebsch, and Borel [36] used it, under the name "forme binaire", to represent the newly discovered P1 and P2 equations.

one first computes the singular part operator $\mathscr{D}$ of each family, introduced in Sect. 2.1.2 (not to be confused with the above Hirota D operator!). In case the PDE admits more than one family, there exists a privileged family[2], which must be selected. Depending on the structure of singularities, one then introduces one or several fields f, g *via* the singular part transformation. In the simplest case of one field, this transformation is

$$u = \mathscr{D} \log f. \tag{E.4}$$

Then (E.3), which is now a differential polynomial

$$P(\partial_x, \partial_t, f) = 0, \tag{E.5}$$

can be rewritten as a Hirota differential polynomial (bilinear form)[3]

$$P(D_x, D_t) f.f = 0, \tag{E.6}$$

in which P is a polynomial, i.e. without any usual derivative ∂_x, ∂_t.

For instance, the KdV equation

$$bu_t + u_{xxx} - \frac{6}{a} u u_x = 0,\ (a,b) \text{ constant}, \tag{E.7}$$

admits a single family of movable double poles, and the transformation

$$u = -2a\partial_x^2 \log f \tag{E.8}$$

maps it to

$$-2a \left(\frac{\left(bD_xD_t + D_x^4\right) f.f}{f^2} \right)_x = 0, \tag{E.9}$$

which, after taking one primitive, is indeed a Hirota bilinear form

$$\left(bD_xD_t + D_x^4\right) f.f + F(t) f^2 = 0. \tag{E.10}$$

This equation does not depend any more on the coefficient a of the nonlinear term in (E.7), and this is a typical feature of these bilinear representations.

More examples and details on this powerful technique can be found in various summer school lecture notes, e.g. [205, 384].

---

[2] Examples are (5.204) and (5.206), respectively for the SK and KK equations. Most of the time, this privileged family only has positive Fuchs indices.

[3] Such a form only implies the existence of a two-soliton solution, and there do exist [204] PDEs which admit three-soliton solutions but not four-soliton solutions.

E.2 Bilinear Representation of a Bäcklund Transformation

Knowing the bilinear form (E.6) of a soliton PDE, there exists a method [212] to derive a representation of its Bäcklund transformation as a set of two bilinear equations. Like the singular manifold method presented in Sect. 5.5, it requires at some point the input of an order for the spectral problem.

In the example of KdV, the goal is to obtain the two equations (5.101)–(5.102) knowing only (E.10). Consider two different solutions f and F of the bilinear PDE (E.10). One first switches to the variables sum and difference

$$\log f - \log F = V, \ \log f + \log F = W, \tag{E.11}$$

and eliminates the highest derivative in W to obtain

$$bV_{xt} + V_{xxxx} + 6V_{xx}W_{xx} = 0. \tag{E.12}$$

One must generate two bilinear equations from this single non-bilinear equation. The canonical basis $V_{mx,nt}, W_{mx,nt}$ used to represent (E.12) proves inconvenient, and there exists an optimal basis of differential polynomials in two variables known as the *binary Bell polynomials* $Y_{m,n}$ [172] defined as

$$(fF)^{-1} D_x^m D_t^n f.F = Y_{m,n}(V,W), \ V = \log \frac{f}{F}, \ W = \log fF, \tag{E.13}$$

whose first elements are

$$\begin{cases} Y_{0,0}(V,W) = 1, \\ Y_{1,0}(V,W) = V_x, \\ Y_{2,0}(V,W) = W_{xx} + V_x^2, \\ Y_{1,1}(V,W) = W_{xt} + V_x V_t, \\ Y_{3,0}(V,W) = V_{xxx} + 3V_x W_{xx} + V_x^3. \end{cases} \tag{E.14}$$

In this basis, the single equation (E.12) becomes

$$\partial_x (bY_{0,1} + Y_{3,0}) - 3(Y_{1,0}\partial_x Y_{2,0} - Y_{2,0}\partial_x Y_{1,0}) = 0, \tag{E.15}$$

and the assumption which, in the singular manifold method, corresponds to a second order Lax pair is a linear dependence between $Y_{2,0}, Y_{1,0}, Y_{0,0}$,

$$Y_{2,0} + a_1 Y_{1,0} + a_0 Y_{0,0} = 0. \tag{E.16}$$

Equation (E.15) can then be integrated once,

$$bY_{0,1} + Y_{3,0} - 3a_0 Y_{1,0} + KY_{0,0} = 0, \tag{E.17}$$

with K an arbitrary function of t. Since the two equations (E.16)–(E.17) are linear in the binary Bell polynomials and therefore expressible in bilinear form

$$\begin{cases} \left(D_x^2 + a_1 D_x + a_0\right) f.F = 0, \\ \left(bD_t + D_x^3 - 3a_0 D_x + K\right) f.F = 0, \end{cases} \tag{E.18}$$

this is the desired result. A translation $V \to V + c_1 x + c_2 t$ allows one to set the coefficients a_1 and K to zero, leaving a_0 as the only essential parameter, then identified as the spectral parameter λ.

For reference, the main PDEs considered in the text are representable as follows (the representation may not be unique),

$$\text{KdV}: \ \left(bD_xD_t + D_x^4\right) f.f = 0,\ u = -2a\partial_x^2 \log f, \tag{E.19}$$

$$\text{Bq}: \ \left(D_t^2 + (\varepsilon^2\beta^2/3)D_x^4\right) f.f = 0,\ u = 2\beta^2\partial_x^2 \log f, \tag{E.20}$$

$$\text{KP}: \ \left(\frac{\varepsilon^2\beta^2}{3} D_{x_1}^4 + \varepsilon^2 D_{x_1} D_{x_3} + D_{x_2}^2\right) f.f = 0,\ u = 2\beta^2\partial_x^2 \log f, \tag{E.21}$$

$$\text{mKdV}: \ \left(D_x^3 - bD_t\right) g.f = 0,\ D_x^2 g.f = 0,\ u = \alpha\partial_x \text{Log}(f/g), \tag{E.22}$$

$$\text{SG}: \ \begin{cases} (D_xD_t - \mu) f.f + g^2 = 0,\ (D_xD_t - \mu) g.g + f^2 = 0, \\ u = (2/a)\text{Log}(g/f), \end{cases} \tag{E.23}$$

$$\text{NLS}: \ \begin{cases} \left(p_r D_x^2 + iD_t\right) f.g = 0,\ \left(p_r D_x^2 - iD_t\right) h.g = 0, \\ D_x^2 g.g - fh = 0,\ A = f/g,\ \bar{A} = h/g, \end{cases} \tag{E.24}$$

$$\text{SK}: \ \left(D_x(\beta D_t + D_x^5)\right) f.f = 0,\ u = \alpha\partial_x^2 \text{Log} f, \tag{E.25}$$

$$\text{KK}: \ \begin{cases} \left(D_x^6 + 16\beta D_xD_t\right) f.f + 15D_x^2 f.g = 0,\ D_x^4 f.f = fg, \\ u = (\alpha/2)\partial_x^2 \text{Log} f, \end{cases} \tag{E.26}$$

The credits are: KdV [208], Boussinesq [217, 218], KP [382], mKdV [209], SG [210], NLS [211], SK [36, 59, 386], KK [241].

A probably complete list of soliton equations has been produced in this representation [241].

The corresponding bilinear representation of their BT is (the second solution is denoted with a prime)

$$\text{KdV}: \ \left(D_x^2 + \lambda\right) f.f' = 0,\ \left(bD_t + D_x^3 - 3\lambda D_x\right) f.f' = 0, \tag{E.27}$$

$$\text{Bq}: \ \left(\beta\varepsilon D_x^3 + 3D_xD_t - 4\beta\varepsilon\lambda\right) f.f' = 0,\ \left(\beta\varepsilon D_x^2 - D_t\right) f.f' = 0, \tag{E.28}$$

$$\text{KP}: \ \begin{cases} \left(\beta\varepsilon D_{x_1}^2 + D_{x_2}\right) f.f' = 0, \\ \left(3\varepsilon D_{x_3} + \beta^2\varepsilon D_{x_1}^3 - 3\beta D_{x_1} D_{x_2}\right) f.f' = 0, \end{cases} \tag{E.29}$$

$$\text{mKdV}: \ \begin{cases} D_x f'.g = \lambda g f',\ \left(D_x^3 + 3\lambda D_x + bD_t\right) g'.g = 0, \\ D_x g'.f = \lambda f' g,\ \left(D_x^3 + 3\lambda D_x + bD_t\right) f'.f = 0, \end{cases} \tag{E.30}$$

$$\text{SG}: \ \begin{cases} D_x g'.g = \lambda f' f,\ D_t f'.g = \lambda^{-1} g' f, \\ D_x f'.f = \lambda g' g,\ D_t g'.g = \lambda^{-1} f' g, \end{cases} \tag{E.31}$$

$$\text{NLS}: \begin{cases} D_x(g'.f+f'.g)-\mu(f'g-fg')=0\ (p=1,q=1), \\ D_x(g'.h+h'.g)-\mu(h'g-hg')=0, \\ iD_t(g'.f+f'.g)-(D_x^2+\lambda)(f'g-fg')=0, \\ -iD_t(g'.h+h'.g)-(D_x^2+\bar{\lambda})(h'g-hg')=0, \\ (iD_t+D_x^2+2\mu D_x+2\mu^2-\lambda)g'.g-f'.h=0, \\ (-iD_t+D_x^2+2\bar{\mu}D_x+2\bar{\mu}^2-\bar{\lambda})g'.g-h'.f=0, \end{cases} \tag{E.32}$$

$$\text{SK}: \ (D_x^3-\lambda)\,f.f'=0,\ \left(D_x^5+5\lambda D_x^2-(2/3)\beta D_t\right)f.f'=0, \tag{E.33}$$

$$\text{KK}: \begin{cases} (D_x^4-8\lambda D_x)\,f.f'-(f'g+fg')/2=0, \\ \left(18D_x^5-32\beta D_t\right)f.g-15(D_xf'.g-D_xf.g')=0, \\ D_x^4f.f-fg=0,\ D_x^4f'.f'-gg'=0. \end{cases} \tag{E.34}$$

Again, the credits are: KdV [212], Boussinesq [218, 330], KP [218], mKdV [209], SG [210], NLS [328], SK [386], KK [319].

Appendix F
Algorithm for Computing Laurent Series

Such a computation is required in many methods presented in this book, therefore it must be done efficiently. The common way is to substitute u by a finite sum

$$u = \sum_{j=0}^{J} u_j \chi^{j+p}, \tag{F.1}$$

to split the resulting l.h.s. $E(u)$ of the equation into a similar sum

$$E(u) = \sum_{j=0}^{J'} E_j \chi^{j+q}, \tag{F.2}$$

to discard those coefficients E_j in the range $J < j \leq J'$, and to solve the set of $J+1$ equations $E_j = 0$ for the $J+1$ unknowns u_j. This algorithm is time and storage consuming, therefore it should not be implemented.

To avoid generating terms to be discarded later on, there exists a perturbation process equivalent to this computation, which is just an application of the very powerful α-method of Painlevé [348], this is [76, (5.26)]

$$x = x_0 + \varepsilon X,\ u = (\varepsilon X)^p \sum_{n=0}^{+\infty} (\varepsilon X)^n u^{(n)}(x),\ E = (\varepsilon X)^q \sum_{n=0}^{+\infty} (\varepsilon X)^n E^{(n)}(x). \tag{F.3}$$

Its implementation generates the following optimal algorithm[1] (for simplicity, we assume that the leading order $u \sim u_0 \chi^p, E(u) \sim E_0 \chi^q$ has already been computed, that the equation is an autonomous ODE, and we omit the handling of the Fuchs indices and no-log conditions).

```
Input:
  E      some expression, polynomial in u(x),u'(x), etc
  p      singularity degree of u
```

[1] The language used, called *Algebraic manipulation program* [128], has a syntax very close to that of Reduce.

```
  q      singularity degree of E
  u_0    leading coefficient of u
  t[0:J] empty array

Output:
  t[0:J] coefficients of the Laurent series
  zero   a remainder which must be zero.

Program:
  Laurent(u,x,E,p,q,u_0,t) {symbolic zero,j,J;
  declare fr function,t array;
  zero=x**(-q)*let u=(function(x) x**p*(u_0+x*fr(x))) in E;
  t[0]=u_0;J=highbound(t,1);let x**(J+1)=0 active;
  for j=1 to J do {
   t[j]=Solve(fr(x),coeff(zero,x,j));
   zero=let once fr=(function(x) uj+x*fr(x)) in zero;
   zero=let uj=t[j] in poly(zero,uj);};
  let x**(J+1)=0 inactive;write 0==zero; /* Must print "tru
  };
```

List of Figures

3.1 KdV bell-shaped solitary wave. 41
3.2 KS, heteroclinic traveling wave of Kuramoto and Tsuzuki. 56
3.3 CGL3, propagating hole. 59
3.4 KS. Pattern of singularities of the unknown general analytic solution. 75

5.1 KdV. Two-soliton solution 100
5.2 Bianchi diagram 102
5.3 Breather solution of sine-Gordon. 120
5.4 KPP, collision of two fronts. 135

C.1 The copolar trio $(\mathrm{sn}, \mathrm{cn}, \mathrm{dn})$ of Jacobi. 217

List of Tables

2.1 Cubic Hénon–Heiles system. Solution of diophantine conditions. . . . 27
2.2 Cubic Hénon–Heiles system. No-logarithm conditions 27

3.1 Characterization of the general analytic solution of a nonintegrable ODE. 47
3.2 Kuramoto–Sivashinsky equation. The six trigonometric solutions. . . . 55

5.1 Action of reductions on the Lax pair, etc, of a PDE. 144

7.1 Duality of the Dynkin diagrams of the continuous Painlevé equations. 184

B.1 Basic data for Pn equations 202
B.2 Confluence of monodromy exponents of the Pn 203
B.3 The eight inequivalent Painlevé equations 208
B.4 Coalescence of the scalar Lax pairs of the Painlevé equations 211

References

1. M. J. Ablowitz and P. A. Clarkson, *Solitons, nonlinear evolution equations and inverse scattering* (Cambridge University Press, Cambridge, 1991).
2. M. J. Ablowitz, R. Halburd, and B. M. Herbst, On the extension of the Painlevé property to difference equations, Nonlinearity **13** (2000) 889–905.
3. M. J. Ablowitz, D. J. Kaup, A. C. Newell, and H. Segur, The inverse scattering transform-Fourier analysis for nonlinear problems, Stud. Appl. Math. **53** (1974) 249–315.
4. M. J. Ablowitz and J. F. Ladik, Nonlinear differential–difference equations, J. Math. Phys. **16** 598-603 (1975).
5. M. J. Ablowitz, A. Ramani, and H. Segur, Nonlinear evolution equations and ordinary differential equations of Painlevé type, Lett. Nuov. Cimento **23** (1978) 333–338.
6. M. J. Ablowitz, A. Ramani, and H. Segur, A connection between nonlinear evolution equations and ordinary differential equations of P-type. I, J. Math. Phys. **21** (1980) 715–721; II, J. Math. Phys. **21** (1980) 1006–1015.
7. M. J. Ablowitz and H. Segur, Exact linearization of a Painlevé transcendent, Phys. Rev. Lett. **38** (1977) 1103–1106.
8. M. J. Ablowitz and A. Zeppetella, Explicit solutions of Fisher's equation for a special wave speed, Bull. Math. Bio. **41** (1979) 835–840.
9. M. Abramowitz, I. Stegun, *Handbook of Mathematical Functions*, Tenth printing (Dover, New York, 1972).
10. G. P. Agrawal, *Nonlinear fiber optics*, 3rd edition (Academic press, Boston, 2001).
11. A. Amengual, Hernández-García, R. Montagne, and M. San Miguel, Synchrozination of spatiotemporal chaos: the regime of coupled spatiotemporal intermittency, *Phys. Rev. Lett.* **78**, 4379–4382 (1997).
12. R. Anderson, J. Harnad, and P. Winternitz, Systems of ordinary differential equations with nonlinear superposition principles, Physica D **4** (1982) 164–182.
13. A. Ankiewicz and C. Pask, The complete Whittaker theorem for two-dimensional integrable systems and its application, J. Phys. A **16** (1983) 4203–4208.
14. M. Antonowicz and A. Fordy, Coupled KdV equations with multi-Hamiltonian structures, Physica D **28** (1987) 345–357.
15. M. Antonowicz, A. P. Fordy, and Liu Q.-P., Energy-dependent third-order Lax operators, Nonlinearity **4** (1991) 669–684.
16. P. Appell, Sur les équations différentielles algébriques et homogènes par rapport à la fonction inconnue et à ses dérivées, C. R. Acad. Sc. Paris **104** (1887) 1776–1779.
17. P. Appell, Sur les invariants de quelques équations différentielles, J. Math. **5** (1889) 361–423.
18. V.I. Arnol'd, *Les méthodes mathématiques de la mécanique classique* (Nauka, Moscou, 1974) (Mir, Moscou, 1976).
19. M. V. Babich and L. A. Bordag, Projective differential geometrical structure of the Painlevé equations, J. Differential Equations **157** (1999) 452–485.
20. A. V. Bäcklund, Einiges über Kurven- und Flächentransformationen, Lunds Universitets Arsskrift **10** (1875) 1–12.
21. S. Baker, Squared eigenfunction representations of integrable hierarchies, PhD Thesis, University of Leeds (1995).
22. S. Baker, V. Z. Enol'skii, and A. P. Fordy, Integrable quartic potentials and coupled KdV equations, Phys. Lett. A **201** (1995) 167–174.
23. A. P. Bassom, P. A. Clarkson, and A. C. Hicks, Bäcklund transformations and solution hierarchies for the fourth Painlevé equation, Stud. Appl. Math. **95** (1995) 1–71.
24. R. J. Baxter, *Exactly solved models in statistical mechanics* (Associated Press, London, 1982).
25. N. Bekki and K. Nozaki, Formations of spatial patterns and holes in the generalized Ginzburg-Landau equation, Phys. Lett. A **110** (1985) 133–135.

26. V. A. Belinskii, G. W. Gibbons, D. N. Page, and C. N. Pope, Asymptotically Euclidean Bianchi IX metrics in quantum gravity, Phys. Lett. A **76** (1978) 433–435.
27. T.B. Benjamin, Internal waves of permanent form in fluids of great depth, J. Fluid Mech. **29** (1967) 559–592.
28. P. Bergé, Y. Pomeau and C. Vidal, *L'ordre dans le chaos*, 353 pages (Hermann, Paris, 1984). English version (Wiley and sons, New York, 1986).
29. W. Bergweiler, P. Rippon and G. Stallard, Dynamics of meromorphic functions with direct or logarithmic singularities, http://arXiv.org/abs/0704.2712
30. D. Bessis, A new method in the combinatorics of the topological expansion, Commun. Math. Phys. **69** (1979) 147–163
31. L. Bianchi, Sulla trasformazione di Bäcklund per le superficie pseudosferiche, Rend. Accad. Naz. Lincei **1** (1892) 3–12; *Luigi Bianchi, Opere*, vol. V, 163–173 (Edizioni Cremonese (Unione Matematica Italiana), Roma, 1957).
32. G.D. Birkhoff, The generalized Riemann problem for linear differential equations and the allied problems for linear difference and $q-$difference equations, Proc. Am. Acad. Arts and Sciences **49** (1914) 521–568.
33. M. Błaszak and S. Rauch-Wojciechowski, A generalized Hénon–Heiles system and related integrable Newton equations, J. Math. Phys. **35** (1994) 1693–1709.
34. P. Boalch, From Klein to Painlevé via Fourier, Laplace and Jimbo, Proc. London Math. Soc. **90** (2005) 167–208.
35. M. Boiti, V. G. Marikhin, F. Pempinelli, and A. B. Shabat, Self-similar solutions of NLS-type dynamical systems, 18 pages, http://arXiv.org/abs/solv-int/9909021.
36. É. Borel, Remarques sur les équations différentielles dont l'intégrale générale est une fonction entière, C. R. Acad. Sc. Paris **138** (1904) 337–339.
37. F. Boulier, *Étude et implantation de quelques algorithmes en algèbre différentielle* (Thèse, Université des sciences et technologies de Lille, 1994).
38. T. Bountis, L. Drossos, M. Lakshmanan, and S. Parthasarathy, On the nonintegrability of a family of Duffing–van der Pol oscillators, J. Phys. A **26** (1993) 6927–6942.
39. J. Boussinesq, Théorie des ondes qui se propagent le long d'un canal rectangulaire horizontal, en communiquant au liquide contenu dans ce canal des vitesses sensiblement pareilles de la surface du fond, J. de math. pures et appliquées **17** (1872) 55–108.
40. L. Brenig and A. Goriely, Universal canonical forms for time-continuous dynamical systems, Phys. Rev. A **4** (1989) 4119–4122.
41. É. Brézin and V. A. Kazakov, Exactly solvable field theories of closed strings, Phys. Lett. B **236** (1990) 144–150.
42. C. Brezinski, J. van Isegehm, Padé approximations, 47–222, *Handbook of Numerical Analysis*, Vol. III, eds. P.G. Ciarlet and J.-L. Lions (North-Holland, Amsterdam, 1994).
43. C. Briot et J.-C. Bouquet, *Théorie des fonctions elliptiques*, 2ième édition (Gauthier-Villars, 1875, Paris).
44. T. Brugarino, Painlevé property, auto-Bäcklund transformation, Lax pairs, and reduction to the standard form for the Korteweg–de Vries equation with nonuniformities, J. Math. Phys. **30** (1989) 1013–1015.
45. F. J. Bureau, Differential equations with fixed critical points, Annali di Mat. pura ed applicata **LXIV** (1964) 229–364 [abbreviated as M. I].
46. F. J. Bureau, Differential equations with fixed critical points, Annali di Matematica pura ed applicata **LXVI** (1964) 1–116 [abbreviated as M. II].
47. F. J. Bureau, Équations différentielles du second ordre en Y et du second degré en $\ddot{Y}$ dont l'intégrale générale est à points critiques fixes, Annali di Matematica pura ed applicata **91** (1972) 163–281 [abbreviated as M. III].
48. J. Burguete, H. Chaté, F. Daviaud, and N. Mukolobwiez : Bekki–Nozaki amplitude holes in hydrothermal nonlinear waves, Phys. Rev. Lett. **82** (1999) 3252–3255.
49. F. Calogero and A. Degasperis, *Spectral transform and solitons I*, (North Holland, Amsterdam, 1982)
50. R. Camassa and D. D. Holm, An integrable shallow water equation with peaked solitons, Phys. Rev. Lett. **71** (1993) 1661–1664.

51. J. Canosa, Diffusion in nonlinear multiplicative media, JMP **10** (1969) 1862–1868.
52. H. W. Capel, F. W. Nijhoff, and V. G. Papageorgiou, Complete integrability of Lagrangian mappings and lattices of KdV type, Phys. Lett. A **155** (1991) 377–387.
53. F. Cariello and M. Tabor, Painlevé expansions for nonintegrable evolution equations, Physica D **39** (1989) 77–94.
54. F. Cariello and M. Tabor, Similarity reductions from extended Painlevé expansions for nonintegrable evolution equations, Physica D **53** (1991) 59–70.
55. É. Cartan, Sur les variétés à connexion projective, Bull. Soc. Math. France **52** (1924) 205–241.
56. P. J. Caudrey, The inverse problem for the third order equation $u_{xxx} + q(x)u_x + r(x)u = -i\zeta^3 u$, Phys. Lett. A **79** (1980) 264–268.
57. V.K. Chandrasekhar, M. Senthil Velan and M. Lakshmanan, New aspects of integrability of force-free Duffing–van der Pol ocsillator and related nonlinear systems, J. Phys. A **37** (2004) 4527–4534.
58. Chang Y. F., M. Tabor, and J. Weiss, Analytic structure of the Hénon–Heiles Hamiltonian in integrable and nonintegrable regimes, J. Math. Phys. **23** (1982) 531–538.
59. J. Chazy, Sur les équations différentielles du troisième ordre et d'ordre supérieur dont l'intégrale générale a ses points critiques fixes, Thèse, Paris (1910); Acta Math. **34** (1911) 317–385. Table des matières commentée avec index, R. Conte (1991), 6 pages.
60. Hsing-Hen Chen, General derivation of Bäcklund transformations from inverse scattering problems, Phys. Rev. Lett. **33** (1974) 925–928.
61. Hsing-Hen Chen, A Bäcklund transformation in two dimensions, J. Math. Phys. **16** (1975) 2382–2384.
62. Chow K.-w. and R. Conte, Stationary doubly periodic wave of the saturable discrete nonlinear Schrödinger equation, to be submitted (2008).
63. Chow K.-w., R. Conte and Neil Xu, Analytic doubly periodic wave patterns for the integrable discrete nonlinear Schrödinger (Ablowitz–Ladik) model, Phys. Lett. A **346** (2006) 422–429. http://arXiv.org/abs/nlin.PS/0509005
64. D. V. Chudnovsky, G. V. Chudnovsky, and M. Tabor, Painlevé property and multicomponent isospectral deformation equations, Phys. Lett. A **97** (1983) 268–274.
65. P. A. Clarkson, Painlevé analysis of the damped, driven nonlinear Schrödinger equation, Proc. Roy. Soc. Edin. **109** A (1988) 109–126.
66. P. A. Clarkson, A. S. Fokas, and M. J. Ablowitz, Hodograph transformations of linearizable partial differential equations, SIAM J. Appl. Math. **49** (1989) 1188–1209.
67. P.A. Clarkson and P. Winternitz, Symmetry reduction and exact solutions of nonlinear partial differential equations, *The Painlevé property, one century later*, 591–660, ed. R. Conte, CRM series in mathematical physics (Springer, New York, 1999).
68. A. K. Common and M. Musette, Two discretisations of the Ermakov–Pinney equation, Phys. Lett. A **235** (1997) 574–580.
69. R. Conte, Universal invariance properties of Painlevé analysis and Bäcklund transformation in nonlinear partial differential equations, Phys. Lett. A **134** (1988) 100–104.
70. R. Conte, Invariant Painlevé analysis of partial differential equations, Phys. Lett. A **140** (1989) 383–390.
71. R. Conte, Painlevé singular manifold equation and integrability, *Inverse methods in action*, pp. 497–504, ed. P. C. Sabatier, Springer-Verlag series "Inverse problems and theoretical imaging" (Springer-Verlag, Berlin, 1990).
72. R. Conte, Painlevé analysis and Bäcklund transformation of the nonintegrable KPP equation, *Nonlinear evolution equations: integrability and spectral methods*, 187–192, eds. A. Degasperis, A. P. Fordy, and M. Lakshmanan (Manchester University Press, Manchester, 1990).
73. R. Conte, Towards the equivalence between integrability and Painlevé test for partial differential equations, *Nonlinear and chaotic phenomena in plasmas, solids and fluids*, 94–101, eds. W. Rozmus and J. A. Tuszyński (World Scientific, Singapore, 1991).
74. R. Conte, Unification of PDE and ODE versions of Painlevé analysis into a single invariant version, *Painlevé transcendents, their asymptotics and physical applications*, 125–144, eds. D. Levi and P. Winternitz (Plenum, New York, 1992).

75. R. Conte, Singularities of differential equations and integrability, *Introduction to methods of complex analysis and geometry for classical mechanics and nonlinear waves*, 49–143, eds. D. Benest and C. Frœschlé (Éditions Frontières, Gif-sur-Yvette, 1994).
76. R. Conte, The Painlevé approach to nonlinear ordinary differential equations, *The Painlevé property, one century later*, 77–180, ed. R. Conte, CRM series in mathematical physics (Springer, New York, 1999). Solv-int/9710020.
77. R. Conte, Various truncations in Painlevé analysis of partial differential equations, 103–116, *Nonlinear dynamics: integrability and chaos*, eds. M. Daniel, K. M. Tamizhmani, and R. Sahadevan (Narosa publishing house, New Delhi, 2000). solv-int/9812008.
78. R. Conte, Exact solutions of nonlinear partial differential equations by singularity analysis, *Direct and inverse methods in nonlinear evolution equations*, 83 pages, ed. A. Greco (Springer, Berlin, 2003). nlin.SI/0009024. CIME school, Cetraro, 5–12 September 1999.
79. R. Conte, Integration of partially integrable equations, 146–157, *Waves and stability in continuous media*, eds. R. Monaco, G. Mulone, S. Rionero and T. Ruggeri (World Scientific, Singapore, 2006). http://arXiv.org/abs/nlin.SI/0604063
80. R. Conte, On the Lax pairs of the sixth Painlevé equation, RIMS Kôkyûroku Bessatsu **B2** (2007) 21–27. http://arXiv.org/abs/nlin.SI/0701049
81. R. Conte, A closed-form solution in a dynamical system related to Bianchi IX, Phys. Lett. A **372** (2008) 2269–2270. http://arxiv.org/abs/0712.0209
82. R. Conte, A. P. Fordy, and A. Pickering, A perturbative Painlevé approach to nonlinear differential equations, Physica D **69** (1993) 33–58.
83. R. Conte, A. M. Grundland, and M. Musette, A reduction of the resonant three-wave interaction to the generic sixth Painlevé equation, J. Phys. A **39** 12115–12127 (2006), Special issue "One hundred years of Painlevé VI", http://arXiv.org/abs/nlin.SI/0604011.
84. R. Conte and M. Musette, Painlevé analysis and Bäcklund transformation in the Kuramoto-Sivashinsky equation, J. Phys. A **22** (1989) 169–177.
85. R. Conte and M. Musette, Link between solitary waves and projective Riccati equations, J. Phys. A **25** (1992) 5609–5623.
86. R. Conte and M. Musette, Linearity inside nonlinearity: exact solutions to the complex Ginzburg–Landau equation, Physica D **69** (1993) 1–17.
87. R. Conte and M. Musette, Beyond the two–singular manifold method, *Nonlinear physics : theory and experiment*, 67–74, eds. E. Alfinito, M. Boiti, L. Martina, and F. Pempinelli (World Scientific, Singapore, 1996).
88. R. Conte and M. Musette, A new method to test discrete Painlevé equations, Phys. Lett. A **223** (1996) 439–448. Solv-int/9610007.
89. R. Conte and M. Musette, Rules of discretization for Painlevé equations, *Theory of nonlinear special functions : the Painlevé transcendents*, 20 pages, eds. L. Vinet and P. Winternitz (Springer, New York, to appear). http://arXiv.org/abs/solv-int/9803014. Montréal workshop, 13–17 May 1996.
90. R. Conte and M. Musette, Analytic expressions of hydrothermal waves, Reports on mathematical physics **46** (2000) 77–88. nlin.SI/0009022
91. R. Conte and M. Musette, New contiguity relation of the sixth Painlevé equation from a truncation, Physica D **161** (2002) 129–141. nlin.SI/0112007. http://dx.doi.org/10.1016/S0167-2789(01)00372-4
92. R. Conte and M. Musette, First degree birational transformations of the Painlevé equations and their contiguity relations, J. Phys. A **34** (2001) 10507–10522. nlin.SI/0110028.
93. R. Conte and M. Musette, A truncation for obtaining all the first degree birational transformations of the Painlevé transcendents, J. Nonlinear Math. Phys. **9** Supp. 1 (2002) 14–28. http://arXiv.org/abs/nlin.SI/0110031 *Recent advances in integrable systems*, eds. Ma Wenxiu, P. J. Olver, Yan Min, and Hu Ji-shan, First joint meeting of the American mathematical society and the Hong-Kong mathematical society, Hong-Kong, 13–16 December 2000.
94. R. Conte and M. Musette, The Painlevé methods, *Classical and quantum nonlinear integrable systems: theory and application*, 39–63, ed. A. Kundu (IOP Publishing Ltd., Bristol, 2003). http://arXiv.org/abs/nlin.SI/0211048

95. R. Conte, M. Musette, and A. M. Grundland, Bäcklund transformation of partial differential equations from the Painlevé–Gambier classification, II. Tzitzéica equation, J. Math. Phys. **40** (1999) 2092–2106.
96. R. Conte, M. Musette, and C. Verhoeven, Explicit integration of the Hénon–Heiles Hamiltonians, J. Nonlinear Mathematical Physics **12** Supp. 1 (2005) 212–227. Special issue in honor of F. Calogero. http://arxiv.org/abs/nlin.SI/0412057
97. R. Conte, M. Musette, and C. Verhoeven, Hamiltonians with two degrees of freedom admitting a singlevalued general solution, Analysis in theory and applications **21** (2005) 188–200. http://arxiv.org/abs/nlin.SI/0507012
98. G. Contopoulos, B. Grammaticos, and A. Ramani, Painlevé analysis for the Mixmaster Universe model, J. Phys. A **25** (1993) 5795–5799.
99. C. M. Cosgrove, A new formulation of the field equations for the stationary axisymmetric vacuum gravitational field I. General theory, J. Phys. A **11** (1978) 2389–2404.
100. C. M. Cosgrove, Painlevé classification of all semilinear partial differential equations of the second order I. Hyperbolic equations in two independent variables, Stud. Appl. Math. **89** (1993) 1–61.
101. C. M. Cosgrove, Painlevé classification of all semilinear partial differential equations of the second order II. Parabolic and higher dimensional equations, Stud. Appl. Math. **89** (1993) 95–151.
102. C. M. Cosgrove, All binomial-type Painlevé equations of the second order and degree three or higher, Stud. Appl. Math. **90** (1993) 119–187.
103. C. M. Cosgrove, Painlevé classification problems featuring essential singularities, Stud. Appl. Math. **98** (1997) 355–433.
104. C. M. Cosgrove, Higher order Painlevé equations in the polynomial class, I. Bureau symbol *P*2, Stud. Appl. Math. **104** (2000) 1–65.
105. C. M. Cosgrove, Chazy classes IX–XI of third-order differential equations, Stud. Appl. Math. **104** (2000) 171–228.
106. C. M. Cosgrove, Higher order Painlevé equations in the polynomial class, II. Bureau symbol *P*1, http://www.maths.usyd.edu.au:8000/res/Nonlinear/Cos/2000-6.html 113 pages, preprint 2000–6.
107. C. M. Cosgrove and G. Scoufis, Painlevé classification of a class of differential equations of the second order and second degree, Stud. Appl. Math. **88** (1993) 25–87.
108. M. M. Crum, Associated Sturm–Liouville systems, Quart. J. Math. Oxford **6** (1955) 121–127.
109. G. Darboux, Sur la théorie des coordonnées curvilignes et des systèmes orthogonaux, Ann. Éc. Norm. **7** (1878) 101–150.
110. G. Darboux, Sur une proposition relative aux équations linéaires, C. R. Acad. Sc. Paris **94** (1882) 1456–1459.
111. G. Darboux, Sur les surfaces dont la courbure totale est constante, C. R. Acad. Sc. Paris **97** (1883) 848–850.
112. G. Darboux, Sur les surfaces à courbure totale constante, C. R. Acad. Sc. Paris **97** (1883) 892–894.
113. G. Darboux, Sur l'équation aux dérivées partielles des surfaces à courbure constante, C. R. Acad. Sc. Paris **97** (1883) 946–949.
114. G. Darboux, *Leçons sur la théorie générale des surfaces et les applications géométriques du calcul infinitésimal* (Gauthier–Villars, 4 volumes, Paris, 1894). Reprinted, *Théorie générale des surfaces* (Chelsea, New York, 1972).
115. E. Date, M. Jimbo, M. Kashiwara, and T. Miwa, KP hierarchies of orthogonal and symplectic type – Transformation groups for soliton equations VI –, JPSJ **50** (1981) 3813–3818.
116. H. T. Davis, *Introduction to nonlinear differential and integral equations*, 566 pages, no. O-556037 (U.S. Government Printing Office, Washington D.C., 1961).
117. A. Degasperis, D. D. Holm, and A. N. W. Hone, A new integrable equation with peakon solutions, Theor. Math. Phys. **133**, 1463–1474 (2002). nlin.SI/0205023.
118. A. Degasperis and P. Procesi, Asymptotic integrability, 23–37, *Symmetry and perturbation theory*, eds. A. Degasperis and G. Gaeta (World scientific, Singapore, 1999).

119. J. Demaret and Y. de Rop, The fractal nature of the power spectrum as an indicator of chaos in the Bianchi IX cosmological model, Phys. Lett. B **299** (1993) 223–228.
120. P. Di Francesco, 2-D quantum and topological gravities, matrix models, and integrable differential systems, 229–285. ed. R. Conte, CRM series in mathematical physics (Springer, New York, 1999).
121. R. Dickson, F. Gesztesy and K. Unterkofler, A new approach to the Boussinesq hierarchy, Math. Nachr. **198** (1999) 51–108.
122. R. K. Dodd and J. D. Gibbon, The prolongation structure of a higher order Korteweg–de Vries equation, Proc. Roy. Soc. London A **358** (1977) 287–296.
123. M. R. Douglas and S. H. Shenker, Strings in less than one dimension, Nucl. Phys. B **335** (1990) 635–654.
124. J. Drach, Sur l'intégration par quadratures de l'équation $\frac{d^2y}{dx^2} = [\varphi(x) + h]y$, C. R. Acad. Sc. Paris **168** (1919) 337–340.
125. P. G. Drazin and R. S. Johnson, *Solitons: an introduction* (Cambridge University Press, Cambridge, 1989).
126. V. G. Drinfel'd and V. V. Sokolov, Equations of Korteweg–de Vries type and simple Lie Algebras, Soviet Math. Dokl. **23** (1981) 457–462.
127. V. G. Drinfel'd and V. V. Sokolov, Lie algebras and equations of Korteweg–de Vries type, Itogi Nauki i Tekhniki, Seriya Sovremennye Problemy Matematiki **24** (1984) 81–180 [English: Journal of Soviet Math. **30** (1985) 1975–2036].
128. J.-M. Drouffe, AMP reference manual, version 9 (1993), SPhT, CEA-Saclay, F-91191 Gif-sur-Yvette Cedex. `Jean-Michel.Drouffe@normalesup.org`
129. V.S. Dryuma, Analytic solution of the two-dimensional Korteweg-de Vries (KdV) equation", Pis'ma Zh. Eksp. Teor. Fiz. **19** (1974) 753–755 [English: Soviet Physics JETP Letters **19** (1974) 387–388].
130. B. Dubrovin, Geometry of 2D topological field theories, Lecture notes in mathematics **1620** (1995) 120–348.
131. B. Dubrovin and M. Mazzocco, Monodromy of certain Painlevé-VI transcendents and reflection groups, Invent. math. **141** (2000) 55–147. math.AG/9806056.
132. A. E. Eremenko, Meromorphic traveling wave solutions of the Kuramoto–Sivashinsky equation, J. of mathematical physics, analysis and geometry **2** (2006) 278–286. http://arxiv.org/abs/nlin.SI/0504053.
133. V. P. Ermakov, Équations différentielles du deuxième ordre. Conditions d'intégrabilité sous forme finale. Univ. Izv. Kiev (1880) Ser. 3, No. 9, 1–25. [English translation by A. O. Harin, 29 pages].
134. P. G. Estévez, P. R. Gordoa, L. Martínez Alonso, and E. Medina Reus, Modified singular manifold expansion: application to the Boussinesq and Mikhailov–Shabat systems, J. Phys. A **26** (1993) 1915–1925.
135. H. Exton, On nonlinear ordinary differential equations with fixed critical points, Rendiconti di Matematica **4** (1971) 385–448.
136. L. Faddeev and L. Takhtajan, Liouville model on the lattice, *Field theory, quantum gravity and strings*, 166–179, eds. H.J. de Vega and N. Sánchez, Lecture notes in physics **246** (Springer-Verlag, Berlin, 1986).
137. G. Falqui and C.-M. Viallet, Singularity, complexity and quasi-integrability of rational mappings, Commun. Math. Phys. **154** (1993) 111–125.
138. N. Farkas and E. Kra, *Riemann surfaces* (Springer, Berlin, 1980).
139. M.J. Feigenbaum, Quantitative universality for a class of nonlinear transformations, J. Stat. Physics **19** (1978) 25–52.
140. R. A. Fisher, The wave of advance of advantageous genes, Ann. Eug. **7** (1936) 355–369.
141. S. Flach, Y. Zolotaryuk and K. Kladko, Moving lattice kinks and pulses: An inverse method, Phys. Rev. E **59** (1999) 6105–6115.
142. H. Flaschka and A. C. Newell, Monodromy– and spectrum–preserving deformations I, Commun. Math. Phys. **76** (1980) 65–116.

143. A. S. Fokas and M. J. Ablowitz, Linearization of the Korteweg–de Vries and Painlevé II equations, Phys. Rev. Lett. **47** (1981) 1096–1100.
144. A. S. Fokas and M. J. Ablowitz, On a unified approach to transformations and elementary solutions of Painlevé equations, J. Math. Phys. **23** (1982) 2033–2042.
145. A. S. Fokas, B. Grammaticos, and A. Ramani, From continuous to discrete Painlevé equations, Journal of mathematical analysis and applications **180** (1993) 342–360.
146. A. P. Fordy (ed.), *Soliton theory: a survey of results* (Manchester University Press, Manchester, 1990).
147. A. P. Fordy, The Hénon–Heiles system revisited, Physica D **52** (1991) 204–210.
148. A. P. Fordy and J. Gibbons, Some remarkable nonlinear transformations, Phys. Lett. A **75** (1980) 325–325.
149. A. P. Fordy and A. Pickering, Analysing negative resonances in the Painlevé test, Phys. Lett. A **160** (1991) 347–354.
150. A. P. Fordy and A. Pickering, Generalised Painlevé truncation : expansions in Riccati pseudopotentials, *Symmetries and Integrability of Difference equations*, eds. P. A. Clarkson and F. W. Nijhoff, London Mathematical Society LNS **255** (1999) 287–298, Cambridge University Press.
151. B. Fornberg and G. B. Whitham, A numerical and theoretical study of certain nonlinear wave phenomena, Phil. Trans. Roy. Soc. **A 289** (1978) 373–404.
152. A. R. Forsyth, *Theory of differential equations.* (Cambridge University Press, Cambridge, 1906). Reprinted (Dover, New York, 1959).
153. J.-D. Fournier, E. A. Spiegel, and O. Thual, Meromorphic integrals of two nonintegrable systems, *Nonlinear dynamics*, 366–373, eds. G. Servizi and G. Turchetti (World Scientific, Singapore, 1989).
154. I.B. Frenkel and V.G. Turaev, Elliptic solutions of the Yang–Baxter equation and modular hypergeometric functions, 171–204, *Arnol'd–Gelfand mathematical seminars*, eds. V.I. Arnol'd, I.M. Gel'fand, V.S. Retakh and M. Smirno (Birkhäuser, Boston, 1997).
155. G. Freud, On the coefficients in the recursion formulae of orthogonal polynomials, Proc. Roy. Irish Acad. A **76** (1976) 1–6.
156. R. Fuchs, Sur quelques équations différentielles linéaires du second ordre, C. R. Acad. Sc. Paris **141** (1905) 555–558.
157. R. Fuchs, Über lineare homogene Differentialgleichungen zweiter Ordnung mit drei im Endlichen gelegenen wesentlich singulären Stellen, Math. Annalen **63** (1907) 301–321.
158. B. Fuchssteiner, Some tricks from the symmetry-toolbox for nonlinear equations: Generelizations of the Camassa–Holm equation, Physica D **95** (1996) 229–243.
159. B. Fuchssteiner and A. S. Fokas, Symplectic structures, their Bäcklund transformations and hereditary symmetries, Physica D **4** (1981) 47–66.
160. A. Fujimoto and Y. Watanabe, Classification of fifth-order evolution equations with nontrivial symmetries, Math. Japonica **28** (1983) 43–65.
161. L. Gagnon and P. Winternitz, Symmetry classes of variable coefficient nonlinear Schrödinger equations, J. Phys. A **26** (1993) 7061–7076.
162. B. Gambier, Sur les équations différentielles dont l'intégrale générale est uniforme, C. R. Acad. Sc. Paris **142** (1906) 1403–1406.
163. B. Gambier, Sur les équations différentielles du second ordre et du premier degré dont l'intégrale générale est à points critiques fixes, Thèse, Paris (1909); Acta Math. **33** (1910) 1–55.
164. T. I. Garagash, On a modification of the Painlevé test for the BLP equation, *Nonlinear evolution equations and dynamical systems*, 130–133, eds. V. G. Makhankov, I. V. Puzynin, and O. K. Pashaev, (World Scientific, Singapore, 1993).
165. C. S. Gardner, J. M. Greene, M. D. Kruskal, and R. M. Miura, Method for solving the Korteweg–de Vries equation, Phys. Rev. Lett. **19** (1967) 1095–1097.
166. R. Garnier, Sur une classe de systémes différentiels abéliens déduits de la théorie des équations linéaires, Rendiconti del Circolo Matematico di Palermo **43** (1919) 155–191.

167. R. Garnier, Sur des équations différentielles du troisième ordre dont l'intégrale générale est uniforme et sur une classe d'équations nouvelles d'ordre supérieur dont l'intégrale générale a ses points critiques fixes, Thèse, Paris (1911); Ann. Éc. Norm. **29** (1912) 1–126.
168. R. Garnier, Sur un théorème de Schwarz, Comment. Math. Helvetici **25** (1951) 140–172.
169. R. Garnier, Sur des systèmes différentiels du second ordre dont l'intégrale générale est uniforme, Ann. Éc. Norm. **77** (1960) 123–144.
170. J. D. Gibbon, P. Radmore, M. Tabor, and D. Wood, The Painlevé property and Hirota's method, Stud. Appl. Math. **72** (1985) 39–63.
171. G. W. Gibbons and C. N. Pope, The positive action conjecture and asymptotically Euclidean metrics in quantum gravity, Commun. Math. Phys. **66** (1979) 267–290.
172. C. Gilson, F. Lambert, J. Nimmo, and R. Willox, On the combinatorics of Hirota $D-$operators, Proc. Roy. Soc. London A **431** (1996) 361–369.
173. C. Gilson and A. Pickering, Factorization and Painlevé analysis of a class of nonlinear third-order partial differential equations, J. Phys. A **28** (1995) 2871–2888.
174. A.A. Goldberg and I.V.Ostrovskii, *Distribution of values of meromorphic functions* (Nauka, Moscow, 1970) (in Russian).
175. P. Gordoa, N. Joshi, and A. Pickering, Mappings preserving locations of movable poles: a new extension of the truncation method to ordinary differential equations, Nonlinearity **12** (1999) 955–968.
176. P. Gordoa and A. Pickering, Nonisospectral scattering problems: a key to integrable hierarchies, J. Math. Phys. **40** (1999) 5749–5786.
177. A. Goriely, *Integrability and nonintegrability of dynamical systems*, Advanced series in nonlinear dynamics **19** (World Scientific, Singapore, 2001).
178. E. Goursat, *Cours d'analyse mathématique* (Gauthier-Villars, Paris, 1924). [English translation: *A course in mathematical analysis* (Dover, New York, 1956)]
179. B. Grammaticos, B. Dorizzi, and A. Ramani, Integrability of Hamiltonians with third- and fourth-degree polynomial potentials, J. Math. Phys. **24** (1983) 2289–2295.
180. B. Grammaticos, B. Dorizzi, and A. Ramani, Hamiltonians with higher order integrals and the "weak-Painlevé" concept, J. Math. Phys. **25** (1984) 3470–3473.
181. B. Grammaticos, F. W. Nijhoff, V. Papageorgiou, A. Ramani, and J. Satsuma, Linearization and solutions of the discrete Painlevé III equation, Phys. Lett. A **185** (1994) 446–452.
182. B. Grammaticos, F. W. Nijhoff, and A. Ramani, Discrete Painlevé equations, *The Painlevé property, one century later*, 413–516, ed. R. Conte, CRM series in mathematical physics (Springer, New York, 1999).
183. B. Grammaticos and A. Ramani, Discrete Painlevé equations: a review, 245–321, in *Discrete integrable systems* Lecture notes in physics **644** (Springer Verlag, Berlin, 2004).
184. B. Grammaticos, A. Ramani, and V. Papageorgiou, Do integrable mappings have the Painlevé property? Phys. Rev. Lett. **67** (1991) 1825–1828.
185. V. I. Gromak, Theory of Painlevé's equation, Differentsial'nye Uravneniya **11** (1975) 373–376 [English : Diff. equ. **11** (1975) 285–287].
186. V. I. Gromak, Solutions of Painlevé's fifth problem, Differentsial'nye Uravneniya **12** (1976) 740–746 [English : Diff. equ. **12** (1976) 519–521].
187. V. I. Gromak, On the algebraic solutions of the third Painlevé equation, Doklady Akad. Nauk Bielorussian SSR **23** (1979) 499–502.
188. V. I. Gromak and G. Filipuk, On the transformations of the sixth Painlevé equation, J. Nonlinear Mathematical Physics **10**, Supp. 2, 57–67 (2003).
189. V.I. Gromak, I. Laine and S. Shimomura, *Painlevé differential equations in the complex plane*, viii+303 pages (de Gruyter, Berlin, 2002).
190. V. I. Gromak and N.A. Lukashevich, *The analytic solutions of the Painlevé equations*, 157 pages (in Russian), (Universitetskoye Publishers, Minsk, 1990). ISBN 5–7855–0319–0.
191. D. J. Gross and A. A. Migdal, Nonperturbative tw-dimensional quantum gravity, Phys. Rev. Lett. **64** (1990) 127–130.
192. G.-H. Halphen, Sur un système d'équations différentielles, C. R. Acad. Sc. Paris **92** (1881) 1101–1103. Reprinted, *Œuvres*, Gauthier-Villars, Paris, tome 2, 475–477 (1918).

193. G.-H. Halphen, *Traité des fonctions elliptiques et de leurs applications*, Gauthier-Villars, Paris. Première partie, Théorie des fonctions elliptiques et de leurs développements en série, 492 pages (1886). Deuxième partie, Applications à la mécanique, à la physique, à la géodésie, à la géométrie et au calcul intégral, 659 pages (1888). Troisième partie, Fragments, 272 pages (1891). `http://historical.library.cornell.edu/math/math_H.html` (3 volumes).
194. W.K. Hayman, *Meromorphic functions* (Clarendon Press, Oxford, 1964).
195. He Yuzan, Yuan Wenjun and Li Yezhou, *Lecture notes on the analytic theory of Painlevé equations*, 232 pages (Science press, Beijing, 2005). ISBN 7-03-014596-8. http://www.sciencep.com. [in Chinese].
196. M. van Hecke, Building blocks of spatiotemporal intermittency, Phys. Rev. Lett. **80** (1998) 1896–1899.
197. M. van Hecke, C. Storm, and W. van Saarlos, Sources, sinks and wavenumber selection in coupled CGL equations and experimental implications for counter-propagating wave systems, Physica D **133** (1999) 1–47. Patt–sol/9902005.
198. M. Hénon and C. Heiles, The applicability of the third integral of motion: some numerical experiments, Astron. J. **69** (1964) 73–79.
199. W. Hereman, P. P. Banerjee, A. Korpel, G. Assanto, A. Van Immerzeele, and A. Meerpoel, Exact solitary wave solutions of non-linear evolution and wave equations using a direct algebraic method, J. Phys. A **19** (1986) 607–628.
200. W. Hereman, P. P. Banerjee and M.R. Chatterjee, Derivation and implicit solution of the Harry-Dym equation and its connections with the Korteweg–de Vries equation, J. Phys. A **22** (1989) 241–255.
201. W. Hereman and A. Nuseir, Symbolic methods to construct exact solutions of nonlinear partial differential equations, Mathematics and Computers in Simulation **43** (1997) 13–27.
202. E. Hernández-García, A. Amengual, R. Montagne, M. San Miguel, P. Colet, and M. Hoyuelos: Moving pictures, *Europhysics News* **29**, 184–187 (1998).
203. J. Hietarinta, Classical versus quantum integrability, J. Math. Phys. **25** (1984) 1833–1840.
204. J. Hietarinta, Direct method for the search of the second invariant, Phys. Rep. **147** (1987) 87–154.
205. J. Hietarinta, Painlevé equations in terms of entire functions, *The Painlevé property, one century later*, 661–686, ed. R. Conte, CRM series in mathematical physics (Springer, New York, 1999).
206. J. Hietarinta and C.-M. Viallet, Singularity confinement and chaos in discrete systems, Phys. Rev. Lett. **81** (1998) 325–328.
207. E. Hille, *Ordinary differential equations in the complex domain* (J. Wiley and sons, New York, 1976).
208. R. Hirota, Exact solutions of the KdV equation from multiple collisions of solitons, Phys. Rev. Lett. **27** (1971) 1192–1194.
209. R. Hirota, Exact solution of the modified Korteweg–de Vries equation from multiple collisions of solitons, JPSJ **33** (1972) 1456–1458.
210. R. Hirota, Exact solution of the sine-Gordon equation from multiple collisions of solitons, JPSJ **33** (1972) 1459–1463.
211. R. Hirota, Exact envelope-soliton solutions of a nonlinear wave equation, JMP **14** (1973) 805–809.
212. R. Hirota, A new form of Bäcklund transformations and its relation to the inverse scattering problem, Prog. Theor. Phys. **52** (1974) 1498–1512.
213. R. Hirota, Direct methods of finding exact solutions of nonlinear evolution equations, *Bäcklund transformations, the inverse scattering method, solitons and their applications*, 40–68, ed. R. M. Miura, Springer lecture notes in mathematics **515** (Springer Verlag, Berlin, 1976).
214. R. Hirota, Nonlinear partial difference equations. V. Nonlinear equations reducible to linear equations, JPSJ **46** (1979) 312–319.

215. R. Hirota, Fundamental properties of the binary operators in soliton theory and their generalization, *Dynamical problems in soliton systems*, 42–49, ed. S. Takeno, (Springer, Berlin, 1985)
216. R. Hirota and J. Satsuma, N-soliton solutions of model equations for shallow water waves, J. Phys. Soc. Japan Letters **40** (1976) 611–612.
217. R. Hirota and J. Satsuma, Nonlinear evolution equations generated from the Bäcklund transformation for the Boussinesq equation, Prog. Theor. Phys. **57** (1977) 797–807.
218. R. Hirota and J. Satsuma, A simple structure of superposition formula of the Bäcklund transformation, JPSJ **45** (1978) 1741–1750.
219. R. Hirota, Soliton solutions to the BKP equations, I. The Pfaffian technique, JPSJ **58** (1989) 2285–2296.
220. N. J. Hitchin, Poncelet polygons and the Painlevé equations, 151–185, *Geometry and analysis*, ed. Ramanan (Oxford university press, Oxford, 1995).
221. L. Hlavatý, On the Painlevé classification of partial differential equations, J. Math. Phys. **31** (1990) 605–609.
222. Mark van Hoeij, package "algcurves", Maple V (1997). `http://www.math.fsu.edu/~hoeij/algcurves.html`
223. A. N. W. Hone, Exact discretization of the Ermakov–Pinney equation, Phys. Lett. A **263** (1999) 347–354.
224. A.N.W. Hone, Non-existence of elliptic travelling wave solutions of the complex Ginzburg–Landau equation, Physica D **205** (2005) 292–306.
225. A. N. W. Hone, V. Novikov and C. Verhoeven, An integrable hierarchy with a perturbed Hénon–Heiles system, Inverse Problems **22** (2006) 2001–2020.
226. A.N.W. Hone and J-P. Wang, Prolongation algebras and Hamiltonian operators for peakon equations, Inverse Problems **19** (2003) 129–145.
227. Paul Hoyer, Über die Integration eines Differentialgleichungssystems von der Form

$$\frac{dx_1}{dt} = a_1x_2x_3 + a_2x_3x_1 + a_3x_1x_2$$
$$\frac{dx_2}{dt} = b_1x_2x_3 + b_2x_3x_1 + b_3x_1x_2$$
$$\frac{dx_3}{dt} = c_1x_2x_3 + c_2x_3x_1 + c_3x_1x_2$$

durch elliptische Funktionen, Dissertation Königl. Friedrich-Wilhelms Universität in Berlin, 1–36 (1879).
228. L. Hsu and N. Kamran, Classification of second-order ordinary differential equations admitting Lie groups of fiber-preserving point symmetries, Proc. London Math. Soc. **58** (1989) 387–416.
229. Hu Xing-biao and Li Yong, Some results on the Caudrey–Dodd–Gibbon–Kotera–Sawada equation, J. Phys. A **24** (1991) 3205–3212.
230. A. Iatrou and J.A.G. Roberts, Integrable mappings of the plane preserving biquadratic invariant curves, J. Phys. A **34** (2001) 6617–6636.
231. M. Inaba, K. Iwasaki and M-H. Saito, Dynamics of the sixth Painlevé equation, 103–167, *Théories asymptotiques et équations de Painlevé*, eds. E. Delabaere and M. Loday, Séminaires et congrès **14** (Société mathématique de France, Paris, 2006).
232. E. L. Ince, *Ordinary differential equations* (Longmans, Green, and co., London and New York, 1926). Reprinted (Dover, New York, 1956). Russian translation (GTIU, Khar'kov, 1939).
233. A. R. Its, A. V. Kitaev, and A. S. Fokas, The isomonodromy approach in the theory of two–dimensional quantum gravitation, Usp. Matem. Nauk **45** (1990) 135–136 [English : Russ. Math. Surveys **45** (1990) 155–157].
234. C. Itzykson and J.-B. Zuber, The planar approximation. II, J. Math. Phys. **21** (1980) 411–421.

235. K. Iwasaki, H. Kimura, S. Shimomura, and M. Yoshida, *From Gauss to Painlevé: a modern theory of special functions*, 347 pages (Vieweg, Braunschweig, 1991).
236. A. Jeffrey and Xu S., Travelling wave solutions to certain non-linear evolution equations, Int. J. Non-Linear Mechanics **24** (1989) 425–429.
237. M. Jimbo (ed.), *Yang–Baxter equation in integrable systems* (World Scientific, Singapore, 1990).
238. M. Jimbo, M. D. Kruskal, and T. Miwa, Painlevé test for the self-dual Yang–Mills equation, Phys. Lett. A **92** (1982) 59–60.
239. M. Jimbo and T. Miwa, Studies on holonomic quantum fields. XVII, Proc. Japan Acad. **56** (1980) 405–410.
240. M. Jimbo and T. Miwa, Monodromy preserving deformations of linear ordinary differential equations with rational coefficients. II, Physica D **2** (1981) 407–448.
241. M. Jimbo and T. Miwa, Solitons and infinite dimensional Lie algebras, Publ. RIMS, Kyoto **19** (1983) 943–1001.
242. M. Jimbo and H. Sakai, A q-analog of the sixth Painlevé equation, Lett. Math. Phys. **38** (1996) 145–154.
243. B. B. Kadomtsev and V. I. Petviashvili, Dokl. Akad. Nauk SSSR **192** (1970) 753–756 [English : Sov. Phys. Dokl. **15** (1970) 539–541].
244. K. Kajiwara, T. Masuda, M. Noumi, Y. Ohta and Y. Yamada, ${}_{10}E_9$ solution to the elliptic Painlevé equation, J. Phys. A 36 (2003) L263–L272.
245. K. Kajiwara, T. Masuda, M. Noumi, Y. Ohta and Y. Yamada, Point configurations, Cremona transformations and the elliptic difference Painlevé equation, 169–198, *Théories asymptotiques et équations de Painlevé*, eds. E. Delabaere and M. Loday, Séminaires et congrès **14** (Société mathématique de France, Paris, 2006).
246. D. J. Kaup, On the inverse scattering problem for cubic eigenvalue problems of the class $\psi_{xxx}+6Q\psi_x+6R\psi=\lambda\,\psi$, Stud. Appl. Math. **62** (1980) 189–216.
247. T. Kawahara and M. Tanaka, Interactions of traveling fronts: an exact solution of a nonlinear diffusion equation, Phys. Lett. A **97** (1983) 311–314.
248. Shunji Kawamoto, An exact transformation from the Harry–Dym equation to the modified KdV equation, J. Phys. Soc. Japan **54** (1985) 2055–2056.
249. L. R. Keefe, Integrability and structural stability of solutions to the Ginzburg-Landau equation, Phys. Fluids **29** (1986) 3135–3141.
250. A. Khare, K.Ø. Rasmussen, M.R. Samuelsen and A. Saxena, Exact solutions of the saturable discrete nonlinear Schrödinger equation, J. Phys. A **38** (2005) 807–814.
251. K. Kimura, H. Yahagi, R. Hirota, A. Ramani, B. Grammaticos and Y. Ohta, A new class of integrable discrete systems, J. Phys. A **35** (2002) 9205–9212.
252. A. V. Kitaev, On similarity reductions of the three-wave resonant system to the Painlevé equations, J. Phys. A **23** (1990) 3453–3553.
253. A. V. Kitaev, Caustics in 1+1 integrable systems, J. Math. Phys. **35** (1994) 2934–2954.
254. A. V. Kitaev, Special functions of the isomonodromy type, rational transformations of spectral parameter, and algebraic solutions of the sixth Painlevé equation, Algebra i Analiz **14** (2002) 121–139. English translation: St. Petersburg Math. J. **14** (2003) 453–465. http://arXiv.org/abs/nlin.SI/0102020
255. A. N. Kolmogorov, I. G. Petrovskii and N. S. Piskunov, The study of a diffusion equation, related to the increase of the quantity of matter, and its application to one biological problem, Bulletin de l'Université d'État de Moscou, série internationale, section A Math. Méc. **1** (1937) 1–26.
256. D. J. Korteweg and G. de Vries, On the change of form of long waves advancing in a rectangular canal, and on a new type of long stationary waves, Phil. Mag. **39** (1895) 422–443.
257. Sophie Kowalevski, Sur le problème de la rotation d'un corps solide autour d'un point fixe, Acta Math. **12** (1889) 177–232.
258. M. Kruskal, Nonlinear wave equations, 310–354, *Dynamical systems, theory and applications*, ed. J. Moser, Lecture notes in physics **38** (Springer-Verlag, Berlin, 1975).

259. N. A. Kudryashov, Exact soliton solutions of the generalized evolution equation of wave dynamics, Prikladnaia Matematika i Mekhanika **52** (1988) 465–470 [English : Journal of applied mathematics and mechanics **52** (1988) 361–365]
260. N. A. Kudryashov, Exact solutions of a generalized equation of Ginzburg–Landau, Matematicheskoye modelirovanie **1** (1989) 151–158 [in Russian].
261. N.A. Kudryashov, The first and second Painlevé equations of higher order and some relations between them, Phys. Lett. A **224** (1997) 353–360.
262. N.A. Kudryashov and M. B. Soukharev, Uniformization and transcendence of solutions for the first and second Painlevé hierarchies, Phys. Lett. A **237** (1998) 206–216.
263. Y. Kuramoto and T. Tsuzuki, Persistent propagation of concentration waves in dissipative media far from thermal equilibrium, Prog. Theor. Phys. **55** (1976) 356–369.
264. M. Kuś, Integrals of motion for the Lorenz system, J. Phys. A **16** (1983) L689–L691.
265. S. Labrunie, Contribution à l'étude de l'intégrabilité des équations différentielles et des équations aux différences finies non-linéaires, Thèse (Université Paris VII, 1996).
266. E. Laguerre, Sur la réduction en fractions continues d'une fraction qui satisfait à une équation différentielle linéaire du premier ordre dont les coefficients sont rationnels, J. de math. pures et appliquées **1** (1885) 135–165. *Oeuvres*, vol. II, 685–711.
267. I. Laine, *Nevanlinna theory and complex differential equations*, (de Gruyter, Berlin and New York, 1992).
268. G. L. Lamb Jr, Propagation of ultrashort optical pulses, Phys. Lett. A **25** (1967) 181–182.
269. G. L. Lamb Jr, Bäcklund transformations for certain nonlinear evolution equations, J. Math. Phys. **15** (1974) 2157–2165.
270. L. D. Landau et E. M. Lifshitz, *Théorie classique des champs*, chapitre "Problèmes cosmologiques" (Éditions Mir, Moscou, 3ième édition et suiv., 1971).
271. G. L. Lamb Jr, *Elements of soliton theory* (Wiley and sons, New York, 1980).
272. S. Lang, *Introduction to complex hyperbolic spaces* (Springer-Verlag, New York, 1987).
273. A. Latifi, M. Musette, and R. Conte, The Bianchi IX (mixmaster) cosmological model is not integrable, Phys. Letters A **194** (1994) 83–92; **197** (1995) 459–460. chao-dyn/9409002.
274. P. D. Lax, Integrals of nonlinear equations of evolution and solitary waves, Comm. Pure Appl. Math. **21** (1968) 467–490.
275. J. Lega, Défauts topologiques associés à la brisure de l'invariance de translation dans le temps, Thèse (Université de Nice, 28 mars 1989).
276. J. Lega, Traveling hole solutions of the complex Ginzburg–Landau equation: a review, Physica D **152–153** (2001) 269–287.
277. J. Lega, J. V. Moloney, and A. C. Newell, Swift–Hohenberg equation for lasers, Phys. Rev. Lett. **73** (1994) 2978–2981.
278. D. Levi, On a new Darboux transformation for the contsruction of exact solutions of the Schrödinger equation, Inverse Problems **4** (1988) 165–172.
279. D. Levi and O. Ragnisco, Non-isospectral deformations and Darboux transformations for the third-order spectral problem, Inverse Problems **4** (1988) 815–828.
280. G. Levine and M. Tabor, Integrating the nonintegrable: analytic structure of the Lorenz system revisited, Physica D **33** (1988) 189–210.
281. Lin Run-liang, R. Conte, and M. Musette, On the Lax pairs of the continuous and discrete sixth Painlevé equations, J. Nonlinear Mathematical Physics **10**, Supp. 2, 107–118 (2003).
282. R. Liouville, Sur les invariants de certaines équations différentielles et sur leurs applications, Journal de l'École Polytechnique **59** (1889) 7–76.
283. J. Llibre and Zhang Xiang, Invariant algebraic surfaces of the Lorenz system, J. Math. Phys. **43** (2002) 1622–1645.
284. E. N. Lorenz, Deterministic nonperiodic flow, J. Atmos. Sci. **20** (1963) 130–141.
285. I. Loris, On reduced CKP equations, Inverse Problems **15** (1999) 1099–1109.
286. N.A. Lukashevivh, Solutions of the fifth equation of Painlevé, Differentsial'nye Uravneniya **4** (1968) 1413–1420 [English : Diff. equ. **4** (1968) 732–735].
287. N. A. Lukashevich, Theory of the fourth Painlevé equation, Differentsial'nye Uravneniya **3** (1967) 771–789 [English : Diff. equ. **3** (1967) 395–399].

288. N. A. Lukashevich, The second Painlevé equation, Differentsial'nye Uravneniya **7** (1971) 1124–1125 [English : Diff. equ. **7** (1971) 853–854].
289. W. Magnus, F. Oberhettinger and R.P. Soni, *Formulas and theorems for the special functions of mathematical physics*, third enlarged ed. (Springer, Berlin, 1966).
290. F. Magri, G. Falqui, and M. Pedroni, The method of Poisson pairs in the theory of nonlinear PDEs, *Direct and inverse methods in nonlinear evolution equations*, 85–136, ed. A. Greco, Lecture notes in physics **632** (Springer Verlag, Berlin, 2003). http://xxx.lanl.gov/abs/nlin.SI/0002009
291. G. Mahoux, Introduction to the theory of isomonodromic deformations of linear ordinary differential equations with rational coefficients, *The Painlevé property, one century later*, 35–76, ed. R. Conte, CRM series in mathematical physics (Springer, New York, 1999).
292. B. Malgrange, Sur les déformations isomonodromiques. I. Singularités régulières, 401–426, *Mathematics and physics* (Paris, 1979/1982), Progr. Math. **37** (Birkhäuser, Boston, 1983).
293. J. Malmquist, Sur les équations différentielles du second ordre dont l'intégrale générale a ses points critiques fixes, Arkiv för Math. Astr. Fys. **17** (1922–23) 1–89.
294. S.V. Manakov, On the theory of two-dimensional stationary self-focusing of electromagnetic waves, Zh. Eksp. Teor. Fiz. **65**, 505–516 (1973) [JETP **38**, 248–253 (1974)].
295. Yu. I. Manin, Sixth Painlevé equation, universal elliptic curve, and mirror of P^2, *Geometry of differential equations*, eds. A. Khovanskii, A. Varchenko, and V. Vassiliev, AMS Transl., ser. 2, **186 (39)** (1998) 131–151. Alg-geom/9605010.
296. P. Manneville, *Dissipative structures and weak turbulence* (Academic Press, Boston, 1990). French adaptation: *Structures dissipatives, chaos et turbulence* (Aléa-Saclay, Gif-sur-Yvette, 1991).
297. E. Mansfield, *Differential Groebner bases* (Ph D Thesis, University of Sydney, 1991).
298. L. Martina and P. Winternitz, Analysis and applications of the symmetry group of the multidimensional three-wave resonant interaction problem, Annals of Physics **196** (1989) 231–277.
299. I. P. Martynov, Differential equations with stationary critical singularities, Differentsial'nye Uravneniya **9** (1973) 1780–1791 [English : Diff. equ. **9** (1973) 1368–1376
300. L. J. Mason and N. M. J. Woodhouse, Self-duality and the Painlevé transcendents, Nonlinearity **6** (1993) 569–581.
301. V. B. Matveev and M. A. Salle, *Darboux transformations and solitons*, Springer series in nonlinear dynamics (Springer, Berlin, 1991).
302. E. M. McMillan, A problem in the stability of periodic systems, *Topics in modern physics*, 219–244, eds. W. E. Brittin and H. Odabasi (Colorado Associated University Press, Boulder, 1971).
303. D. M. Michelson, Steady solutions of the Kuramoto–Sivashinsky equation, Physica D **19** (1986) 89–111.
304. A. V. Mikhailov, A. B. Shabat, and V. V. Sokolov, The symmetry approach to classification of integrable equations, *What is integrability?*, 115–184, ed. V. E. Zakharov (Springer, Berlin, 1991).
305. W.E. Milne, The numerical determination of characteristic numbers, Physical Review 35 (1930) 863–867.
306. R.M. Miura, Korteweg–de Vries equation and generalizations. I. A remarkable explicit nonlinear transformation, J. Math. Phys. **9** (1968) 1202–1204.
307. H. C. Morris, Prolongation structures and a generalized inverse scattering problem, J. Math. Phys. **17** (1976) 1867–1869.
308. D. Mumford, *Tata lectures on theta, II* (Birkhäuser, Basel, 1983).
309. G. Muntingh and M. van der Put, Order one equations with the Painlevé property, Indag. Mathem., N.S. **18** (2007) 83–95.
310. Y. Murata, Rational solutions of the second and the fourth equations of Painlevé, Funkc. Ekvacioj. Ser. Int. **28** (1985) 1–32.
311. Y. Murata, Classical solutions of the third Painlevé equation, Nagoya Math. J. **139** (1995) 37–65.

312. M. Musette, Painlevé–Darboux transformation in nonlinear PDEs, *Inverse methods in action*, 570–577, ed. P. C. Sabatier, Springer-Verlag series "Inverse problems and theoretical imaging" (Springer-Verlag, Berlin, 1990).
313. M. Musette, Painlevé analysis for nonlinear partial differential equations, *The Painlevé property, one century later*, 517–572, ed. R. Conte, CRM series in mathematical physics (Springer, New York, 1999).
314. M. Musette, Nonlinear superposition formulae of integrable partial differential equations by the singular manifold method, *Direct and inverse methods in nonlinear evolution equations*, 137–170, ed. A. Greco, Lecture notes in physics **632** (Springer Verlag, Berlin, 2003).
315. M. Musette and R. Conte, Algorithmic method for deriving Lax pairs from the invariant Painlevé analysis of nonlinear partial differential equations, J. Math. Phys. **32** (1991) 1450–1457.
316. M. Musette and R. Conte, Solitary waves and Lax pairs from polynomial expansions of nonlinear differential equations, *Nonlinear evolution equations and dynamical systems*, 161–170, eds. M. Boiti, L. Martina, and F. Pempinelli (World Scientific, Singapore, 1992).
317. M. Musette and R. Conte, The two–singular manifold method, I. Modified KdV and sine-Gordon equations, J. Phys. A **27** (1994) 3895–3913.
318. M. Musette and R. Conte, Non-Fuchsian extension to the Painlevé test, Phys. Lett. A **206** (1995) 340–346. Solv-int/9508003.
319. M. Musette and R. Conte, Bäcklund transformation of partial differential equations from the Painlevé–Gambier classification, I. Kaup–Kupershmidt equation, J. Math. Phys. **39** (1998) 5617–5630.
320. M. Musette and R. Conte, Analytic solitary waves of nonintegrable equations, Physica D **181** (2003) 70–79. http://arXiv.org/abs/nlin.PS/0302051
321. M. Musette and C. Verhoeven, Nonlinear superposition formula for the Kaup–Kupershmidt partial differential equation, Physica D **144** (2000) 211–220.
322. A. C. Newell and J. A. Whitehead, Finite bandwidth, finite amplitude convection, J. Fluid Mech. **38** (1969) 279–303.
323. R. Nevanlinna, *Le théorème de Picard–Borel et la théorie des fonctions méromorphes* (Gauthier-Villars, 1929, Paris).
324. R. Nevanlinna, *Eindeutige analytische Funktionen, 2 Aufl.* (Springer-Verlag, Berlin, 1955).
325. A. Newell, M. Tabor, and Zeng Y.-b., A unified approach to Painlevé expansions, Physica D **29** (1987) 1–68.
326. F. W. Nijhoff and V. G. Papageorgiou, Similarity reductions of integrable lattices and discrete analogues of the Painlevé II equation, Phys. Lett. A **153** (1991) 337–344.
327. F. W. Nijhoff, A. Ramani, B. Grammaticos, and Y. Ohta, On discrete Painlevé equations associated with the lattice KdV systems and the Painlevé VI equation, Stud. Appl. Math. , **106** (2001) 261–314. solv-int/9812011.
328. J. J. C. Nimmo, A bilinear Bäcklund transformation for the nonlinear Schrödinger equation, Phys. Lett. A **99** (1983) 279–280.
329. J. J. C. Nimmo, Darboux transformations from reductions of the KP hierarchy, *Nonlinear evolution equations and dynamical systems*, 168–177, eds. V. G. Makhankov, A. R. Bishop, and D. D. Holm (World Scientific, Singapore, 1995).
330. J. J. C. Nimmo and N. C. Freeman, A method of obtaining the $N-$soliton solution of the Boussinesq equation in terms of a Wronskian, Phys. Lett. A **95** (1983) 4–6.
331. M. Noumi, *Painlevé equations through symmetry*, Translations of Mathematical Monographs **233** (American Mathematical Society, Providence, RI, 2004), translated from the 2000 Japanese original by the author.
332. M. Noumi, Special functions arising from discrete Painlevé property, Journal of Computational and Applied Mathematics **202** (2002) 48–55.
333. K. Nozaki and N. Bekki, Pattern selection and spatiotemporal transition to chaos in the Ginzburg-Landau equation, Phys. Rev. Lett. **51** (1983) 2171–2174.
334. K. Nozaki and N. Bekki, Exact solutions of the generalized Ginzburg-Landau equation, J. Phys. Soc. Japan **53** (1984) 1581–1582.

335. W. Oevel and W. Schief, Darboux theorems and the KP hierarchy, *Applications of analytic and geometric methods to nonlinear differential equations*, 193–206, ed. P. A. Clarkson (Plenum, New York, 1993).
336. Y. Ohta, A. Ramani and B. Grammaticos, Elliptic discrete Painlevé equations, J. Phys. A **35** (2002) L653–L659.
337. Y. Ohta, J. Satsuma, D. Takahashi and T. Tokihiro, An elementary introduction to Sato theory, Prog. Theor. Phys. Suppl. **94** (1988) 210–241; addenda Prog. Theor. Phys. **80** (1988) 742.
338. Y. Ohyama and S. Okumura, A coalescent diagram of the Painlevé equations from the viewpoint of isomonodromic deformations, J. Phys. A: Math. Gen. **39** (2006) 12129-12151. Special issue "One hundred years of Painlevé VI".
339. K. Okamoto, Sur les feuilletages associés aux équations du second ordre à points critiques fixes, Japan. J. Math. **5** (1979) 1–79.
340. K. Okamoto, Isomonodromic deformation and Painlevé equations, and the Garnier system, J. Fac. Sci. Univ. Tokyo, Sect. IA **33** (1986) 575–618.
341. K. Okamoto, Studies on the Painlevé equations, I, Sixth Painlevé equation, Ann. Mat. Pura Appl. **146** (1987) 337–381.
342. K. Okamoto, Studies on the Painlevé equations, II, Fifth Painlevé equation, Japan. J. Math., **13** (1987) 47–76.
343. P. J. Olver, *Applications of Lie groups to differential equations* (Springer, Berlin, 1986).
344. H. Ono, Algebraic solitary waves in stratified fluids, J. Phys. Soc. Japan **39** (1975) 1082–1091.
345. L. V. Ovsiannikov, *Group analysis of differential equations* (Academic Press, New York, 1982).
346. P. Painlevé, *Leçons sur la théorie analytique des équations différentielles* (Leçons de Stockholm, 1895) (Hermann, Paris, 1897). Reprinted, *Oeuvres de Paul Painlevé*, vol. I (Éditions du CNRS, Paris, 1973). `http://historical.library.cornell.edu/math/math_P.html`
347. P. Painlevé, Sur les équations différentielles du second ordre à points critiques fixes, C. R. Acad. Sc. Paris **126** (1898) 1697–1700.
348. P. Painlevé, Mémoire sur les équations différentielles dont l'intégrale générale est uniforme, Bull. Soc. Math. France **28** (1900) 201–261.
349. P. Painlevé, Sur les équations différentielles du second ordre et d'ordre supérieur dont l'intégrale générale est uniforme, Acta Math. **25** (1902) 1–85.
350. P. Painlevé, Sur les équations différentielles du second ordre à points critiques fixes, C. R. Acad. Sc. Paris **143** (1906) 1111–1117.
351. *Oeuvres de Paul Painlevé*, 3 volumes (Éditions du CNRS, Paris, 1973, 1974, 1976).
352. N. R. Pereira and L. Stenflo, Nonlinear Schrödinger equation including growth and damping, Phys. Fluids **20** (1977) 1733–1743.
353. V. Periwal and D. Shevitz, Unitary-matrix models as exactly solvable string theories, Phys. Rev. Lett. **64** 1326–1329 (1990).
354. É. Picard, Mémoire sur la théorie des fonctions algébriques de deux variables, J. Math. Pures Appl. **5** (1889) 135–319.
355. É. Picard, Remarques sur les équations différentielles, Acta Math. **17** (1893) 297–300.
356. A. Pickering, A new truncation in Painlevé analysis, J. Phys. A **26** (1993) 4395–4405.
357. A. Pickering, The singular manifold method revisited, J. Math. Phys. **37** (1996) 1894–1927.
358. E. Pinney, The nonlinear differential equation $y''(x) + p(x)y(x) + c/y^3(x) = 0$, Proc. Amer. Math. Soc. **1** (1950) 681–681.
359. Lev P. Pitaevskii and S. Stringari, *Bose-Einstein condensation* (Clarendon Press, Oxford, 2003).
360. H. Poincaré, Sur les groupes des équations linéaires, Acta mathematica **4** (1883) 201–312. Reprinted, *Oeuvres* (Gauthier-Villars, Paris), tome II, 300–401.
361. *Oeuvres de Poincaré*, 12 volumes (Gauthier-Villars, Paris, 1951–1956).
362. Y. Pomeau and P. Manneville, Stability and fluctuations of a spatially periodic flow, J. Physique Lett. **40** (1979) L609–L612.

363. J.-F. Pommaret, *Lie pseudogroups and mechanics* (Gordon and Breach science publishers, New York, 1988).
364. R. B. Potts (1981), Exact solution of a difference approximation to Duffing's equation, J. Austral. Math. Soc. B **23** (1981) 64–77.
365. R. B. Potts, Ordinary and partial difference equations, J. Austral. Math. Soc. B **27** (1986) 488–501.
366. R. B. Potts (1987), Weierstrass elliptic difference equations, Bull. Austral. Math. Soc. **35** (1987) 43–48.
367. G. R. W. Quispel, J. A. G. Roberts, and C. J. Thompson, Integrable mappings and soliton equations, Phys. Lett. A **126** (1988) 419–421.
368. G. R. W. Quispel, J. A. G. Roberts, and C. J. Thompson, Integrable mappings and soliton equations II, Physica D **34** (1989) 183–192.
369. A. Ramani, B. Dorizzi, and B. Grammaticos, Painlevé conjecture revisited, Phys. Rev. Lett. **49** (1982) 1539–1541.
370. A. Ramani and B. Grammaticos, Discrete Painlevé equations : coalescences, limits and degeneracies, Physica A **228** (1996) 160–171.
371. A. Ramani, B. Grammaticos and J. Hietarinta, Discrete versions of the Painlevé equations, Phys. Rev. Lett. **67** (1991) 1829–1832.
372. V. Ravoson, L. Gavrilov, and R. Caboz, Separability and Lax pairs for Hénon–Heiles system, J. Math. Phys. **34** (1993) 2385–2393.
373. C. Rogers and S. Carillo, On reciprocal properties of the Caudrey–Dodd–Gibbon and Kaup–Kupershmidt hierarchies, Physica Scripta **36** (1987) 865–869.
374. C. Rogers and W. F. Shadwick, *Bäcklund transformations and their applications* (Academic press, New York, 1982).
375. F. J. Romeiras, A note on integrable two-degrees-of-freedom Hamiltonian systems with a second integral quartic in the momenta, J. Phys. A **28** (1995) 5633–5642.
376. W. van Saarloos, Front propagation into unstable states, Physics reports **386** 29–222 (2003).
377. H. Sakai, Rational surfaces associated with affine root systems and geometry of the Painlevé equations, Comm. Math. Phys. **220** (2001) 165–229.
378. S. Yu. Sakovich, On integrability of differential constraints arising from the singularity analysis, J. Nonlinear Mathematical Physics **9** (2002) 21–25. nlin.SI/0004037.
379. M. Salerno, V. Z. Enol'skii, and D. V. Leykin, Canonical transformation between integrable Hénon–Heiles systems, Phys. Rev. E **49** (1994) 5897–5899.
380. M. San Miguel, Phase instabilities in the laser vector complex Ginzburg–Landau equation, Phys. Rev. Lett. **75**, 425–428 (1995).
381. M. Sato and Y. Sato, Soliton equations as dynamical systems on infinite dimensional Grassmann manifolds, Lecture Notes in Num. Appl. Anal. **5** (1982) 259–271.
382. J. Satsuma, N-soliton solution of the two-dimensional Korteweg–de Vries equation, J. Phys. Soc. Japan **40** (1976) 286–290.
383. J. Satsuma, Exact solutions of Burgers' equation with reaction terms, *Topics in soliton theory and exact solvable nonlinear equations*, 255–262, eds. M. J. Ablowitz, B. Fuchssteiner, and M. D. Kruskal (World Scientific, Singapore, 1987).
384. J. Satsuma, Hirota bilinear method for nonlinear evolution equations, 171–222, ed. A. Greco, Lecture notes in physics **632** (Springer Verlag, Berlin, 2003).
385. J. Satsuma and R. Hirota, A coupled KdV equation is one case of the four-reduction of the KP hierarchy, J. Phys. Soc. Japan **51** (1982) 3390–3397.
386. J. Satsuma and D. J. Kaup, A Bäcklund transformation for a higher order Korteweg–de Vries equation, J. Phys. Soc. Japan **43** (1977) 692–697.
387. K. Sawada and T. Kotera, A method for finding N-soliton solutions of the K.d.V. equation and K. d. V.-like equation, Prog. Theor. Phys. **51** (1974) 1355–1367.
388. F. Schwarz, An algorithm for determining polynomial first integrals of autonomous systems, Journal of symbolic computation **1** (1985) 229–233.
389. J. Segert, Frobenius manifolds from Yang–Mills instantons, Math. Res. Lett. **5** (1998) 327–344.

390. H. Segur, Solitons and the inverse scattering transform, *Topics in ocean physics*, 235–277, eds. A. R. Osborne and P. Malanotte Rizzoli (North-Holland publishing co., Amsterdam, 1982).
391. T. Sen and M. Tabor, Lie symmetries of the Lorenz model, Physica D **44** (1990) 313–339.
392. J. Shohat, A differential equation for orthogonal polynomials, Duke Math. J. **5** (1939) 401–417.
393. E. K. Sklyanin, Separation of variables – New trends –, Prog. Theor. Phys. Suppl. **118** (1995) 35–60.
394. V.P. Spiridonov, A.S. Zhedanov, Orthogonal polynomials and combinatorics, 365–388, *Special functions 2000*, eds. J. Bustoz, M.E.H. Ismail and S.K. Suslov (Kluwer, Dordrecht, 2001).
395. J. Springael, R. Conte, and M. Musette, On exact solutions of the Bianchi IX cosmological model, Regular and chaotic dynamics **3** (1998) 3–8. Solv-int/9804008.
396. P. Stäckel, Sur une classe de problèmes de Dynamique, C. R. Acad. Sc. Paris **116** (1893) 485–487. Sur une classe de problèmes de Dynamique, qui se réduisent à des quadratures, C. R. Acad. Sc. Paris **116** (1893) 1284–1286.
397. P. Stäckel, Sur l'intégration de l'équation différentielle de Hamilton, C. R. Acad. Sc. Paris **121** (1895) 489–492.
398. J. Swift and P. C. Hohenberg, Hydrodynamic fluctuations at the convective instability, Phys. Rev. A **15** (1977) 319–328.
399. K.M. Tamizhmani and R. Conte, unpublished (1988).
400. K.M. Tamizhmani, A. Ramani, B.G. Grammaticos and T. Tamizhmani, Discrete integrability, *Classical and quantum nonlinear integrable systems: theory and application*, 64–94, ed. A. Kundu (IOP Publishing Ltd., Bristol, 2003).
401. A. H. Taub, Empty space-times admitting a three-parameter group of motions, Annals of Math. **53** (1951) 472–490.
402. O. Thual and U. Frisch, Natural boundary in the Kuramoto model, *Combustion and nonlinear phenomena*, 327–336, eds. P. Clavin, B. Larrouturou, and P. Pelcé (Éditions de physique, Les Ulis, 1986).
403. S. Toh, Statistical model with localized structures describing the spatio-temporal chaos of Kuramoto–Sivashinsky equation, JPSJ **56** (1987) 949–962.
404. A. Tresse, Détermination des invariants ponctuels de l'équation différentielle ordinaire du second ordre $y'' = \omega(x,y,y')$. Leipzig 1896. 87 S. gr. 8°. Fürstl. Jablonowski'schen Gesellschaft zu Leipzig. Nr. **32** (**13** der math.-naturw. Section). Mémoire couronné par l'Académie Jablonowski; S. Hirkel, Leipzig (1896).
405. C. Tresser et P. Coullet, Itérations d'endomorphismes et groupe de renormalisation, C. R. Acad. Sc. Paris **287** A (1978) 577–580.
406. T. Tsuda, K. Okamoto and H. Sakai, Folding transformations of the Painlevé equations, Math. Annalen **331** (2005) 713–738.
407. H. Umemura, Birational automorphic groups and differential equations, Nagoya Math. J. **119** (1990) 1–80.
408. G. Valiron, *Cours d'analyse mathématique*, 2ième éd. (Masson, Paris, 1950).
409. P. Vanhaecke, Integrable systems in the realm of algebraic geometry, 218 pages, Lecture notes in mathematics **1638** (Springer, Berlin, 1996).
410. C. Verhoeven, Van Bäcklund-transformatie tot niet-lineaire superpositieformule voor een aantal vijfde orde partieeldifferentiaalvergelijkingen, Licenciaatsverhandeling, (VUB, Brussels, July 1999).
411. C. Verhoeven, Integration of Hamiltonian systems of Hénon–Heiles type and their associated soliton equations, PhD thesis, Vrije Universiteit Brussel (May 2003).
412. C. Verhoeven and M. Musette, Grammian N-soliton solutions of a coupled KdV system, J. Phys. A **34** (2001) L721–L725.
413. C. Verhoeven, M. Musette and R. Conte, Integration of a generalized Hénon–Heiles Hamiltonian, J. Math. Phys. **43** (2002) 1906–1915. http://arXiv.org/abs/nlin.SI/0112030.

414. C. Verhoeven, M. Musette and R. Conte, General solution for Hamiltonians with extended cubic and quartic potentials, Theor. Math. Phys. **134**, 128–138 (2003). http://arXiv.org/abs/nlin.SI/0301011
415. C. Verhoeven, M. Musette and R. Conte, On reductions of some KdV-type systems and their link to the quartic Hénon–Heiles Hamiltonian, 12 pages, *Bilinear integrable systems - from classical to quantum, continuous to discrete*, ed. P. van Moerbeke (Kluwer, Dordrecht, 2004). http://arXiv.org/abs/nlin.SI/0405034
416. S.Yu. Vernov, Proof of the absence of elliptic solutions of the cubic complex Ginzburg-Landau equation, Teoreticheskaya i Matematicheskaya Fizika **146** (2006) 161–171 Theor. Math. Phys. **146** (2006) 131–139. http://arXiv.org/abs/nlin.PS/0503009.
417. S. Yu. Vernov, Elliptic solutions of the quintic complex one-dimensional Ginzburg-Landau equation, J. Phys. A **40** 9833–9844 (2007). http://arXiv.org/abs/nlin.PS/0602060
418. A. P. Veselov, Growth and integrability in the dynamics of mappings, Commun. Math. Phys. **145** (1992) 181–193.
419. A.P. Vorobev, On the rational solutions of the second Painlevé equation, Differentsial'nye Uravneniya **1** (1965) 79–81 [English : Diff. equ. **1** (1965) 58–59].
420. M. Wadati, Bäcklund transformations for solutions of the modified Korteweg–de Vries equation, JPSJ **36** (1974) 1498–1498.
421. M. Wadati, H. Sanuki, and K. Konno, Relationships among inverse method, Bäcklund transformation and an infinite number of conservation laws, Prog. Theor. Phys. **53** (1975) 419–436.
422. H. D. Wahlquist, Bäcklund transformation of potentials of the Korteweg–de Vries equation and the interaction of solitons with cnoidal waves, *Bäcklund transformations, the inverse scattering method, solitons and their applications*, 162–183, ed. R. M. Miura, Springer lecture notes in mathematics **515** (Springer Verlag, Berlin, 1976).
423. H. D. Wahlquist and F. B. Estabrook, Bäcklund transformation for solutions of the Korteweg–de Vries equation, Phys. Rev. Lett. **31** (1973) 1386–1390.
424. H. D. Wahlquist and F. B. Estabrook, Prolongation structures of nonlinear evolution equations, J. Math. Phys. **16** (1975) 1–7.
425. J. Weiss, The Painlevé property for partial differential equations. II: Bäcklund transformation, Lax pairs, and the Schwarzian derivative, J. Math. Phys. **24** (1983) 1405–1413.
426. J. Weiss, Bäcklund transformation and linearizations of the Hénon–Heiles system, Phys. Lett. A **102** (1984) 329–331.
427. J. Weiss, On classes of integrable systems and the Painlevé property, J. Math. Phys. **25** (1984) 13–24.
428. J. Weiss, The sine-Gordon equations: Complete and partial integrability, J. Math. Phys. **25** (1984) 2226–2235.
429. J. Weiss, The Painlevé property and Bäcklund transformations for the sequence of Boussinesq equations, J. Math. Phys. **26** (1985) 258–269.
430. J. Weiss, Modified equations, rational solutions, and the Painlevé property for the Kadomtsev–Petviashvili and Hirota–Satsuma equations, J. Math. Phys. **26** (1985) 2174–2180.
431. J. Weiss, M. Tabor, and G. Carnevale, The Painlevé property for partial differential equations, J. Math. Phys. **24** (1983) 522–526.
432. H. Wittich, *Neuere Untersuchungen über eindeutige analytische Funktionen* (Springer-Verlag, Berlin, 1955).
433. S. Wojciechowski, Separability of an integrable case of the Hénon–Heiles system, Phys. Lett. A **100** (1984) 277–278.
434. S. Wojciechowski, Integrability of one particle in a perturbed central quartic potential, Physica Scripta **31** (1985) 433–438.
435. Wikipedia project, articles on "Affine Weyl groups", "Coxeter groups", etc `http://en.wikipedia.org/wiki/Coxeter_group`
436. R. Willox, private communication (1995).
437. A.I. Yablonskii, On rational solutions of the second Painlevé equation, Vesti Akad. Nauk Bielorussian SSR, Ser. Fiz. Tekh. Nauk **3** (1959) 30–35.

438. Yee T.-l. and R. Conte, Another integrable case in the Lorenz model, J. Phys. A **37** (2004) L113–L115. http://arXiv.org/abs/nlin.CD/0402033
439. Yee T.-l., R. Conte, and M. Musette, Sur la "solution analytique générale" d'une équation différentielle chaotique du troisième ordre, 195–212, *From combinatorics to dynamical systems*, eds. F. Fauvet and C. Mitschi, IRMA lectures in mathematics and theoretical physics **3** (de Gruyter, Berlin, 2003). http://arXiv.org/abs/nlin.PS/0302056 Journées de calcul formel, Strasbourg, IRMA, 21–22 mars 2002.
440. V. E. Zakharov (ed.), *What is integrability?*, 321 pages (Springer, Berlin, 1991).
441. V.E. Zakharov and S. V. Manakov, Resonant interaction of wave packets in nonlinear media, Pis'ma Zh. Eksp. Teor. Fiz. **18** (1973) 413–417 [English : Soviet Physics JETP Letters **18** (1973) 243–245].
442. V. E. Zakharov and A. B. Shabat, Exact theory of two-dimensional self-focusing and one-dimensional self-modulation of waves in nonlinear media, Zh. Eksp. Teor. Fiz. **61** (1971) 118–134 [English : Soviet physics JETP **34** (1972) 62–69].
443. V. E. Zakharov, On stochastization of one-dimensional chains of nonlinear oscillators, Zh. Eksp. Teor. Fiz.**65** (1973) 219–225 [English : Soviet Physics JETP **38** (1974) 108–110].
444. V. E. Zakharov and A. B. Shabat, Interaction between solitons in a stable medium, Zh. Eksp. Teor. Fiz. **64** (1973) 1627–1639 [English : Soviet Physics JETP **37** (1973) 823–828].
445. V. E. Zakharov and A. B. Shabat, A scheme for integrating the nonlinear equations of mathematical physics by the method of the inverse scattering problem. I, Funktsional'nyi Analiz i Ego Prilozheniya **8** (1974) 43–53 [English : Funct. Anal. Appl. **8** (1974) 226–235].
446. Zhao J-X, Elliptic deformation of the travelling wave of the Kuramoto–Sivashinsky equation, private communication (2007).

Index

N-soliton solution, 82, 99, 122, 130
P1 equation, 45, 170, 188, 193, 220, 223
P2 equation, 37, 79, 137, 138, 140, 141, 170, 187, 220, 223
P3 equation, 37, 38, 194
P4 equation, 220
P6 Picard case, 9, 199
P6 equation, 139, 142, 143, 166, 183, 184, 186, 192, 194, 195, 197, 199, 200, 203, 207, 209, 210, 213, 220
Pn equation, 8, 184, 187, 195, 199–204, 207–210, 212
α-method, 28, 30, 182, 229
singular part transformation, 101, 105

addition formula, 180, 216, 218
affine
  transformation, 204, 207
  Weyl group, 207, 208
Airy equation, 212
Arnol'd–Liouville integrability, xiv
Askey–Wilson polynomial, 186

Bäcklund
  parameter, 83, 106
  transformation, 81, 106, 112, 114, 121, 122, 126
  transformation (auto–), 81, 121
background, 39, 80
Bell polynomials, 225
Bessel equation, 212
Bianchi
  diagram, 101, 128
  permutability theorem, 101
Bianchi IX cosmological model, xii
bilinear
  Bäcklund transformation, 225
  form, 223
  formalism, 129, 223
birational transformation, 140, 141, 178, 192
Boussinesq equation, xii, 95, 113
branch point, x
breather solution, 120
Burgers equation, 83, 198

Camassa–Holm equation, 84
Cartan equivalence classes, 193
Cartan matrix, 208
CGL5 equation, 71
chaos, 18, 45
characteristic
  manifold, 78, 84, 85
  reduction, 79, 136, 144
classical
  function, 187, 193
  solution, 186, 187, 193, 199, 212
classification, 38, 85, 123, 152, 167, 182, 191, 194
cnoidal waves, 215
coalescence, 183, 186, 200, 204, 210
complete equation, 123, 162
complex Ginzburg–Landau (CGL3) equation, x, xi, 19, 47, 55, 64, 69, 82
confluence, 200
conservation law, 82, 85, 113–116
contiguity relation, 142, 178
continuum limit, 164, 166, 167, 169, 172, 176, 179, 180, 182, 184, 186
  perturbation of the, 172
covariance, 94
Coxeter group, 208
criterium of polynomial growth, xiv
critical, 9, 29
critical point, x

crossratio, 102, 192
Crum transformation, 94, 99

d'Alembert equation, 81
Darboux
  coordinates, 145
  transformation, 94, 97–99, 101, 123, 144, 178
  transformation (binary), 98
Darboux–Halphen system, 73
dedication, v
deformation parameters, 139
Degasperis–Procesi equation, 84
degeneracy, 200
degree, 6, 165, 170, 219
determining equations, 50, 53, 57, 62, 106, 110, 111, 114, 117, 120, 124, 132, 135, 143
differential part, 97
diophantine condition, 21, 25, 26, 90
discrete
  d – P1 equation, 166, 174
  d – P2 equation, 166
  d – P6 equation, 184, 186
  d – Pn equation, 183–185
  e – P6 equation, 186
  e – Pn equation, 184
  q – P6 equation, 183, 186
  q – Pn equation, 183, 184
  elliptic equation, 183
  elliptic equation of Jacobi, 180
  elliptic equation of Weierstrass, 180
  equation, 163
  Lax pair, 176
  Painlevé property, 165, 169, 179
  Painlevé test, 165
  Riccati equation, 174
discretization, 164
distance, 46, 47, 63
dominant
  behavior, 16
  terms, 13, 16, 19
double scaling limit, 164
Duffing–van der Pol oscillator, x, 21, 71

elliptic
  coordinates, 154
  equation, 9, 36, 63, 67–69, 183, 187
  equation of Halphen, 180
  equation of Jacobi, 43, 48, 141, 180
  equation of Weierstrass, 7, 39, 40, 42, 43, 48, 141, 180, 189, 195, 199, 216
  function, xi, xv, 23, 33, 43, 47–49, 76, 134, 165, 185, 187, 197, 215, 221
  function of Halphen, 217
  function of Jacobi, 7, 216
  function of Weierstrass, 194, 216
  integral, 7
  second order PDE, 198
  solution, 52, 63, 74, 132, 134, 136, 182
Ermakov–Pinney equation, xv, 43, 177
exact discretization, 179
exact solution, 6
expansion variable, 11, 85, 87, 88, 103

fiber-preserving transformations, 193
finite-difference equation, 164
Fisher equation, xiii, 93, 130
fixed singularity, 2
Fuchs indices, 14, 18
Fuchsian perturbative method, 28, 73
Fuchsian type, 17

Gambier, 11
Gauss hypergeometric equation, 31, 142, 200, 209, 212
  elliptic, 186
general analytic solution, xi, 33, 46, 47, 51, 55, 58, 63, 74
general solution, 6, 15, 78, 188
genus, 60, 64, 67, 68, 70, 148, 151, 152, 162, 185, 195, 212, 213, 215
Groebner basis
  differential, 106, 132
group of reflections, 208

Hénon–Heiles Hamiltonian
  cubic, x, 23, 24, 146
  quartic, xiii, 24, 153
Hamilton–Jacobi equation, xiv, 145, 147, 149, 150, 155, 157, 159
Harry–Dym equation, 84
heat equation, 78, 83
heteroclinic, 47, 58, 59, 132
hierarchy, 82, 122, 125, 130, 160, 197
Hirota operator, 129
hodograph transformation, 71, 72, 84, 85
homoclinic, 47, 60
homographic
  group, xiv, 7, 8, 86, 123, 144, 167, 192, 194–197, 203
  invariant, 86, 104, 110, 126, 210
  transformation, 7, 36, 42, 67, 85, 89, 141, 174, 181, 183, 191, 207, 212, 216
Hoyer, 11, 73
hyperbolic
  second order PDE, 198
hyperelliptic

curve, 148
system, 148, 152

indicial equation, 14, 89
integrability, 6, 18, 45, 80, 83, 84, 94, 106, 144, 165, 171, 187, 193
Arnol'd–Liouville, 145, 147, 154
by IST, 19, 78, 79, 85
Liouville, 145, 146, 153
partial, 44, 45, 82, 89, 91, 93, 94, 130
inverse spectral transform, xii, 19, 78, 79, 85
irreducibility, 183, 191, 193, 202
isomonodromic deformation, 139
isospectral, 96

Kadomtsev–Petviashvili equation, xiii, 95
Kaup–Kupershmidt equation, xiii, 26, 122, 124, 126–128
Korteweg–de Vries equation, xi, 78–80, 87, 94, 110, 122, 125
fifth order, 26, 122
Kowalevski, 11, 28
KPP equation, 89, 133
Kuramoto–Sivashinsky (KS) equation, ix, xi, 16, 45, 47, 51, 52, 63, 68, 82, 188

Laurent series, 11, 65
Lax pair, 95, 210
Lax representation, 96
order, 96
Riccati representation, 97
scalar representation, 95
string representation, 97
zero-curvature representation, 96
linear part, 1
linear Schrödinger equation, xv, 43, 177
Liouville equation, 81, 83, 198
Liouville integrability, xiv
logarithmic derivative, 4
logistic map, xiv, 163, 164, 173, 175
Lorenz model, ix, 12, 34

meromorphic, 219
function, 219
globally, 219
locally, 219
microdifferential operator, 97
Miura transformation, 112
modified
Boussinesq equation, 115
equation, 112
Korteweg–de Vries equation, 83, 112, 116, 120
movable singularity, ix, 2

Nevanlinna
characteristic function, 220
Nevanlinna theory, 166
no-logarithm condition, 15, 90
nonFuchsian perturbative method, 31
nonlinear Schrödinger equation, x, xi, 19, 40, 80, 82
nonlinear superposition formula, 94, 102

one-soliton solution, 82, 119
order, 165
differential, 5, 8
Lax pair, 96, 103
leading, 125
matrix, 20
of a determinant, 66
of a pole, 53
of discretization, 165, 180
of growth, 166, 220
of perturbation, 3, 18, 29
scattering, 105

Padé approximant, 74
Painlevé
analysis (invariant), 86
equations, 8
function, xi, 8, 33
property, 8, 189
discrete, xiv, 166, 172
weak, 22, 23
property for ODEs, ix, 1, 11
property for PDEs, 83
test, 189
discrete, 167
weak, 22
test for ODEs, ix, 11
test for PDEs, 85
parabolic
second order PDE, 198
particular solution, ix, 188
perturbation of the continuum limit, xiv
Pfaffian, 223
plane wave solution, 82, 181
point transformations, 192
projective Riccati system, 33, 60, 97, 104, 128
pseudopotential, 103, 104, 121, 127, 144
psi-series, 15, 45
Puiseux series, 22, 45

QRT mapping, 183
quasilinear equation, 84, 89

ramification point, x
reduction, xii, 78

residue criterium, 48
resultant, 66
resummation, 3, 87
Riccati equation, 1, 33, 52, 60, 71, 72, 86, 102, 103, 111, 117, 118, 121, 123, 128, 138, 143, 164, 177, 188, 192, 212
  discrete, 174

Sawada–Kotera equation, xiii, 26, 122, 124, 126–128
scattering operator, 96, 116
Schwarzian
  derivative, 86
  equation, 178
  equation, discrete, 179
self-adjoint, 98
self-dual Yang–Mills, 79
semilinear equation, 84
separating variables, xiv, 145, 147, 154, 162
sine-Gordon equation, 116, 117, 198
singular manifold
  equation, 111, 114, 119, 122, 123
  method, xii, 105, 107
  variable, 77
singular part
  operator, 18, 53, 54, 100, 105, 224
  transformation, 5, 94, 124, 178, 224
singular solution, 174, 188
singularity confinement method, xiv, 167
solitary wave, 40, 47, 55, 59, 69, 80, 82, 93, 182, 187
  bright, 43, 115
  dark, 43
soliton, 82
soliton equation, 80, 82
spectral parameter, 95, 106
Sturm–Liouville equation, 43, 98, 177
subequation, 34
Swift–Hohenberg equation, 46

three-soliton solution, 224
three-wave resonant interaction, 80
traveling wave, xi, 80
traveling wave reduction, 79, 134
truncation
  equations, 107
  method, xi, 54, 105, 107
  variable, 107
two-soliton solution, xiii, 119, 130, 134, 224
Tzitzéica equation, 198

unreachable, 46

vacuum, 81, 94
van der Pol oscillator, 21

wave vector, 97
weak Painlevé property, x, 84, 147
weak Painlevé test, x
Weber–Hermite equation, 212
Whittaker equation, 212
Wronskian, 99

GPSR Compliance
The European Union's (EU) General Product Safety Regulation (GPSR) is a set of rules that requires consumer products to be safe and our obligations to ensure this.

If you have any concerns about our products, you can contact us on

ProductSafety@springernature.com

In case Publisher is established outside the EU, the EU authorized representative is:

Springer Nature Customer Service Center GmbH
Europaplatz 3
69115 Heidelberg, Germany

www.ingramcontent.com/pod-product-compliance
Ingram Content Group UK Ltd.
Pitfield, Milton Keynes, MK11 3LW, UK
UKHW021319160826
14477UKWH00023B/496
9789400796270